LIVING WITH GEOMETRY

Coming to an understanding with God, Life and the Universe...

MICHAEL A. GREEN

Outskirts Press, Inc.
Denver, Colorado

Outskirts Press, Inc.
http://www.outskirtspress.com

ISBN: 978-1-4327-2818-2

Outskirts Press and the "OP" logo are trademarks belonging to Outskirts Press, Inc.

PRINTED IN THE UNITED STATES OF AMERICA

Foreword

The story unfolds during the latter half of the 20 th. century and on into the earlier stages of the 21 st. Observations of the human condition, its perceptions of life, its state of being and notes on its social and physical environment coupled with studies and practices involving the subject of geometry and the disciplines of its applications in the professions of surveying and engineering along with a parallel interest in the ancient sciences led to a series of long standing queries resulting in ruminations that provided profound insights. When it took only a twirl of the compasses to bisect a one unit square, then make a line join an invaluable set of truths emerged supported by firm mathematical principles. The memory of an ancient, grand science based on simplicity itself made itself known and the most astounding revelations of our times were at hand. When the foundations of the Establishment began to crack, the trumpets sounded, hence, the game was afoot. From a hands on, down to earth practical approach in the study I found myself in possession of an array of cutting edge information that provided the solutions to a number of timeless riddles. It was the forgotten knowledge that enabled a traverse of the heavens to be conducted by way of theorems that provided the square root values for 2, 3, 5, Ancient Pi and Phi. The hidden workings of life and the Universe were revealed after having been lost for thousands of years, that which the world populace is presently unaware of. There was no need of telescopes, space vehicles or electronic gadgetry of any sort to perform the survey as it was discovered, ancient astronomy is founded on a "Phi Ratio Code". Amazingly it was the unique components of Golden Section Geometry used in harmony with elemental arithmetic that proved to be instrumental in manifesting a trustworthy and accurate outcome in the endeavor. Living with Geometry was generated in order to place the keys that unlock the secrets of the lost sciences into the care of the hands it belongs to, therefore, a thorough review of the manuscript along with many drawings will assist in the object of the exercise. The details of the decoding procedure are somewhat intense and it will require close attention to master them. The promise is, a new outlook and direction will be found within the contents, wherein answers to the mysteries that have tormented the mind of man since time immemorial are now yours to treasure. For an introduction, please see five self explanatory drawings following the dedications. In time the reader will become very familiar with just what the "Phi Proportion and the Geometry of Nature and Life" is all about…

Living with Geometry offers answers and insights into The Mysteries of all times:

Who, or what is God? What is the evidence that supports this notion? What is it that runs the machinery of life and the Universe? What are the origins of life? What is the most fundamental miracle of life we need to be acquainted with? Why hasn't the Universe just exploded and vanished, what makes it work? The Big Bang Theory is challenged and put on hold. What does infinity mean and why is the concept of it important to us? How can the domain of nature tell us what we need to know? Life after death? Life elsewhere in the Universe? How did ancient peoples view life and operate, and what was the nature of their successes? What did they accomplish or know about in their time that we should know about today? What can help us answer these questions? The Emerald Tablet and its meaning. The true meaning of the Number of the Beast, what really was the "Serpent" in the Tree of Knowledge, what is the true story behind Noah's Ark and the Tower of Babel? The Seven Hermetic Principles or Laws of Nature which unlock the doorway to the Temple of Supreme Knowledge and Wisdom. The true dimensions of the Great Pyramid, how to use them, so that the code that has been locked up within it for thousands of years tells us what we need to know. Full details of the Golden Section, the Geometry of Nature, God's Ratio, the Divine Proportion put forward in the scriptures by scribes who had no real understanding of it, and other resources that have provided the keys to understanding the workings of life, the Universe, God and who we are.

The proposal of traveling throughout the infinite expanses of space to find out what makes life and the Universe tick probably wouldn't get you burned at the stake today but it might make you a social outcast or get you locked up in a nut house if there is no basis, object or agreeable result from the exercise to be had. Therefore, it was decided to put my feet up for a time and privately travel that inner universe at the speed of thought to discover the workings of the outer one, meaning finding reliable reference material, making use of constructive thought, some practical applications and applying natural intuition to seek out what can be rediscovered about the ancient sciences and its attached wisdom, and what can be learned from the geometry of nature that will help us function properly today and in future. Based on what was discovered by delving into the mysteries of life and the Universe it was concluded there is a complete link and equality between the inner universe, the mind of Man, and the outer one which was found to be very much alive. It was also found that the truth is stranger and far more delightful than fiction once it was determined the Golden Section, or Geometry of nature is the Language of the Living Universe. What has been determined after taking a long hard look at what we have going for us and what is working against us is, we as a species, who are totally dependant on others on this planet need to reconnect with the natural domain in order to understand the true processes of life, the Cosmos and God to be free from our material confines and this will happen when we learn to speak and think in terms of this very special language, because herein lies the freedom and understanding Man has sought after for eons. There is a true rich vein of metaphysical gold within us all that needs to be mined and refined in order to develop our spiritual understanding and sense of "being". A cordial invitation is extended to those who wish to join with me in the pursuit of truth, the joy of enlightenment and the freedom these valuable and all too often elusive commodities provide.

Table of Contents

The shape of the Universe. Golden Egg Theorem. Why an egg is uniquely shaped the way it is. Egg dome structures. How to design and build them and why they should be built.

Bridging the gap between the physical and metaphysical, dealing with the difference between the mundane and the spiritual. Why instruction in the Golden Section should be forthwith introduced into the education system at all levels. How we can adjust our thoughts and actions and work practices with the Golden Section, the key to the ancient sciences, to deal constructively with today and tomorrow. More theorems. *Children of the stars ~ Space travel in an ancient era ~ Where we are from and where we are going.* 8

and the 12 Planets of the Solar System. Final analysis of the Masonic Symbol. The The Big Bang Number Theorem. A Labor of Love Project. Review of the mystic, or metaphysical meanings of numbers. Some closing comments on the future of Living with Geometry.

Drawings in support of and in conjunction with the manuscript are presented at the end of each chapter. Each image is a story within itself and are worth many words and impressions. A number of stunning theorems never seen before developed as a result of the research that took place in chapters 3 to 8, and these will surely capture the interest of the reader.

Not unlike a masterpiece of art, one ill placed brush stroke on the canvas will spoil the effect intended by the artist, and the same can be said for a composition of words if just a few are used incorrectly, when the author wishes to convey important facts and information to the reader. Words, combined with images, drawings, pictures or certain symbols say much more, hence, "Living with Geometry" is dedicated to those who care to make notes on the details of geometry and numbers because they are very important subjects to be acquainted with for our development as thinking beings. Within these contents there are answers for those who question the mysteries that life presents but find no firm answers to them by studying the rough approximations, partial truths and primitive interpretations offered by modern science, or by congregating in the cathedrals of religion to engage with the unknowable, pretense and meaningless dogma, or by attending the class rooms of our inefficient learning institutions under the directions of a system that presents but a shadow of what true education is all about. To complete the list of foes in our midst that hinder our progress, Big Business gobbles up our pocket books and freedom of spirit because it cares for nothing more than material gain.

What is available in Living with Geometry are tangible ways to think and study through our problems in order to find solutions to them, because not all is what it appears to be, and the game is always afoot when there are mysteries to be solved and clear answers to be had. In the following pages there will be no disappointments because the Universe provides, truth and enlightenment on the most important matters in life will find you…

Michael A. Green ~ Applications Geometer

First of all it needs to be explained, this is not a traditional mathematics course, though it might appear as such to begin with. The basics are reviewed in a somewhat unconventional but amusing manner and are presented in a certain sequence unlike a typical mathematics program in order that we end up on the same page and speaking the same "special language in pursuit of the truth", and become familiar with the thought processes that helped find the much cherished answers needed. Abstract, mind numbing complexities are not offered in this review, therefore, there is no need to shy away from what, to many, appears an intimidating subject. There are a number of surprises in store for those who have reached the higher levels in the subject by way of the limitations of the present day education system. I am certified at the professional level in the areas of surveying and engineering, and I might be known as an applications geometer possessing a record of over forty years practice, with a hands on, down to earth practical approach in many aspects of mathematical applications. I am not unlike a teacher it could be said, but my true position is more in line with the required mind set on the subject's attached philosophies, uses and applications for material needs and playful mental stimulation for the sheer joy of riding on the wave of the learning curve in research mode to see where constructive thoughts might lead to. If the reader allows it, simplicity and enlightenment from the study will be theirs. Beginning in chapter 3 a state of superconsciousness will take hold then peak out near the ending in chapter 8. Within this presentation are cutting edge revelations never seen before in this time era which promises to bring the reader to an astounding level of understanding with God, Life and the Universe. It will become known as a journey along the Golden Spiral Road of Life. With that being said, please join with me on an intriguing adventure, the results of which will rock the foundations of the Establishment and provide us and our fellows with a new outlook and direction in life. Drawings with the lettering MGTS in the corners of the drawings are my business trade mark initials, which means I have looked into solutions of various mathematical concepts and present them to the attention of the reader's scrutiny. Let us roll up our sleeves and get to work as a team, there will be no dull moments or regrets. Be prepared for some number crunching. I trust the information herein, turns out to be one of your better investments. Thank you, yours truly...

Michael Green ~ MGTS Technologies Inc...

ARCHITECTURE OF THE LIVING UNIVERSE:

Half buried in the sands of time, the ancient sciences were lost...

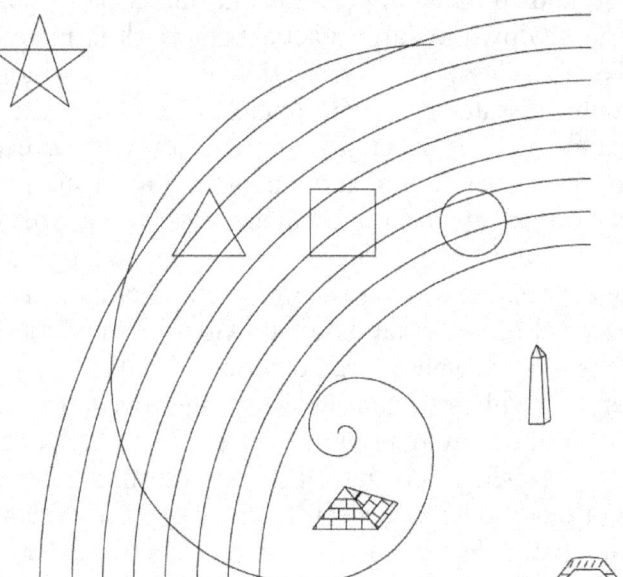

until the Great Pyramid spoke, in the Language of the Universe, and the most astounding revelations of our time were at hand...

MGTS

ELEMENTS OF THE GOLDEN SECTION / Phi RATIO (a)

$\underline{\sqrt{3}}$ = ONE ROYAL CUBIT

a+b is to a as a is to b

$\dfrac{a+b}{a} = \dfrac{a}{b}$

$a + b = \sqrt{5}$

$a - b = 1.0$

$a \times b = 1.0$

ANGLE ABC = 63° 26' 03"

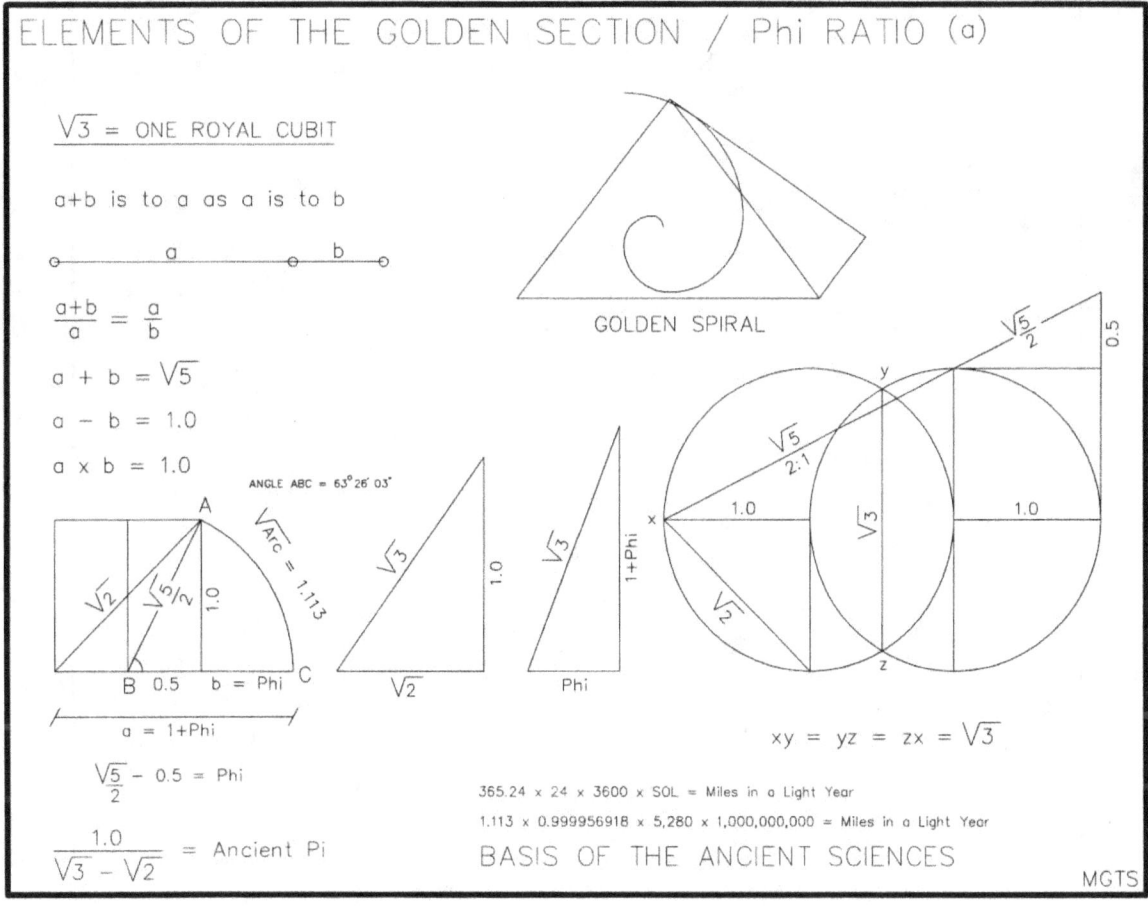

GOLDEN SPIRAL

$xy = yz = zx = \sqrt{3}$

$\dfrac{\sqrt{5}}{2} - 0.5 = Phi$

$\dfrac{1.0}{\sqrt{3} - \sqrt{2}}$ = Ancient Pi

365.24 x 24 x 3600 x SOL = Miles in a Light Year

1.113 x 0.999956918 x 5,280 x 1,000,000,000 = Miles in a Light Year

BASIS OF THE ANCIENT SCIENCES

MGTS

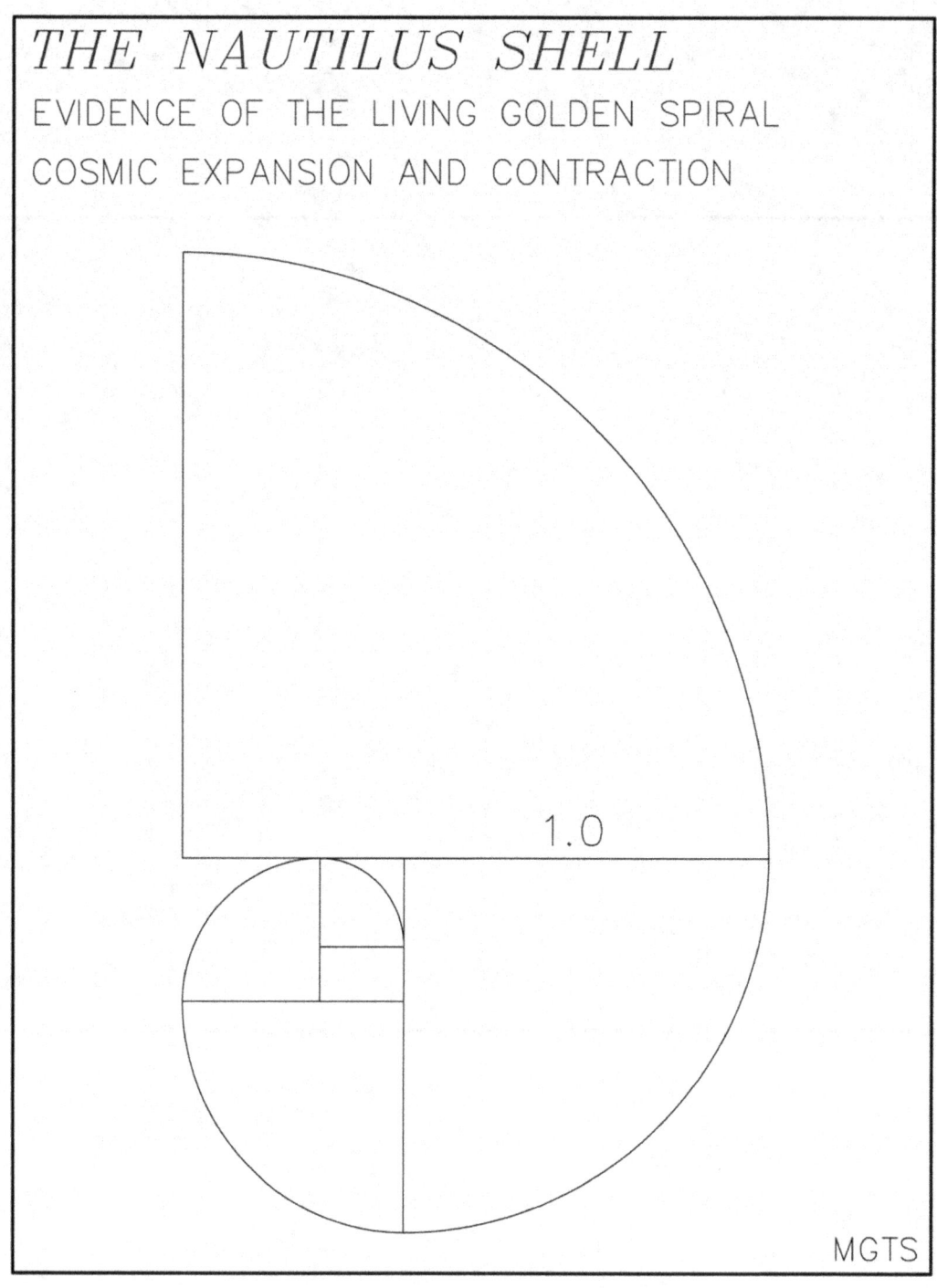

THE NAUTILUS SHELL

EVIDENCE OF THE LIVING GOLDEN SPIRAL

COSMIC EXPANSION AND CONTRACTION

1.0

MGTS

Royal Cubit Theorem — Based on Ancient Earth Dimensions

One Mile = 5,280 ft.

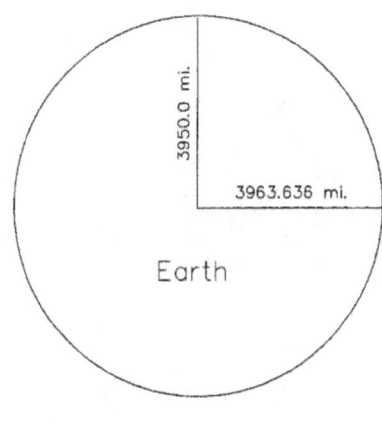

Earth

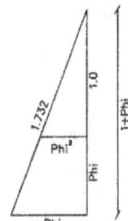

Equatorial Radius = 3,963.636 mi.

Polar Radius = 3,950.000 mi.

3,963.636 − 3,950.000 = 13.636 mi.

3,950.000 − 13.636 = 3,936.364 mi.

$$\frac{3,936.364 \times 5,280}{12,000,000} = \sqrt{3} = 1.732 \text{ ft.}$$

Note:

1.732/2 = 0.866 x 1,000,000

= 866,000 mi. = Diameter of the Sun

1.0/1.732 x 100,000 x 2 = 115,473.4411 lt. yr.

= Diameter of the Milky Way

Statement:

One Royal Cubit = $\sqrt{3}$ = 1.732 ft.

to three places of decimal

1.732 ft. = Capstone Height Great Pyramid

1.732/20 = 0.0866 = Pyramid Inch

= Ten Millionth Part of the Sun's Diameter

MGTS

ROYAL CUBIT THEOREM MGTS

BASED ON ANCIENT EARTH MEASURES

EQUATORIAL RADIUS 3963.636 mi.

LESS POLAR RADIUS 3950.0 mi.

DIFFERENCE = 13.636 mi.

POLAR RADIUS − 13.636 = 3936.3636 mi.

$$\frac{3936.36 \times 5280}{12,000,000} = \sqrt{3} \ (1.732 \ \text{ft.})$$

STATEMENT:

TRUE ROYAL CUBIT = $\sqrt{3}$

$$\frac{\sqrt{3}}{20} = 0.0866 \ / \ \text{PYRAMID INCH}$$

TEN MILLIONTH PART OF THE SUN'S DIAMETER

GOLDEN SECTION

GENESIS OF DESIGN FOR THE GREAT PYRAMID

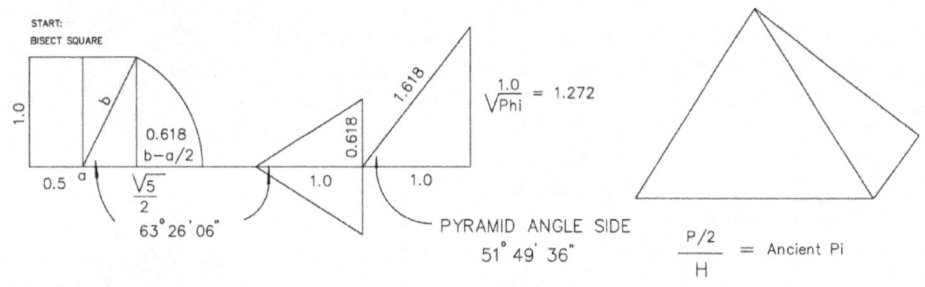

START:
BISECT SQUARE

1.0

b

0.618
$b-a/2$

0.5 a

$\dfrac{\sqrt{5}}{2}$

63° 26' 06"

0.618

1.0 1.0

1.618

$\dfrac{1.0}{\sqrt{Phi}} = 1.272$

PYRAMID ANGLE SIDE
51° 49' 36"

$\dfrac{P/2}{H}$ = Ancient Pi

NOTE:

$b = \sqrt{1.0^2 + 0.5^2} = 1.118$

$b - 0.5 = 0.618$

$a + 0.618 = 1.618$

MGTS

CHAPTER 1

Introduction - Form and Space

The word geometry is rather dryly defined in the dictionary as being a branch of mathematics dealing with the relations, properties and measurements of solids, surfaces, lines and angles. This is quite true in physical terms yet unlike any other subject it literally and uniquely encompasses and expresses the finite in relationship with the infinite simultaneously. Furthermore, it is from a certain type of geometry that the thought processes for art, science, language, numbers and philosophy are derived and unfortunately, for the most part western civilization put this valuable information to the side and has forgotten about how it works. Originally, or historically speaking, it is said, geometry was one of the two fields of pre-modern mathematics, the other being the study of numbers. We might think geometry came first because it stimulates the sense of sight, then numbers would play a roll in activating the analytical mind in order to evaluate it. Actually, the two interact with each other in a splendid way. We think about what we see. In terms of simple geometry everyone knows at first glance what a point, triangle, circle and a square is in finite terms but in this branch of mathematics at the elevated level we will be discussing shortly, its associated parts have far more to offer in terms of deeper meanings with regards to the *metaphysical* aspects of life if one is so inclined. According to the dictionary metaphysics is that part of philosophy concerned with the study of the ultimate causes and the underlying nature of how "things" work. "Things" such as the influences, reasons for and the meaning of life and existence, to be more precise. An area that in reality, modern science and religion, who, historically speaking, haven't been much help to us on at all. There is a natural desire to learn and understand life and the Cosmos in all of us. To begin with look at a circle, its radius is a finite dimension, yet its arc is based on the unique irrational number Pi, the value of which repeats itself to infinity. This is rather remarkable in itself, anyone with compasses in hand can draw an arc or circle and should acclaim himself a genius but for the most part this exercise is likened to simple child's play and taken for granted, therefore it goes by unnoticed without much thought attached to it.

The basic geometry in the two dimensional plane we work with is somewhat effective because it can be seen, or visually equated. Endeavors to design, build and construct would never get off the ground if it wasn't for the rules of geometry. Without geometry there would be no language and means to communicate. Without geometry there would be no sense of physical sequence, progression and order. The true definition of a point becomes important in the following interpretation of true meanings. A point is an infinitesimal reference anywhere in time and space. It could be a dot on a sheet of paper being the center of a circle or a sphere, the corner of a triangle or square, it could be Earth's location in the solar system, it could be the Sun in reference to its position in the Milky Way and the infinite realm of the Universe, or philosophically speaking it could be the psyche of you or me, relative to a process known as the progression of conscious life on Earth at any time or anywhere in the Universe.

When a triangle becomes a tetrahedron, a square becomes a cube and a circle becomes a sphere the study becomes much more interesting and the limitations of the two dimensional plane are transcended. Then we have the third dimension, form and space. To serve as an introduction to the remarkable qualities of this subject, following are some graphics which make this exercises possible.

Many intensive and creative hours can be spent constructing three dimensional space from the known two dimensional planar dimensions. For starters all you need is a decent compass set, a supply of 32" x 24" bristle board paper, white glue, masking tape, scissors, a razor knife, a triangular scale for folding and a couple of set squares or one of those rolo rulers for drawing parallel lines. The regular 5 platonic solids are the tetrahedron, cube or hexahedron, octahedron, dodecahedron and icosahedron are shown in terms of their isometric projections. These are good ones to start with. A sphere is a special case. One can be constructed by cutting out three equal sized circles and using some ingenuity. Usually there is a cutting blade included in a quality compass set that makes it possible to cut out circles in paper. Add a spool of nylon thread and thumbtacks to the list so your models can be suspended from the ceiling and be more fully appreciated. What needs to be done to construct a sphere one particular way is cut out three equal radius circles each having approximately 3/8 " parallel widths. On one circle cut half way through its width top and bottom from the inside on its vertical at center, on the second cut halfway through its width top and bottom from the outside on its vertical at center. On the third circle repeat the step for circle one and fit all three together like a simple three dimensional puzzle and voila, a sphere has been constructed. Attach a thread to it and hang it up for viewing. You will never tire of glancing at it as the air currents move and spin it around. Included in this presentation are construction drawings for the above mentioned solids including a Phi based pyramid. This will be explained in the following chapter. The learning curve is very brief for this exercise. All that is needed is a basic knowledge of geometry and drawing procedures. Study the following construction drawings to get acquainted with the intentions of the process. The first step is to draw the lines on the dull side of the bristle board paper so your finished form will be shiny side out. There are no calculations to be made, only a basic skill using the compasses, straight edge and knowing about the principle of parallel lines is required. Bristle board paper has substantial tensile strength so quite large models can be made. Make a judgment on this based on the sheet size. If you have a Cad drawing program and printer or plotter it is an easy matter to tape your drawing atop the bristle board paper and punch thumbtack holes at the drawing points. If you don't have a light table tape your sheet in a sunny window and join the points using pencil and straight edge. You might first wish to use the razor knife and scale edge to cut along the lines. Once you have more confidence, use the scissors with care on the perimeter lines. In the folding and creasing operation an example of the congruency factor will be seen as one triangular or pentagonal area matches the other. To get good creases stroke the scissor handles over the folds using some pressure. To fold and crease along the glue tab lines fit a smaller piece of paper under the line and holding the scale on the line with one hand then fold by firmly yet gently pulling upwards with your other fingers. The mucilage can be applied to the glue tabs with an artists brush, a pointy scrap of the paper, or your finger. What to do next will fall into place, as the matching pieces fit together in a natural sequence. Give the glue time to set as you fit the form together in stages. You are bound to get some dried glue on your models. This will disappear easily by rubbing it off using a damp cloth. At a more advanced level another suggestion is to draw windows inside and parallel to the design lines and cut them out accordingly. By doing this the space inside the finished forms will be even more fully appreciated. It will resemble what Kepler in the 17 th. Century and many others before, and since have been spending a lot of time on. This is especially where the razor knife can be used for cutting and actually the models are easier to put together by doing this. Ensure the glue tabs are slightly narrower than the frames. In putting the dodecahedron and icosahedron together you might feel as though you need a second pair of hands because it almost takes the dexterity of a brain surgeon to do it. Partner up with a friend to work on them if possible. By the time your models are constructed you have become the mad scientist or like Edward Scissor Hands, ankle deep in paper scraps with blobs of glue here and there and on yourself, but by then it is time to stand back and admire your handiwork. One more suggestion is to give your models some color. Try red spray paint on the tetrahedron, perhaps black for the cube and orange for the octahedron or any that suit your choices. Use bright gold on the dodecahedron and icosahedron for reasons that will become clear further along on in this presentation. Use the speckle stone spray paint if possible, it will give texture to the surfaces of the forms. To give your models luster and a longer lifetime spray them with Varathane or clear Krylon. There are the thirteen other platonic solids by Archimedes if you are looking for other challenges. Over 2000 years ago one of his famous quotes in his lecture hall was, " Let no person destitute of geometry enter these doors ". What he was saying it might be assumed is, the subject of geometry is very important, so don't show up for classes unless you know something about it and are serious. You might wish to try some platonic solids of your own. Nice ones to make are crystal and diamond shapes. You can teach yourself how to make these with practice.

Further along is a two dimensional drawing of a quartz crystal and the associated construction drawing that shows how this can be done. Give the sphere or the atomic energy sign a try, these ones are gratifying. The delightful aspect of this exercise is the learning gains and hands on feel for geometry that is experienced. No matter what your age, it might occur this type of thing is not done in our school programs in any big way yet it should be. If you are enrolled in a math program be encouraged to participate in this exercise and if you have young people attending school show them how much fun it is and talk to their math teachers about it. The other more modern term for platonic solid is symmetrical polyhedron. Once the construction of the simpler solids is conquered there are many star formations to work on. There is another way to express the platonic solids and star forms using bristle board paper. This is accomplished by drawing out three shapes of equal size and cutting them out then folding each one at the center vertical line. Then by gluing one face to the other three angles of 120 degrees equaling 360 degrees have been formed in a natural manner. After the glue has dried and a little hand manipulation is made with the fins of the model to balance the interior angles a form has been constructed that deserves to be painted and hung up. Construct a three dimensional atomic sign this way. Go to the drawing that shows three ellipsoids. If you cannot draw ellipsoids by hand and if you don't have a cad program an ellipsoid template can be picked up at a stationary or drafting supply store. You might want to spray paint this one silver and splash on some glue here and there and sprinkle on some silver glitter. Then a model of the molecular theory becomes available. This will be very attractive as it floats and spins around with the air currents in your working area. It would be so great to make all these forms out of 1/8 "silicone bronze, gold and silver rods using colored plastic sheeting and all that but not everyone has a budget for this. The idea at this stage is to gain an appreciation for geometric form and become acquainted with its unique qualities in three dimensions. If the interest is keen enough the sky is the limit. An unfolding of the conscious state takes place while involved in these practices.

Looking at the broader view on this topic there are a number of different geometries available to us. Due to technological advances these include hyperbolic geometry which help us understand maneuvering mechanics for operations in space and spherical geometry which assist our navigators and aviators to circumnavigate the globe. These geometries are beyond the scope of Euclidean geometry that has been available to us in our secondary education system, however, everything has its purpose and place. The Mercator projection is an innovation and adaptation from the 16 th. century which equates the earth's curved surface into the plane and allows us to work in the plane then equate to a curved surface when needed and this is why Euclidean geometry is of value in the fields of surveying and map making. References to known or astronomically derived points along the meridians and latitudes of the globe relative to the equator are made to make this work possible. Another type of geometry used in business and commerce as well as science and even politics etc. is graph types. We have the pie graph which shows what portion of a circle a part makes up, then the bar graph which shows how variables differ from each other and there is the line graph which shows how a variable changes over time periods. These are examples of how, because of geometry, an assessment of a state or condition can be made by the sense of sight and the analytical power of the mind. The seductive logos for sport, business, political causes etc. based on geometry cannot go by unnoticed. The abundance of geometry for roads, bridges, parking lots, athletic facilities and equipment cannot go ignored either. The machinery and tools we use to build, kitchen utensils such as cutlery, pots, pans, plates, cups etc. all have specific dimensions and geometric design relative to their uses. Music is the sound of geometry, or the geometry of sound embraced by art. Art is an expression of geometry, an octave consists of 8 notes, the number of infinity. The other type of note that can be made is, the words geometry and infinity each consist of 8 letters and the list goes on and on. Depressing as it is, mention of the war machinery or abuse of geometry can not be overlooked either. Never the less, if it wasn't for geometry we would have difficulty realizing our needs in the physical plane but the importance of it on the spiritual and mental levels should not be overlooked because that, as it will be discovered, is very important to us. It is fair to say, to a very large extent we all live with geometry in one way or the other. In truth, as originally stated, the understanding of this subject puts us in a good position being that it is the basis for all that is of value to us in the material plane and for study purposes it shows how the complimentary subjects such as algebra, trigonometry and the mathematics of motion are interrelated, but something in our repertoire of mathematics has been missing for a long while. To be concise, we all live with another type of geometry not so well understood by us today. In reaching back into the ancient past we have the opportunity to make a study of this area of debate and reconnect with the fundamentals of life, the language of the Universe and a whole lot more. There is much more following…

THE PLATONIC SOLIDS

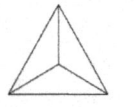

TETRAHEDRON

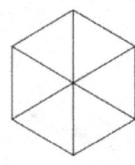

HEXAHEDRON
CUBE

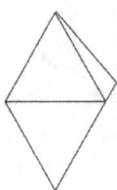

OCTAHEDRON

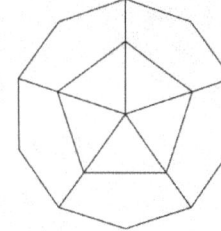

DODECAHEDRON

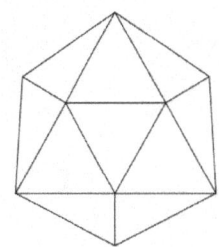

ICOSAHEDRON

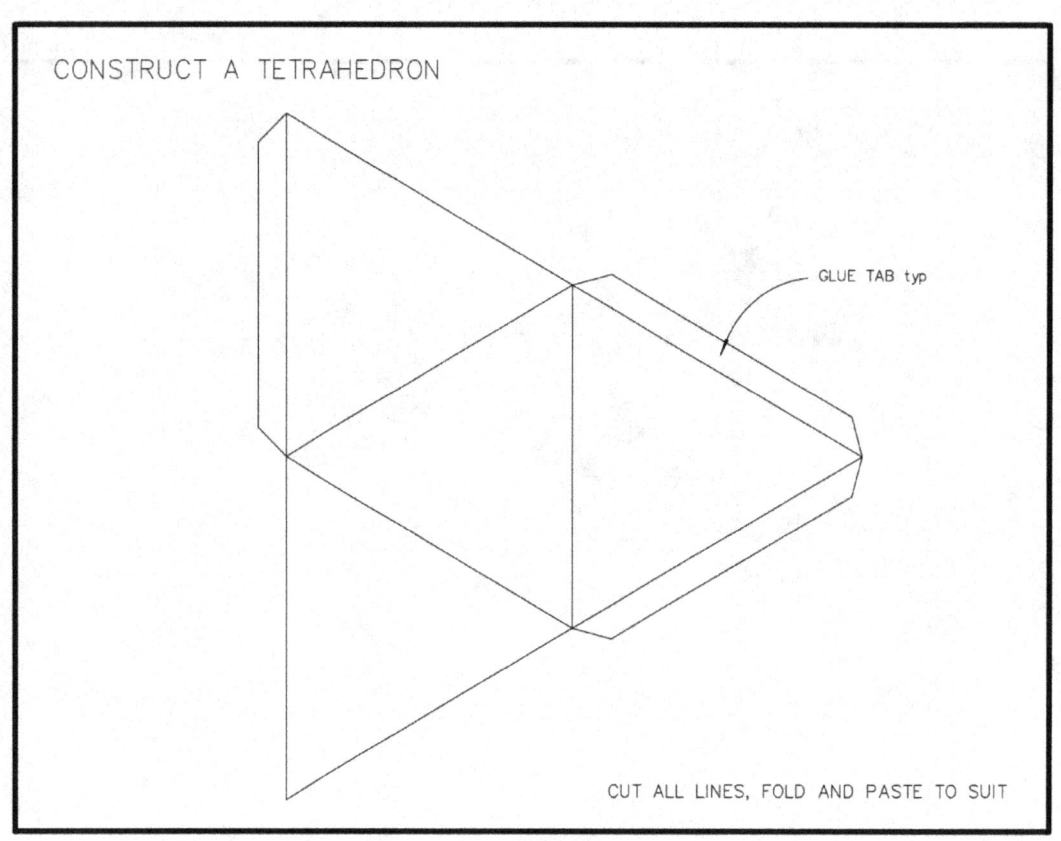

CONSTRUCT A TETRAHEDRON

GLUE TAB typ

CUT ALL LINES, FOLD AND PASTE TO SUIT

CONSTRUCT A CUBE

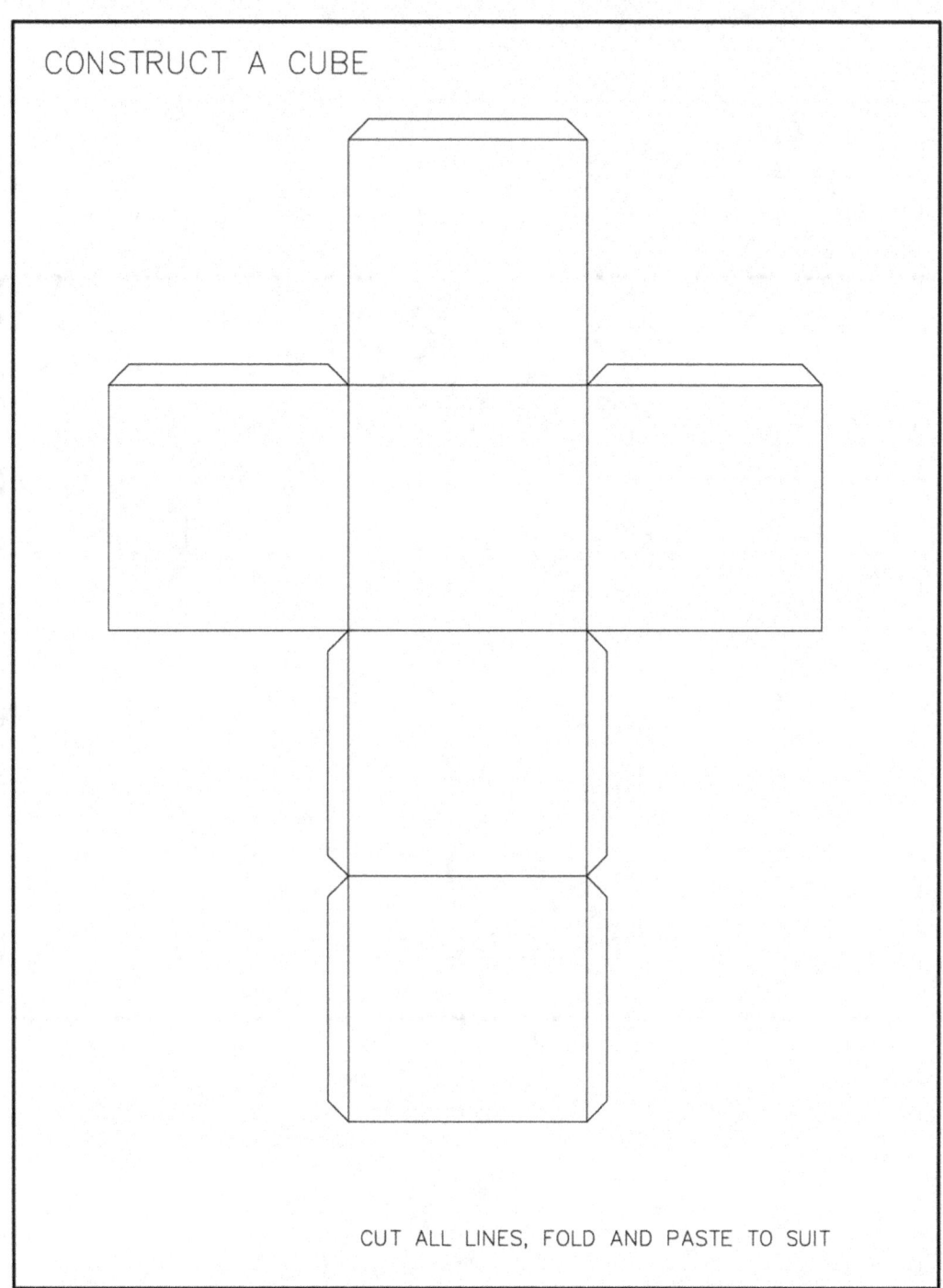

CUT ALL LINES, FOLD AND PASTE TO SUIT

CONSTRUCT AN OCTAHEDRON

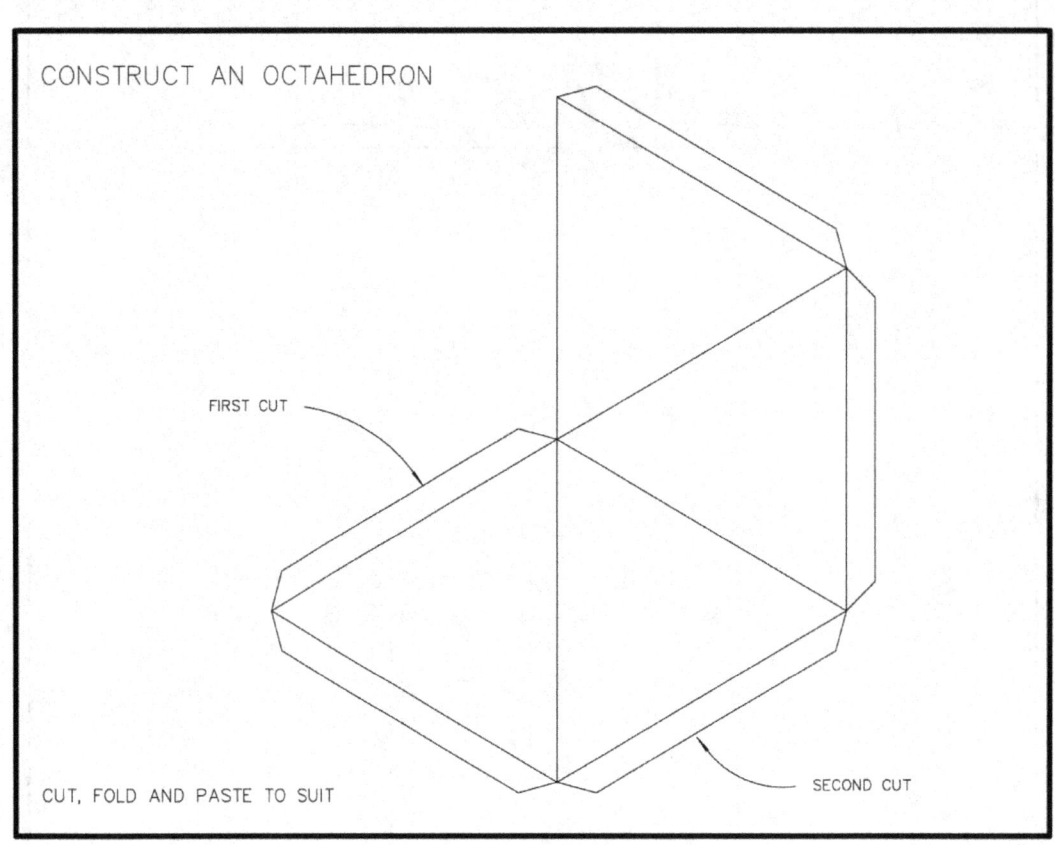

FIRST CUT

SECOND CUT

CUT, FOLD AND PASTE TO SUIT

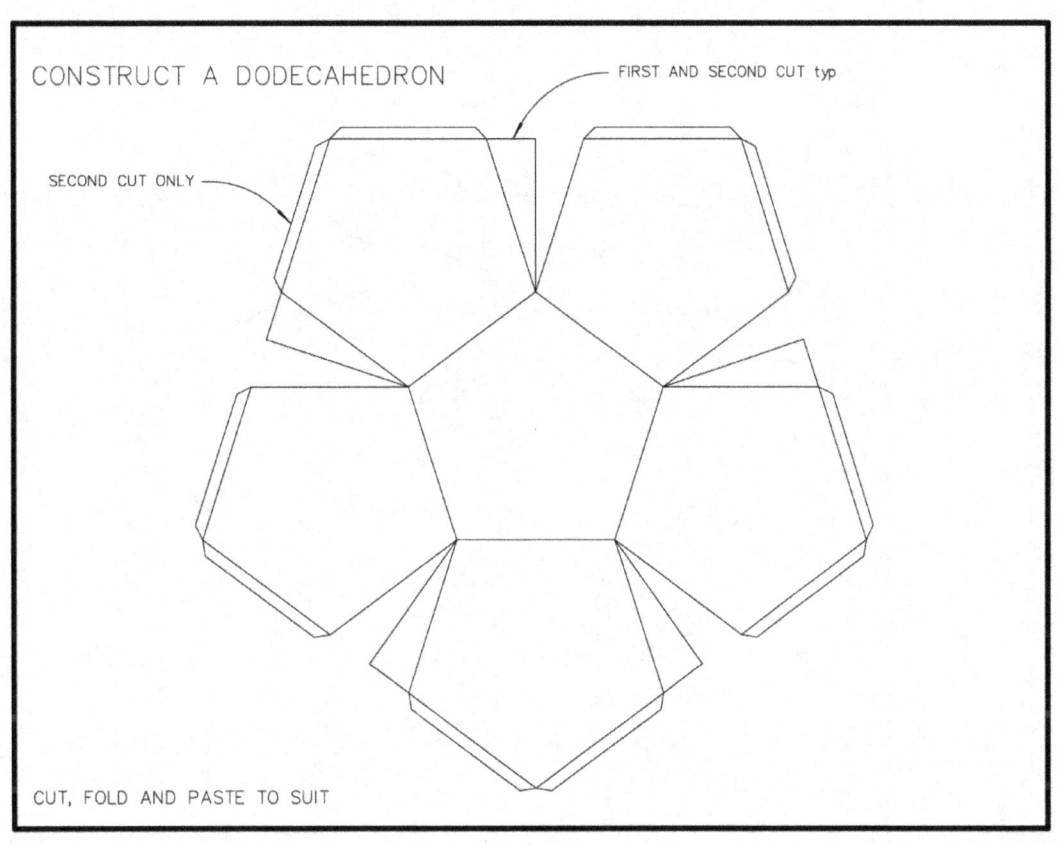

CONSTRUCT A DODECAHEDRON

FIRST AND SECOND CUT typ

SECOND CUT ONLY

CUT, FOLD AND PASTE TO SUIT

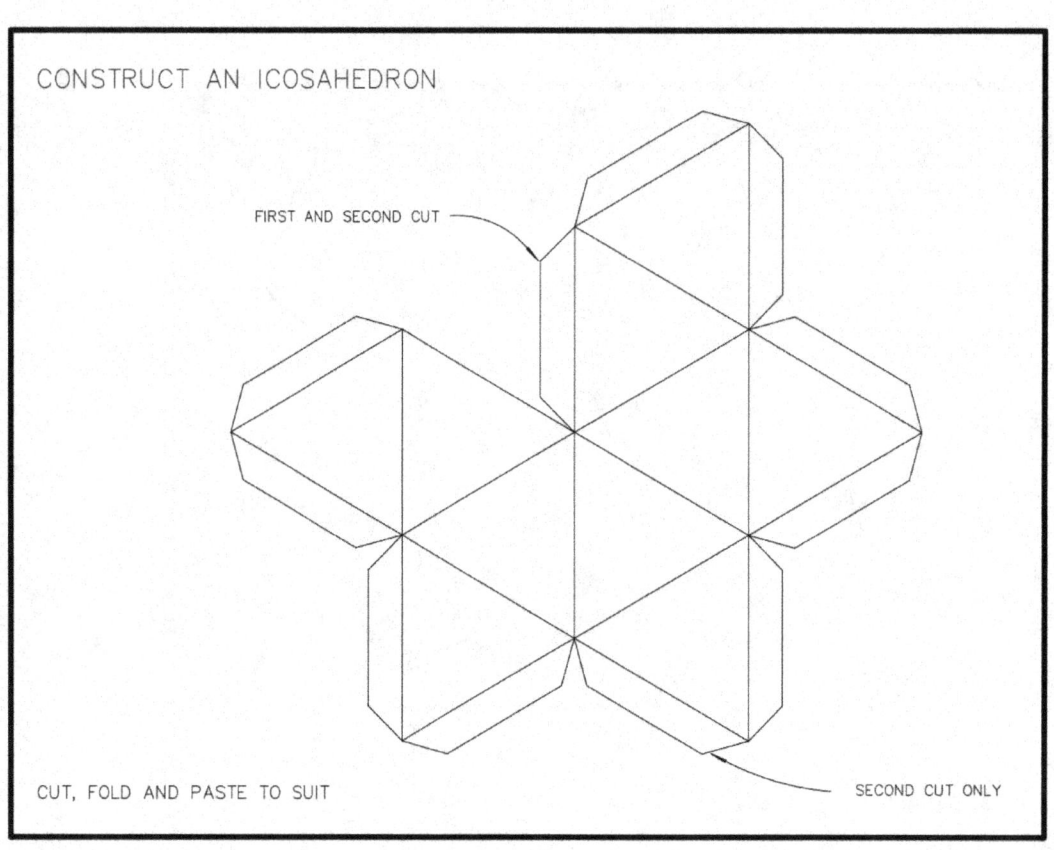

CONSTRUCT AN ICOSAHEDRON

FIRST AND SECOND CUT

CUT, FOLD AND PASTE TO SUIT

SECOND CUT ONLY

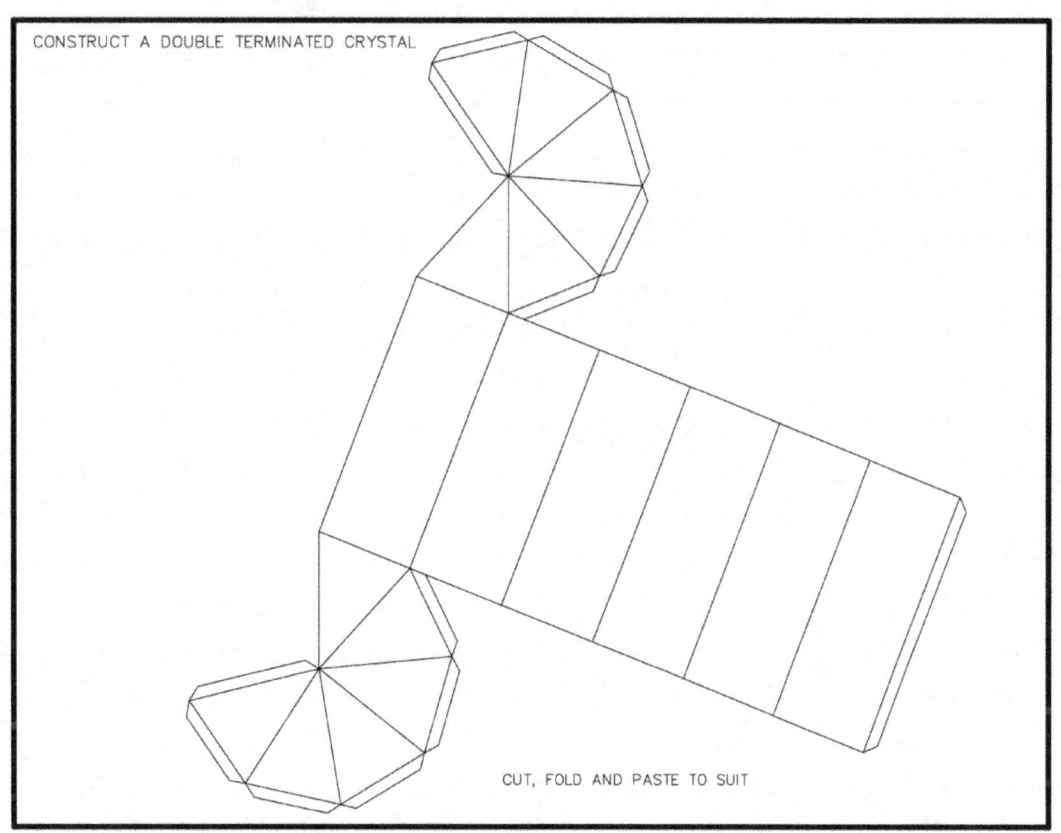

CONSTRUCT A DOUBLE TERMINATED CRYSTAL

CUT, FOLD AND PASTE TO SUIT

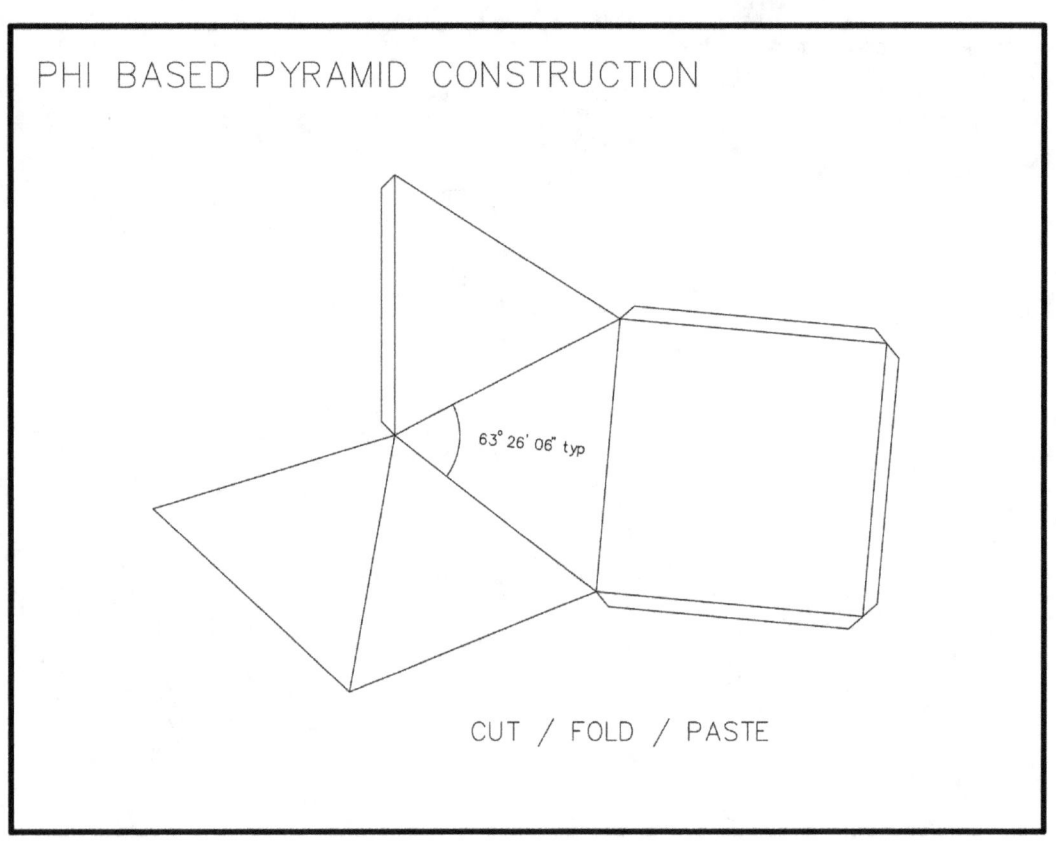

PHI BASED PYRAMID CONSTRUCTION

63° 26' 06" typ

CUT / FOLD / PASTE

CHAPTER 2

Introduction to the Golden Section ~ The Geometry of Nature

In the following presentation the natural laws of the microcosm and macrocosm will be explored using that above mentioned very special brand of geometry which was used extensively in more ancient times before the use of its applications disappeared. It appeared in early Greek culture and died out, then it underwent a brief revival in the 12 th. century, then again during the 14 th. to 17 th. centuries in Europe, and somehow its popularity again came to an end in the mainstream and the subject matter went into obscurity and was lost to the general public until more recent times. Once the secrets of the Golden Section are revealed the true beauty of geometry goes far beyond its regular definition and the doorway to understanding the workings of the natural laws of life, the cosmos and access to understanding the God force is reopened. In the process we will examine the human condition and its state of "being" today, within what will surely come to be known in endearing terms as, "The Architecture of the Living Universe ".

This segment of geometry is based on another irrational number value known as Phi. Somehow during our development period since ancient times appreciation of this function was left behind and only the value of Pi survived, when in fact it will be discovered it is these ratios, or functions working together which say it all, with regards to what we need to know, and are therefore, very dear to us.

An introduction to Phi ~ the Golden Section ~ God's Ratio ~ the Divine Proportion

Phi is an ancient Greek word said to have acknowledged their 5 th. Century B.C. sculptor Phidias who worked extensively with the Golden Section Proportion in his many projects. The early Greeks were by no means the inventors of this Ratio, as it came from old Egypt after they cultivated much knowledge from there, but we know of no word in this culture that stands for the word Phi. One look at their hieroglyph for the number one which is two vertical parallel lines of one unit with a space between them of one half unit tells us the hypotenuse distance equals the square root of 5, divided by 2 = 1.118 - 0.5 = 0.618, the number value for Phi. Therefore, their number one says it all, it explains how the processes of life and the Cosmos work. Another look at their hieroglyph for one hundred which bears an uncanny resemblance to a * golden spiral * increasing in magnitude to the left, counter clockwise, leads to the conclusion that these symbols best represent the meaning of that word expressed in number, or proportion values, not words. Reference to the Golden Spiral will be made shortly and as for the number value definition of Phi, that is coming up next.

Note: All Golden Spirals shown in the following when it comes to planet orbits etc. increase to the left in magnitude, counterclockwise, just as the ancient Egyptian hieroglyph for 100 indicates, because the planets revolve around the sun in this direction. To draw one clockwise isn't necessarily wrong as this expression of geometry is appealing either way. The key is knowing when counter clockwise or clockwise is applicable. No harm will be done as long as this is understood and any misunderstandings are quickly overcome once the reverse image of the graphics are viewed in a

13

mirror. Different ways of perceiving and studying various concepts will be presented along the way in this discussion.

Phi in Number Value Form:

Phi = One half the square root of 5 less 1/2 or 0.5 = 0.6180339... to infinity
1+Phi = I.0 over Phi, or 1+ Phi = 1.6180339... to infinity
Phi + 1+Phi = square root of 5... The square root of the sum of the squares of Phi and 1+Phi = square root of 3...
Another formula for 1+Phi is: Square root of 5 + 1, divided by 2.
A way to see the relationship of Phi and Ancient Pi: 1+Phi + 1.0 x 6, divided by 5 = Pi or: Square root of Phi x 4 = Ancient Pi. Oh, oh, what are we getting ourselves into here, a blind date coupled with confusion? Not really, give it a chance and stay with it a while and it will all fall into place and make perfect sense…

Uniquely, one is to the whole as the whole is to the one. By way of an interesting coincidence the motto for the trio of French Musketeers, "One for all, and all for One", isn't so far off the mark of the Golden Section Trinity meaning. It is at least one way to remember its terms.

This might all look a little scary but it will be found these values or ratios occur naturally with a few strokes of a compass and the wonderful part of it is no involved calculations are needed, all you need to do is engage with it and let the power of the mind take over. The greatest things known are based on simplicity itself. In today's circle of inquiry the Phi ratio is often referred to as that integral part of sacred geometry. Well, yes of course this proportion is certainly worthy of veneration if one infers its use by ancient peoples and what it has to do with life but to avoid associations with religion for now it might also be better known as simply, the geometry of the natural world and the * Language of the Universe *. It will be seen how it relates to religion later on. Foot long equations are not required, however some basic algebra and descriptive geometry is involved and of course interest will be a key part of the study. The best thing to do is dive in with your compasses and straight edge or a coordinate geometry drawing program if you have the use of one and pursue the Gold of Enlightenment using the ratios of Phi and Pi. It could be said, simply stated, working with the Golden Section is the journey of a lifetime based on 1.0, as in one unit. For drawing purposes the information will fit your hand like a tailor made glove. The most probable reason for this is, the proportions of the human form to the last detail like all other living things on this planet exist in terms of the Golden Section or Golden Ratio. Indeed, this is why your body is not unlike a sacred temple and should be regarded as such. Throughout the ages many a scholar and intellectual have been captivated by this awesome subject. One such person was Italian mathematician, Leonardo Pisano Fibonacci during the 12 th, century to whom we are so grateful for his research and records on the subject. His was a numbers approach rather than a geometric one which came to the same conclusions about the Golden Section to determine its special qualities. The concept is within any ones grasp and a university degree in mathematics is not required. Actually, that might get in the way for reasons that will be explained later on as we travel along the Golden Spiral Road of life. Proceed from the drawings entitled Elements of the Golden Section and find the Golden Ratio which is based on "simplicity" itself. This is where the Phi takes form in terms of a number equaling a ratio = the Golden Section. Hang on to this drawing because it is very important. It will serve as a reminder on how it works during those fleeting moments of forgetfulness. That is the problem, modern civilizations somehow forgot about it, or lost its grounding in the subject a long time ago and the cost has been high. Pin it on a wall near your work station or in an area you most frequent during the day and soon enough it will become part of the way you think. I hereby make the solemn promise, this is the key to that which has eluded man for ages gone by, that which is understanding how "things" work. Starting with a square of one unit bisected, let the secrets of life and the Universe be known, the natural world and the secrets of the cosmos itself will unfold in your mind and before your eyes. As an aide, shown along with these elements are the golden number value keys of geometry such as the square roots of 2, 3 and 5 and the unique pentagram and five pointed star with which you can draw your first golden spiral. The addition that can be made to this discussion is, a cube with the dimensions of one unit also provides the square root of 3 when a right angle is formed, having a length of the sq. rt. 2, the diagonal distance of the square, and a height of 1.0. When the sq. rt. of 3 is used as the vertical and the adjacent side remains as the sq. rt. of 2 the hypotenuse length = sq. rt. 5.

Following is discussed the way to draw the nautilus shell which is evidence of the Living Golden Spiral. Other terms for the Golden Spiral today are the logarithmic spiral, equiangular spiral, growth spiral and " Spira Mirabils ", or Marvelous Spiral as named by the 17 th. Century mathematician, Jakob Bernoulli. Professor Bernoulli saw the importance of this spiral and requested it to be engraved in his tomb stone because what he saw in it was the presence of God, like a signature on life, like an endorsement on the " Living Universe".

As an experiment to see the Golden Spiral in action fill the kitchen sink and sprinkle bread crumbs on the water, then pull the plug after the oxygen bubbles have dissipated. The bread crumbs will swirl counterclockwise as the sink drains because the center of the earth's gravitation is offset by the center of its magnetism. Incidentally, in spite of a popular misconception, sinks drain counter clockwise in Australia as well. Here is seen the force of gravity operating in terms of the Golden Ratio as the velocity of the water accelerates into a vortex where it exits the drain. The spiral form in the clouds seen in a satellite view of a high or low pressure zone and in the sweeping arms of a spiral galaxy are none other than the Golden Spiral. The Golden Spiral is not to be confused with, for instance the Archimedean Circular Spiral, which is the locus of points corresponding to the location over time of a point moving away from a fixed point with a constant speed along a line which rotates with constant angular velocity. Some uses of this spiral are seen in the scroll pump and early gramophone records. A drawing of this type of spiral is included. This spiral has some visual appeal but it isn't what we are looking for.

To be clear on this the Golden Spiral Arc radii decrease and increase specifically in terms of the Golden Ratio proportion. For example, to draw the nautilus shell use 1.0 x Phi (0.618) x 0.618 and so on for decreasing the arc radii values of the Pi based circles, and 1.0 x 1+ Phi (1.618) x 1.618 and so on for increasing the arc radii values of the Pi based circles. A thought worth pondering over is that this arrangement increases to infinity both ways. To avoid confusion at this stage just employ the right angles as seen in the drawing of the nautilus shell. A classic example of how Pi and Phi are inseparable in nature is seen in this geometric form as described above. Included is the Pythagoren Theorem which plays an important role in the investigation. Have a scientific calculator on hand for back checking on square root values and those square roots of the sums of square roots of numbers etc. as part of the learning process. For a warm up please see the sheet titled Given: Phi / 1.0 / 1+Phi - Compute all Distances Using the Pythagorean Theorem. This one will be fun and good practice for what lies ahead in this presentation. There are a number of drawings, or graphics to view in the following, as words alone are not enough to explain the details. These provide the reader with that extra dimension of understanding. After all a picture says a 1,000 words.

Then, especially have a look at the sheet titled, What Happened When Pi and Phi Were

Separated. This was a critical development in our history that led to our misinterpretation and lack of knowledge and understanding about life's processes. At the same time view the drawing titled Phi and a Circle / D = 1.0. This provides some perspective on the issue before us. Using Pi by itself enables us to compute and draw arcs and circles and twirl the compasses on the level of child's play and this might be acceptable for our material needs but limits us in terms of our capabilities to understand the workings of the natural world, life and the Cosmos. When Pi and Phi are used together the great expanse of the opposite ends of the two infinities, Cosmic Expansion and Contraction, are embodied or expressed by the growth spiral or Golden Ratio. In this the workings of the natural world, life and the Cosmos are seen instantaneously. An enchanting trinity is seen in the values 1.0, Phi and 1+Phi ~ "One is to the Whole as the Whole is to One". The basis of a living organism, or the natural machinery of life is seen in the drawing and that is what it is all about. It is understandable why these two values should never have been separated in study or in practice and when this happened, so called rational science went in the wrong direction and the Church was already there. It is seen the square root of Phi = Ancient Pi over 4 and this shows how these two ratios are interconnected and were inseparable. Now that the secret is out and we see what we have been missing out on, but it is possible to reconnect with their combined uses in our technology today in many effective ways. To begin with we can gain a much better understanding of the natural processes of life, the Cosmos and ourselves. At the same time view the drawing entitled Symmetry in the Golden Ratio and let this settle into your thought processes for the time being. Basically it shows that everything that exists in the universe such as a force or energy vibration is dual in nature and has an equal and opposite value. A full clamshell with its two sides intact found down at the beach verifies this law. Nature and the Universe speak to us at all times. Included is a drawing titled Angel's Wings", showing a pair of Golden Spirals back to back which appears not unlike a butterfly or an angel expressing the universal duality factor. In fact, the proportions or geometry of the wings of any living creature such as a bird, insect or certain mammals and reptiles are uniquely proportioned in the Phi Ratio and since this is the case the fins and body proportions of all fishes share the same attribute. The Holy Bible makes reference to angels on many occasions and the assumption is these also have wings but this inquiry is not about the paranormal or supernatural aspects of life. However, if there are any angels out there, we would want them to be on our side. One might assume, their wings are in the proportions of what is also known as God's ratio. It will be discovered nature tells us many important things about the workings of life and the Universe by way of this constant known as the Golden Ratio and this is one of the grandest things to be acquainted with. From this study all the potential of 1.0, the all number, the Universe can be determined and the Gold of Enlightenment is at hand. The information on the other drawings at the end of this chapter is related to the topic and is self explanatory, the rest is up to you. Have some fun, learning is the greatest adventure, the power of thought and intelligence, our most sagacious gifts. We are on track and headed in the right direction.

Quotes by Notable Individuals Related to Mathematics and Phi

" Geometry has two great treasures: One is the Theorem of Pythagorus: the other, the division of a line into extreme and mean ratio. The first we may compare to a precious jewel ; the second we may name a measure of gold." ~ Johannes Kepler

" The most beautiful thing we can experience is the mysterious. Its is the source of all true art and science." ~ Albert Einstein

"When one sees eternity in things that pass away, then one has pure knowledge.
~ BHAGAVAD GITA

"Without mathematics there is no art." ~ Luca Pacioli

"Like God, the Divine Proportion is always similar to itself." ~ Fra Luca Pacioli

"The good, of course, is always beautiful, and the beautiful never lacks proportion."
~ Plato

"Measure what is measurable, and make measurable what is not so."
~ Galileo Galilei

"The Universe cannot be read until we have learned the "language" and have become familiar with the characters in which it is written. It is written in mathematical language, and the letters are triangles, circles and other geometrical figures, without which means it is humanly impossible to comprehend a single word." ~ Galileo Galilei

"The human mind has first to construct forms, independently, before we can find them in things." ~ Albert Einstein

"Nature hides her secrets because of her essential loftiness, but not by means of ruse." ~ Albert Einstein

"Where there is matter, there is geometry." ~ Johannes Kepler

"Mathematics seems to endow one with something like a new sense." ~ C. Darwin

On the Separation of Pi and Phi

Just how and why Pi and Phi got separated in our studies and influences is a mystery but it must have happened, to our great loss, sometime between the decline of Ancient Egypt and during the emergence of Western Culture. The result of this separation goes back to the old cliché, somewhat similar to, when the left hand didn't know what the right hand was doing. This was the basis of ancient science and Egypt had been operating and prospering with the use of these values and principles for untold millennium. The actual workings of the Great Pyramid which is still a mystery to us shows in its geometry how Pi and Phi work together. And though there are some distasteful aspects about it the following insights might help explain the sequence of events that led to the separation of Pi from Phi and the dark days in our history we would rather forget.

It might be noted that the works of Euclid, Archimedes and Pythagorus, to name a few of their era, which have had a tremendous influence on Western World Culture, however, there are no references to them shedding light on the Golden Ratio and carrying on with the usage and importance of the union between Pi and Phi in the learning process. They did work with it in earlier times when there was an assimilation of Greek and Arabic knowledge but somehow it got lost in the works. Ancient Greece and its scholars studied at Alexandria until its libraries were sacked and had learned much from Ancient Egypt as it went into decline and of course they knew about the Golden

Ratio, their earlier magnificent sculptors and public buildings such as the Parthenon and Olympus were based on this wondrous geometric ratio. At a later time in Greek History the above mentioned began to use the function of Pi extensively but Phi, or the Golden Section was left out of the picture and it might appear with all due respect, the work of these highly intelligent individuals of that time was oriented more toward material needs and the race was on to conquer nature, not work with it, but rather to acquire wealth and power in the material plane. It is as if there was a general consensus within the scientific community of that time that said just forget nature we are taking charge now and will get what we want out of life. We might ask if this period marked the emergence of commercialism, capitalism and material greed. No doubt the state of war was a factor in the events of that time and unfortunately that theme is still with us today. Here is seen the early beginnings of rational science which has led to the study of Modern Physics, Chemistry and Mathematics of which the text books of today are plastered with their names with references to the square root of this and that times modern Pi with no reference to Phi. Perhaps it was since the true understandings of the interaction between ancient Pi and Phi were lost the gap between Man and the natural domain has increasingly widened and the eventual result was a breakdown in the quality of human life. Underlying with this development it may be seen that man of the Western Culture had lost his appreciation for the workings of the natural world and his understanding of the Cosmos starting over some 2500 years ago. As a result the industrialized communities of today have become like heartless robots with a set of inconsistent if not false belief systems and plastic values in which material wealth and the acquirement of money talk loudest. The beginnings of our problems go back a long way and there is no simple way to explain the root cause of them but we can develop some viewpoints on it based on history. What we do know is something went horribly wrong in the affairs of Western Culture after Greece fell into decline. The civilized world was then taken over by an autocratic, corrupt and brutal Roman Empire. By the time its demise was on the horizon the downward spiral of Western Culture was well in progress. It had become mired in a material existence and it might be asked, was this the great offense against God and why the Spirit of Life was pinned to the Cross of Matter? In fact the roadways in Rome were lined with crucified victims at the time. Man of Western Culture endured its own self inflicted punishment because the masses had lost the connection with the natural world, the Cosmos and God all in a predictable sequence. It would seem the laws of karma and cause and effect were collectively in progress at this time and a great myth which at least meant well, was created around it. After that evil and decadence in Rome was done away with a turning point in the history of Western Culture took place when another form of money grubby, power hungry corruption took its place, disguised as Religion, a product of brutality and man's in humanity to man. It became known as the Church, ironically, centered in the capital city of Rome no less, which ended up misleading the confused and suffering populace of which the majority of had no education and were illiterate. Now access to God was being dished out for a price in hard cold cash so the rich could get richer and the poor would stay in the rut of poverty. In the beginning no one was allowed to have a bible and it went to great lengths to snuff out Roman paganism to cover its tracks and destroyed any other opposition to its intents and of course any other religion was classified as heathen to the Church leaders. The many inquisitions against those who opposed Church dogma were the cause of death and hardship to whoever spoke out against it and the crusades to recover Jerusalem from Muslim rule was an infringement on another people and their religion and hard feelings over it linger to this day. We might ask whatever happened to the fine works of Plato, Aristotle and Socrates, the founders of the concepts for philosophy and democracy during the earlier Golden Age in Western culture. All this went out the window when the State and Church got into cahoots with each other and connived a plan for control of the masses to bleed them dry. The Royal Families went along with it as long as they got their cut of the action and of course great efforts were made for conversion to swell the membership in order to promote the wealth and power of the two sides which had their own internal power struggle to deal with. It was a sad set of circumstances when these two influences became the law and rule of the land. By that time the system was already headed for trouble and a third world condition prevailed. This Religion has disintegrated into bits and shards ever since its beginnings but it is still a threat and all this could have been avoided had Western Culture Man not strayed from the pathway and truth about the operations of nature and life. By now we have had a long hard look at the overwhelming historical evidence of the injustices that took place during this destructive period which dragged our forefathers through the dark ages. This scourge was inflicted on the human populace in Western culture because falsity and pretentiousness were allowed to be afoot. The lessons have been harsh, yet we are resilient and have learned from it. Religious persecution and class distinction in a stagnating Europe had much to do with the droves of free thinking people who suffered the hardships in getting to the New World and as we know the story of this epic period unfolded with its own set of problems. As the grab for property and material wealth and the state of war based on empire expansion progressed and carried on in the New World it was the guns, steel, disease and whiskey that led to the demise of the peoples who inhabited it. Then the throngs of missionaries came to beef up the membership of the Church with a bible in one hand and a false bill of goods in the other. As time went on rational science and religion in western culture came to lager heads but neither had the answers and to date it has developed into a standoff. In fact, after a survey of the facts are in it will be seen how religion, rational science and our educational system, through blind ignorance and a set of false values have misled or downright betrayed western world culture. We will be asking ourselves why our studies haven't included the Golden Section and its use for over the past 2500 years, and it is pointed out more than once in this presentation, we would be further ahead had this been the case.

Today the concepts of democracy, free speech and freedom of religion prevail in the complexity of modern day multicultural, industrialized civilization and our society is more or less functional but these ideals have their downsides as we know. We might ask why it took over 2500 years for the works of Plato, Aristotle and Socrates to take hold in our culture. We know what got in the way of it. In the meantime the Church has become toothless menace, surely on its last legs, but ever lurking on the sidelines. The royal families have become more or less figureheads of state who have little to do with the political processes and law structure in society, yet they attend social functions for show, all at the expense of the tax payer. No one has been beheaded by them lately, at least. The Crown Corporations in Canada are run by the government with Big Business in the drivers seat. Life in our social systems is less than perfect but it is salvageable and there are many decent things about it. Efforts are continuously being made to improve the situation but we could use some help and this might be closer at hand than expected. We cannot afford to go backward, or wait another hundred years, because if we don't start to change our views and perspectives now there will be nothing left for our children and their offspring. It is time to move ahead and not wallow in the misery of the past or get tied up by our difficulties today. The challenge is to chart out a more correct pathway for civilization and our fellows. The serious question needs to be asked of ourselves, what do we want our culture and that of others in the world to evolve or re-evolve into, a hopeless case of wreckage and missed opportunities or a success employing the use of mind and actions over obstacles? That first obstacle is us and the way we think or the way we don't think.

The symbolism of the spirit of life on the cross of matter is self explanatory. The six faces of the cross transform to a cube, representing matter, in which the spirit is trapped in a material existence. This was done in chapter one when the cube was created from the planar dimensions. The cube can be likened to a crystal of salt, that all important element to life, and in the Holy Bible there is a metaphorical reference to the story of Lot's wife who was turned into a pillar of salt when she disobeyed the Angels and looked back at the wicked, degenerate city of Sodom, which is to say when an individual, society or culture turns the wrong way it becomes immobilized and stagnant. The answer lies within us, we just need to find it and so often is the case, that which is needed is usually close at hand and can be easily solved.

One of the objects of this presentation is to shrug off the shadow of negativity of past developments through timely enlightenment and put it behind us. Once we reconnect with the relationship between Pi and Phi in our hearts and minds and see the bigger picture about the workings of the Living Universe we can make a new start and move forward in the great circle of life and begin to be more agreeable with the social and physical environment. We can add to and polish what is good about it and move on. In other words we need to come to terms with the natural world and the Cosmos again, to find the true house of God that is within us. We are in the 11 th. hour, therefore it is time to make our move. For now, as the seconds tick away, there is time yet to get back to the basics, unravel a number of mysteries and find what has been lost and rekindle the Human Spirit to get it back on track so it can believe in itself. Let the study continue on without any interference from religion or rational science, as re-acquaintances are made with our trusted friends, the universal principles. Again, this is not a mathematics course, nor is it particularly of a religious nature. As it turns out, the Holy Bible is a valuable historical resource which is full of reference on the Golden Section and wisdom when one reads between the lines. The following is more a search for truths in the higher circles of thought from which a number of conclusions can be made, and though it might come as a surprise, many of the answers will be found in the subject we all live with and in nature, the greatest teacher of all.

Square Root of 3, Vesica Pisces and the Golden Section

It is truly unique how the intersection of two circles of equal radii equaling one unit manifests the value for the square root of 3 and this is a significant number in the Golden Section inquiry. Of course another way to create the square root of 3 is the square root of sum of the squares of Phi and 1+Phi as simply, shown in the drawing titled Elements of the Golden Section. Some interesting vesica pisces values are made available using radii such as Phi, 1+Phi and Pi and what is determined is, these irrational number values are all wholesomely interrelated. These are shown in the included drawings. Commit to memory, sq. rt. 3 = 1.732. It is very important.

Impossible Configurations: Gems of Geometry

A source of history tells of three categories of mathematical problems that are impossible to solve using compasses and straight edge. These are:

Squaring the Circle / Doubling the Cube / Angular Trisection

There are three drawings included which deal with this statement. For a warm up please have a look at them and come to your own conclusions.

The Enneagram

Another rare Gem of Geometry is a beautiful symbol known as the Enneagram, origins unknown, signifying the full number 9. Its theme is 3, 6 and 9, a real collection of trinities. Three for the number of sides in a triangle, six for the number of lines which are unique to the symbol and nine for the numbers from one to that full number 9 before the reset to ten takes place. It most probably comes from ancient Egypt as so many other vital and significant pieces of the knowledge puzzle have come from, but if it didn't originate there it is still worthy of investigation. There are busy associations which employ the use of their claim of wisdom linked with the Enneagram. The contention is the Enneagram is a dynamic system of one to nine personality types that empowers an individual to better understand themselves and others. If a person is interested they are given a personality profile test by one of these groups to determine what their category of personality type is then they are coached by someone in the organization who is allegedly knowledgeable in these matters. To say the least the geometry of this symbol is dynamic, though it doesn't directly correlate with the Golden Ratio. An entertaining drawing is included in this section to show a correspondence between the Enneagram and Golden ratio with regards to the square root of 3 which they have in common. Traces of the ancient wisdom and sciences are with us today in the form of astrology and numerology and so forth but we don't fully understand them. Perhaps the Enneagram fits into this category. For now let us examine how the geometry of this intriguing symbol is derived and see where it goes from there. It never hurts to tinker about and toss ideas around as very often great discoveries are made this way.

As the geometry of this symbol is studied it is seen a circle is divided into nine equal arcs, therefore the angles at center which are subtended by these arcs must equal 40 degrees each. At first glance it looks as though an ancient angular trisection theorem was applied because Angle 9 0 3 equals 120 degrees and one third of this equals 40 degrees. This could be the case, however, it becomes apparent that one half of angle 9 3 6 equals 30 degrees since triangle 9 3 8 is equilateral and Angle 0 4 P equals 10 degrees provides a total of 40 degrees. This enables the figure to be drawn because the chord length subtending by the radius equals an angle of 40 degrees. After this stage, based on close inspection we find that angle 93X equals 10 degrees, therefore Angle 03X equals 40 degrees and by simple construction the intersection at Y on the extension of line 3-X is found by using radius point 4 and Dimension 4-0 making line 3-4 parallel to line 0-Y equaling line 3-4. Now Line 4-0 is parallel and equal to Line 3-y. There are three sides in the equilateral triangle and due to the rules of geometry in the case of this outstanding symbol / number theorem the radius of the circle is one unit or 1.0 and their lengths are the square root of 3 as are the values for dimensions 5 / 8, 8 / 2, 7 / 1 and 1 / 4, therefore we have 3 and the square root of 3 and the digits 1 to 9. By dividing each number value around the circle into 1.0 ie. 1.0 over 2.0, over 3.0, over 4.0, through to over 9.0 then going through that process again 2 over 3, over 4 through to over 9.0 and so on the decimal values for the fundamentals of Cosmic Law are readily made available, or so the claim is.

What is of interest in the Enneagram is, when 1.0 is divided by 7.0 the product is .1428571 and when 3.0 is added the total comes to a value of 3.1428571 which is very close to what we know as Pi. In grade school the short cut to Pi was 22 over 7 which also equals the same. The generally accepted modern day value for Pi is 3.1415927... and this is used to compute the circumference of a circle or an arc part of one using a known diameter of a finite dimension. The difference between this and the one determined from the Enneagram is .0012644. Then we have Ancient Pi, the value given by the ratios of the Great Pyramid when half the perimeter of its base is divided by its height. The square root of Phi to three places of decimal x 4 comes close but doesn't quite make it. Another unique and simpler way of deriving its true value is shown just ahead. To be clear on this the function Ancient Pi is used for astronomical computations in the vertical plane. Several fascinating examples of this application will be shown further along. It is interesting to find a close approximation of Pi as follows:

Square Root of 2 = 1.4142135... + Square Root of 3 = 1.7320508... = 3.1462643 which is quite close to the above. This number value multiplied by modern Pi over Ancient Pi gives a quotient of 3.1432014 which is very close to the Enneagram Pi.

The fact is, in practical terms of today Pi is generally applied using no more than three places of decimal for its finite needs if it is used for example, a machined part of some sort, and it is usually applied using only two or three decimal places in survey and engineering works.

Determine Pi by Direct Measure

Following is an amusing hands on way to determine Pi without extensive calculations. First cut out a circle from the bristle board paper that has an exact diameter of 12" or 1.0 foot. Then make a fine pencil dot at the very edge of the circle. On a flat surface like a counter top covered with a strip of news print place a long straight edge on top of it then set the circle on its edge where the point is and make another point on the paper exactly where that is, Point A. Roll the circle along the edge of the straight edge and make a point at the first revolution on the circle and call it Point B, then roll the circle some more along the straight edge and make another point at the two revolutions point , Point C. To find the Pi value, using a steel measuring tape calibrated in feet, tenths and hundredths of a foot record the measurements from A to B and B to C and add them or just measure from A to C. A magnifying glass can be used to estimate to the thousandths of a foot. If your circle is truly one foot in diameter the measurement between A and B will surely be 3.14 and some thousandths. The workings of a measuring wheel are based on this simple principle. A drawing is provided for this exercise.

What is Modern Pi?

The answer lies in the Golden Numbers...

Ancient Pi. A grand revelation based on simplicity itself is at hand when 1.0 is divided by the sq. rt. of 2, to three places of decimal, 1.414 subtracted from the sq. rt. of 3, to three places of decimal, 1.732 = 0.318, reciprocal = 3.144654088 x 2 = 6.289308176 = (2 x Ancient Pi). To be clear on this very important matter, 1.732 – 1.414 = 0.318, reciprocal = 3.144654088 = Ancient Pi. Note: To 9 decimal places only. By no coincidence: Phi = 0.618 - (3.0, divided by 10)= 0.3 = *0.318, reciprocal = same as above.

Proof: Jumping ahead into the next chapters it it will be seen the exact same value is the result of using the correct dimensions of the Great Pyramid when half of its perimeter is divided by its height. As follows: 880 Royal Cubits, divided by 279.84 Royal Cubits = 3.144654088.

Modern Pi...Modern science has missed the boat on this one as well.

Ancient Pi = 3.144654088 x 0.999
= Pi, 3.141509434, reciprocal = *0.318318318318
Note: To 12 decimal places only and easy to remember.

To carry on: reciprocal of 0.999 = 1.001001001001 x Pi
= 3.144654088 = Ancient Pi. Pi x 2 = 6.283 = (2 x Pi).
Use a calculator that can handle that many decimal places.

This is the short and simple of it. It serves no purpose to know Pi to thousands of places of decimal. It is found that the Ancient Pi function is employed to compute astronomical values in the vertical plane while Pi or, modern Pi is used to determine circumference values of circles with known radii in the horizontal plane including those of planets. Additional input on this fascinating topic is made available further along..

We see a function of Pi in the circle of the moon and sun and Earth from space, in the arc of a rainbow and eagle's wings, in the pupil of the eye and when a dewdrop falls into a still pond the function of Pi emerges in the form of perfectly circular ripples. Take note of the sand dollar or a snail shell on the beach, living creatures that exist and function in terms of Pi and Phi. These inseparable irrational numbers, Pi as well as Phi belong to the infinite Universe, one cannot be fully appreciated without the other, they function together harmoniously and we are amazed to discover the natural world has a total understanding of mathematics. For example a sea mollusk such as the nautilus creates a shell which has an appreciation for those unique ratios of Pi and Phi as well as the right angle. In our domain the rational is determined, paradoxically, by employing the irrational, or in another sense as suggested, the finite is expressed in terms of the infinite.

What is Nature?

Nature is the physical Universe, the greatest teacher of all, a topic of study with hidden lessons that can show us the way. What we need to do is differentiate between human nature and our present day science and the science of nature on its terms.

Nature = the workings of the Physical Universe = God's Presence

The Great Pyramid…that mysterious wonder left behind from ancient times

This topic fits well with the presentation because the Golden Section and its proportions were used in the design for the Great Pyramid of Giza, the obelisks, the Temple of Luxor, the Temple of Isis, the Shrine of Sekhment and the list goes on in ancient Egypt to name just some of these types of constructions. The design details of the Great Pyramid structure can be likened to a bouquet of Golden Section Geometry, a Geometric Expression of Nature, or the Geometry of the Physical Universe. It provides us with the relationship between ancient Pi and Phi. Its dimensions provide the distance from earth to the sun, the speed of light and its relative proportions to earth and sun etc. and we wonder how this could be when modern science tells us this knowledge was unknown until the 20 th. Century. The truth on these matters will be resolved in chapters 3 and 4. One estimate tells us the Great Pyramid was constructed around 5,000 years ago, another reference claims it was built 12,700 years previously, around the time the Sphinx was constructed, but these are but guesses at this time. We are informed its base covers an area of 13 acres and its present height is some 481 ft. It consists of over two million granite and limestone blocks which interlock with great precision. It remains a mystery how these blocks were cut to such high tolerances in the first place and to this day we could not duplicate the construction of such a structure. First we don't fully understand how it works or what it means and our most powerful machines could not lift those enormous granite blocks into place at those heights. There are possibly other simpler ways this could have been done. One way would be to float the stones into place using the buoyancy factor, though this seems unlikely. Enormous coffer dams would have had to be constructed around the site in order to do this. One might think water was present when we learn the level at its base is within one quarter of an inch. It was either that or advanced survey equipment was available at the time. Could the blocks have been levitated into position with a minimal degree of manpower using hydrogen or helium filled balloons? There is also the possibility the granite was somehow reduced to crush form and mixed with water and a limestone additive to make a slurry which was poured into formwork like we do for our concrete towers. To this day scientists cannot figure out how the ancient Egyptians made their mortar. Today concrete slurry is either pumped into the form work or it can be transported in big steel totes into position by large cranes, elevators, conveyor belts or air machines such as helicopters or helium filled balloons. When you come to think of it, how else could those air shafts in the Great Pyramid have been aligned so accurately and how is it that this structure was built exactly at the epicenter of the globe in perfect alignment with the cardinal directions? Since the ancient Egyptians, or their predecessors were very knowledgeable in astronomy they were no doubt equally sophisticated and skilled at all other levels. Electrolysis, even electricity and granite concrete works and air transportation would have been within their capabilities. We may never know what the human sized red granite crystalline sarcophagus in the Kings Chamber was used for or how it worked but it must have been very important to them, the Great Pyramid was much more than an ornament one might think. We don't understand why the pyramids at Giza were arranged with an alignment to the constellation of Orion or how the River Nile was associated by them with the Orion Arm in the Milky Way. This pyramid location in ancient time would have been the equivalent of Greenwich as a reference point for navigation on the globe one would think. There is nothing new under the sun. We are left wondering if the period of the Great Pyramid construction era had something to do with Stone Henge and the many other sacred sites throughout Europe, the British Isles and in other parts of the world. Apparently the Great Pyramid contains coded messages in its dimensions and numerological makeup that prophesize past and future events in the world. Many of the interpretations are quite bizarre and most are only guesses because no one has the right information on it to really understand how this works, therefore it will be best to concentrate on what the Great Pyramid means to us today aside from it being a reminder of a glorious time in the history of the world. What it tells us is the ratios Phi and Pi belong together and understanding the workings of the Golden Section is a very important part of the learning process. The contents of "Living with Geometry " will be an in depth study of this all but forgotten subject and the promise is made, there will be no disappointments when it is seen where it goes. A workable drawing titled, Genesis of Design for the Great Pyramid included at the end of this chapter it offered and it will be seen how Phi and Ancient Pi are part of its workings. The construction drawing for a pyramid at the end of chapter 1 is Phi based and this provides an opportunity for some hands on work with such a form which will become very important as the story unfolds.

An astounding revelation on the global location of the Great Pyramid as it relates to the ancient antipodal sites of Angkor Vihear on the west edge of Asia and the Nazca lines in Peru show a Golden Section relationship between these three sites. There is no coincidence in this yet we are mystified by it and the implications are mind boggling. Obviously the ancient sciences had a handle on the earth's dimensions and a profound knowledge of astronomy and world geography. We wonder how this knowledge got lost, as there was a long period in the dark age history of Western Culture when it was believed the earth was flat and that it was the center of the Universe. The distance between Angkor Vihear and Nazca is one half the circumference of the earth and from this computations and deductions can be made as to where other sacred sites are along these lines. This relates to the Ley line theory discovered in the British Isles earlier in the 20 th. century. It is amazing that the lines and arcs touch every continent in the world. Please see the graphic titled Giza Phi Connection which shows the Golden Ratio relationship between these sites.

The problem with us in determining how these great feats were accomplished, how the knowledge was acquired and the reasoning behind it is, we have made the assumption the people of that time were less developed than we are. In fact 300 miles east of Cairo there are carvings of advanced air machines very similar to ours on a large stone wall 300 miles east of Cairo. We might consider as our technology advances that it might get closer to where these people were some time in the future and the only way we are going to do this is by following their footsteps and the example of their science. If the Greek and Roman Empires got their information from Old Egypt and passed it on to the Western World, we can only speculate where the ancient peoples of Egypt got their information from. There is talk, rumor and conjecture of an Atlantis and a Lemuria when there might have been an Eastern World and a Western World in ancient times but their existence hasn't been fully proven as yet and may never be, therefore the truth of the matter is not known. The most popular theory on where Atlantis was is the Island of Santini near Crete where the Minoan Culture thrived until a fierce volcano erupted with a force 100 times greater than that of Mount St. Helen's. There is some very interesting reading in Plato's Timaeus and Critias and what Egyptian priests told Solon about it, yet the actual reality of Atlantis remains more mythical than fact and perhaps a pipe dream for many. However, it does no harm to keep an open mind on any possibility when it comes to human history. According to fragments of archaeological findings with regards to Atlantis, it might have been a vast empire from the present day western shores of Europe extending to the Caribbean Islands until an enormous cataclysm made it sink deep below the ocean level. Its survivors could only have ended up on land masses east and west of it. This might help to explain the origins of human habitation in various areas of the globe

Human habitation on Earth is one mystery upon another. Originally, it is conceived the Earths surface was one land mass called Pangae then parts of it broke off and drifted into the positions known as the seven continents of present day. After many millions of years while major cataclysms and climate variations of untold magnitude and influence took place any remnants of which might have been glorious civilizations of man are lost forever due to the grinding, crunching and breakdown of the land masses. The popular theory of man's origins today is " Out of Africa ", meaning that is where the first humans appeared and just how this happened is unknown by today's scientists. Geologists can only make intelligent guesses on what has transpired over this vast time period and archeologists can only scratch so deep. The evidence is long buried and inaccessible in the oceans, beneath the sands of time such as in the Sahara Desert and other areas like this where man had once inhabited these areas. There are remnants of long lost civilizations hidden away in the jungles on many of the continents in the world as well. The scholars can go only so far back to the limits of available recorded history. Fossilized sea shells are found at the summit of Mount Everest which had been at one time at the sea bottom and this is only recent history in geological terms on our planet. We might ask if human habitation on earth was probable or not during this time period. We might ask if the surviving members of an advanced civilization that was effected by a cataclysm or drought or worse might not have sought refuge in caves and the like with nothing but the clothes on their backs and went through a degenerative period until their future generations could pick up the pieces and begin new civilizations. There are stone age paintings of now extinct animals found in caves in Europe that require a high degree of intelligence and skill. A question might be, what are the true beginnings and history of peoples in remote areas of the globe? We make the assumption they were always primitive but their true origins are unknown to us. A fossilized human skeleton is found somewhere in Africa and it is believed to be representative of people of that time but what they don't know for sure is whether or not that person suffered down syndrome. Human travel around the globe since ancient times is in evidence. For example, an archeological dig in China unearthed a European people settlement carbon dated 3.000 B.C. Europeans were in North America some 20,000 years ago, and this has been proven by both DNA analysis and archaeology digs on the east side of North America, and this is only part of the story. How can the advanced civilizations in Mezzo American and South America be explained?

When I was quite young I might have got a glance at a pyramid in some book and one of my primary grade school teachers told me when I asked about the pyramids in Egypt, she said, oh there were primitive brown skinned people there who stacked piles of stones to look at the shadows for some reason, then she said I think they were trying to figure out how a sun dial worked or something like that. I wasn't mature enough at the time to understand anything about it and as it turned out this didn't do much for my level of enlightenment on the subject over the next number of years. It was later when I decided to find out about it on my own I got somewhere with the inquiry and discovered there was a wealth of information available and there must have been some incredible things going on in our ancient past. There are ancient pyramids and megalithic structures all over the world. How do we know these were not all connected somehow in some highly sophisticate way and for what reason? World travel and communications services might be a close guess, we just don't know, we haven't been able to put the pieces of the puzzle together yet.

Take note, one the most abundant minerals in the Earth's continental crust is quartz, silicone dioxide. The slope on a quartz crystal is in the Phi, or Golden Section proportion and by no coincidence the slope of the Great Pyramid is the same. The interior granite block core of this megalithic structure consist of quartz and feldspar and was encased by limestone blocks. In fact piezoelectricity is generated when quartz is under pressure, hence, the actual meaning of pyramid is fire in the middle. This tells us the technology of that day was based on what we use for our computers and communication systems today. The faces of the Great Pyramid are concave to a certain degree. Is it possible it made use of energies such as gravity, electromagnetism and piezoelectricity and solar energy no doubt to become a transmitting and receiving crystal for global and interstellar communications with other locations we don't know about or understand today? Our problem is we don't accept the notion that there was a human culture thousands of years before our time who were far more advanced than we are today. The slope on the Great Pyramid and quartz crystal is known as the angle of life because as stated it is found in the proportions of all living things, including the prehistoric fossilized bones of the dinosaurs. What does this mean? For starters it means all living things had or have something in common with life on this planet and the workings of the Universe. In other words all living things are an interrelated part of the living cosmos. An interesting question might be, why does a mineral such as quartz have the life angle when we don't consider a mineral to be a living thing yet sages, mystics and those acquainted with the healing arts have been using quartz crystals and spheres for thousands of years. There was a short lived craze going on in the 80's and 90's about quartz crystals then it went quiet. It claimed a human could communicate with them or vice / versa and they had ancient memories stored in them. Included is a drawing which shows how the elements of the Golden Ratio were employed to design the Great Pyramid. Now that you have constructed models of the basic platonic solids try a Phi based pyramid. This is also a noteworthy platonic solid with the unique Phi ratio form. A construction drawing is made available to get you started.

The Great Pyramid of Giza is the only remaining survivor of the Seven Wonders of the Ancient World. The others are: the Hanging Gardens of Babylon, Temple of Artemis, Statue of Zeus at Olympia. Mausoleum of Maussollos, Colossus of Rhodes and the Light House of Alexandria. What all these projects have in common is geometry in terms of design, surveying, architectural and engineering skills on extraordinary levels, yet the Great Pyramid comes out on top because it's actual workings and use remains a mystery to this day.

UFO's ~ Crop Circles ~ Alien and Ghost Sightings ~ Spontaneous Combustion

Since quartz has come into the discussion this brief insert is justified and timely because it might help to resolve certain mysteries which have to date pervaded our thoughts for quite a long while. It is for some reason an inherent trait in human nature to seek bizarre and irrational explanations for certain conditions and situations that appear to exist without proper analysis of their probable causes by natural phenomenon. There is a distinct possibility quartz and certain other minerals in the ground have a lot to do with the above. When this mineral is under pressure such as in the case of the Great Pyramid or in the Earth which has a radius or depth of nearly 4000 miles quartz and certain other minerals such as rock salt emit piezoelectricity which oscillates at a constant rate like the backward and forward swing of a pendulum. Piezoelectricity is an energy having neither mass or substance though it has a voltage and a physical force. There are many applications for piezoelectricity in today's technology, therefore we know something about it. There are large beds of quartz crystals and most probably very large individual ones deep below the earth's surface and at unpredictable times under pressure the energy is released into the atmosphere especially near fault lines in the form of fiery spheres complete with colors. Hence these are misinterpreted as UFO's. In all cases their descriptions suit the observation of a visible energy phenomenon. Science fiction loves this kind of thing but it might be considered, no one from elsewhere in the Universe has come to visit us. Eyewitness accounts in areas where crop circles have appeared claim to have seen fiery balls of light flitting about in the atmosphere prior to them appearing. It can be assumed this energy rotates at a very fast pace, very possibly that of the speed of light. It is understandable how crop circles could be formed when this energy is close to the ground in a hay field. The pranksters could never do as good a job on these. There are documented reports about lights in the sky and crop circles throughout history. It is as if there is a form of intelligence in the mineral kingdom attached to these puzzling observations we don't fully comprehend, yet we know the geometry of a quartz crystal is an exact hexagonal prism which can refract light and terminates in a six sided pyramid, the slope of which is that of the life angle in the Phi proportion, that interconnecting constant which is found in all other living things. Therefore it is deducted, perhaps contrary to regular belief, that the animal, vegetable and mineral domains on the planet are all alive and well in terms of a relationship we don't quite understand. This piezoelectric energy phenomenon could also explain alien and ghost sightings and spontaneous combustion in cases where people have been burned or caught on fire for no apparent reason and it could be the cause of forest and grass fires in uninhabited wilderness areas. Furthermore, take for instance ghost or spirit appearances in old stone castles and the like. Piezoelectricity would be generated by the pressure created by one stone stacked on the other and there might be anomalous electro magnetic activity present at the same time which would create physical energy apparitions and so forth. These are considerations on the matter at least. Old castles and sacred sites in the British Isles are well known for ghost sightings and this is perhaps understandable because there are many of these in the region which are linked with earth energies but we do not fully understand how this works.

Back to the Main Topic

While Golden Section geometry was applied during the Renaissance in 14 th. to 17 th. century Europe a great period of vigorous artistic and intellectual activity ensued. The works of Fibonacci had resurfaced and the creations of Michelangelo, Leonardo Davinci and many others were in progress but it became heresy to the Church because any relationship with the ancient world and its knowledge was condemned by them. Then the study was forced underground and became obscure except to those who worked with it in private. It was lost to the public and to our great regret it hasn't been taught in the public school system in modern times. Why wouldn't anyone want to see their child in their formative years drawing the Golden Spiral such as the nautilus shell and gaining an insight into the workings of the natural world and life? Because of this we have been deprived of our heritage from ancient times through Greece and Rome and were disconnected from the natural world and the truth about the workings of the cosmos and God and this has been going on for too long now. Around that time a clergy member was publicly burned at the stake for proclaiming the stars had planets revolving around them and Galileo was under house arrest until his death for determining the truth about the workings of our solar system. Here, truly was a period of degeneracy unparalleled in the history of Western World affairs when raw sewage was flowing down the streets and disease was decimating the human population. Social conditions in those days were no different then when the Third Reich establishment was in power in Nazi Germany when neighbors snitched on each other out of fear. When the Church had an iron grip on the populace people were dragged out of their homes to have their tongues carved out, disemboweled, burned at the stake or tortured on the rack for blasphemy or for conversion. As the technology of the industrial revolution advanced in leaps and bounds until present day, as discussed, rational science had very little or no reference to the Golden Ratio to work with and this is very unfortunate for us as we will discover. Fortunately access to this topic is in abundance on the Net today and many dedicated souls are sharing their research to help us get a handle on it. After working out the details on the drawing titled, Elements of the Golden Section / Phi Ratio (a) I saw something of great importance and decided to get in on the act too.

As stated earlier, what is seen engages the mental faculties, because a picture says a thousand words. It was decided to show a sequence of informative drawings at the end of chapter 2 in order to get warmed up to the concept of the Golden Ratio. The images tell a story and serve to answer a few questions at least. The last sheet is my one page business blurb that explains what I have been doing for a living using geometry, while engaged in the professions of survey / engineering for the past 40 years or so, while conducting a sideline research into the mysteries of the Ancient Sciences. After wading through heaps of false and misleading information on the topic for a lengthy period evidence of a grand science emerged when a true assessment of the Great Pyramid, the era when it was built, how it was constructed and by whom and why came to light. This approach will be a fast-track route to getting into the subject in a big way in the following chapter and finding those elusive answers to a number of very important unanswered questions that have boggled the brains of our best scientists for past ages. In the early stages of the inquiry I elected to construct a Phi based pyramid oriented to magnetic north with a chair at its center, one third its height. The object was to use the enclosure as a green house once it was enveloped in clear plastic sheeting and a sitting chamber where I could meditate in my own space over the challenging undertaking that was before me. To make a long story short, It was as if I was connecting with the universal consciousness within the structure, the same proportions as the Great Pyramid, when a premonition cut in and I knew that I would be getting into the topic in the form of a book with substantial meaning. The symbol for infinity is one of great intrigue and its definition plays a huge role in the storyline. It can be seen how the Golden Ratio helps to explain it.

ELEMENTS OF THE GOLDEN SECTION / Phi RATIO (a)

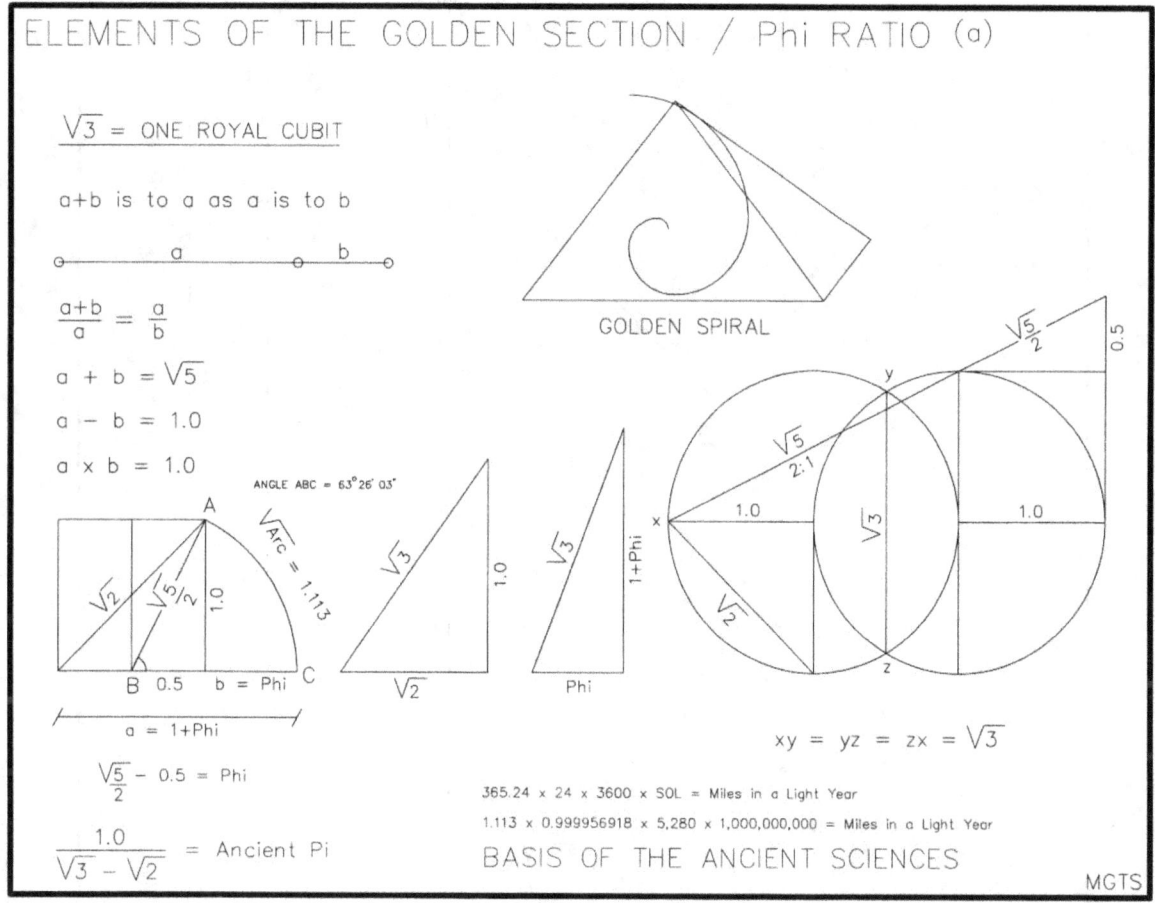

$\underline{\sqrt{3}}$ = ONE ROYAL CUBIT

a+b is to a as a is to b

$$\frac{a+b}{a} = \frac{a}{b}$$

$a + b = \sqrt{5}$

$a - b = 1.0$

$a \times b = 1.0$

ANGLE ABC = 63° 26′ 03″

GOLDEN SPIRAL

$\sqrt{\text{Arc}} = 1.113$

$xy = yz = zx = \sqrt{3}$

$\dfrac{\sqrt{5}}{2} - 0.5 = Phi$

$\dfrac{1.0}{\sqrt{3} - \sqrt{2}}$ = Ancient Pi

365.24 × 24 × 3600 × SOL = Miles in a Light Year

1.113 × 0.999956918 × 5,280 × 1,000,000,000 = Miles in a Light Year

BASIS OF THE ANCIENT SCIENCES

MGTS

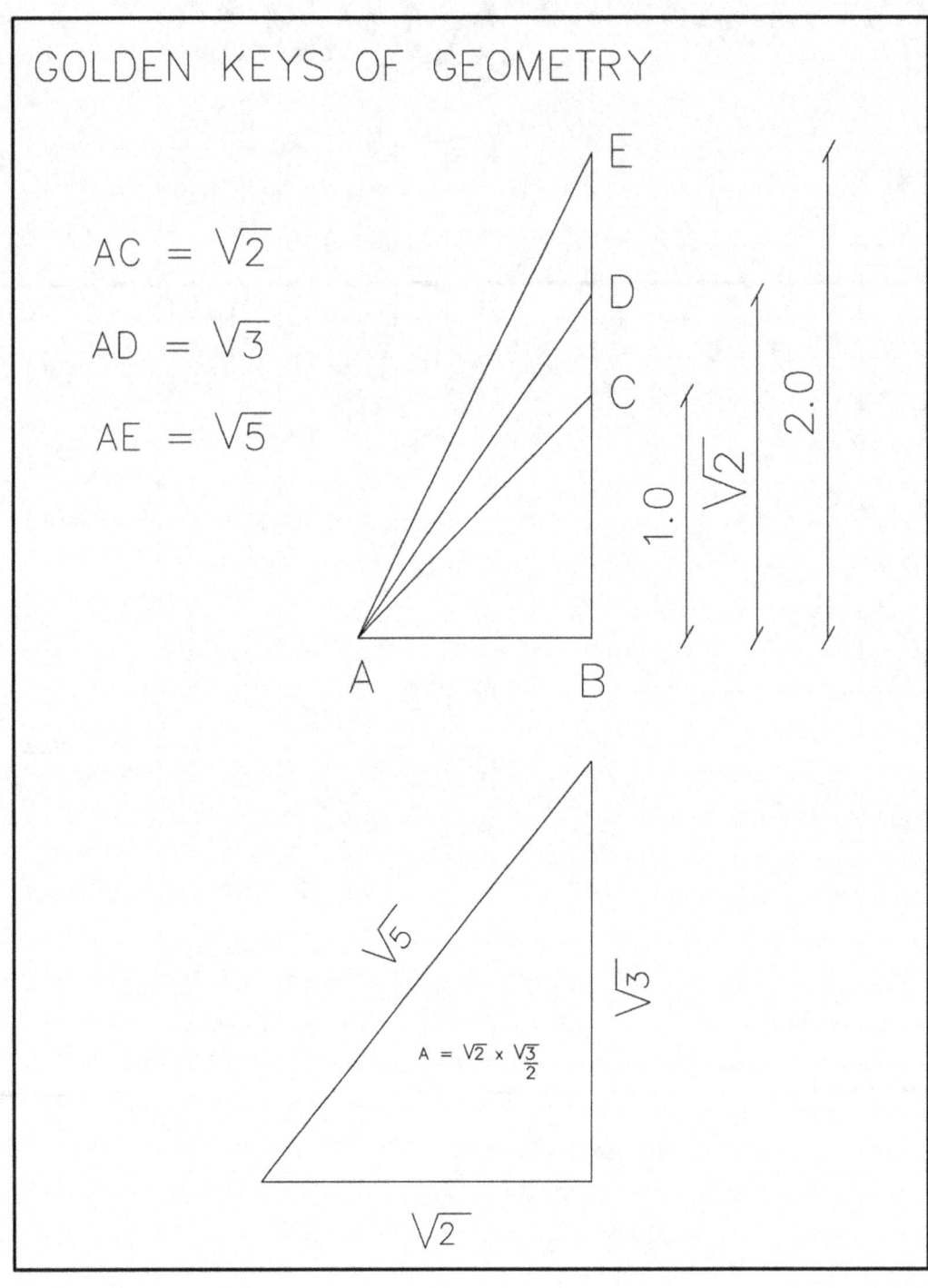

GOLDEN KEYS OF GEOMETRY

$AC = \sqrt{2}$

$AD = \sqrt{3}$

$AE = \sqrt{5}$

E

D

C

A

B

1.0

$\sqrt{2}$

2.0

$\sqrt{5}$

$\sqrt{3}$

$A = \sqrt{2} \times \dfrac{\sqrt{3}}{2}$

$\sqrt{2}$

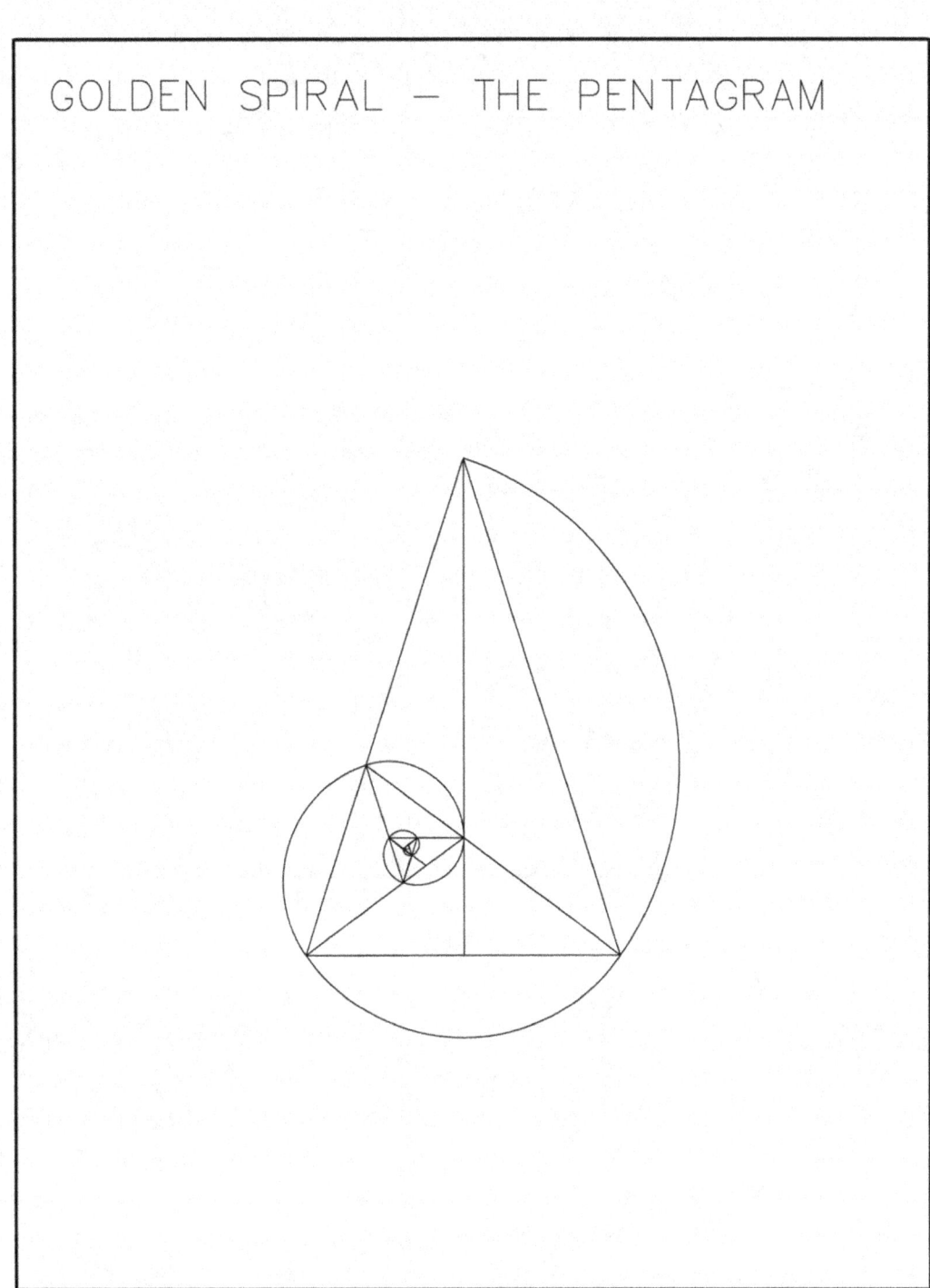

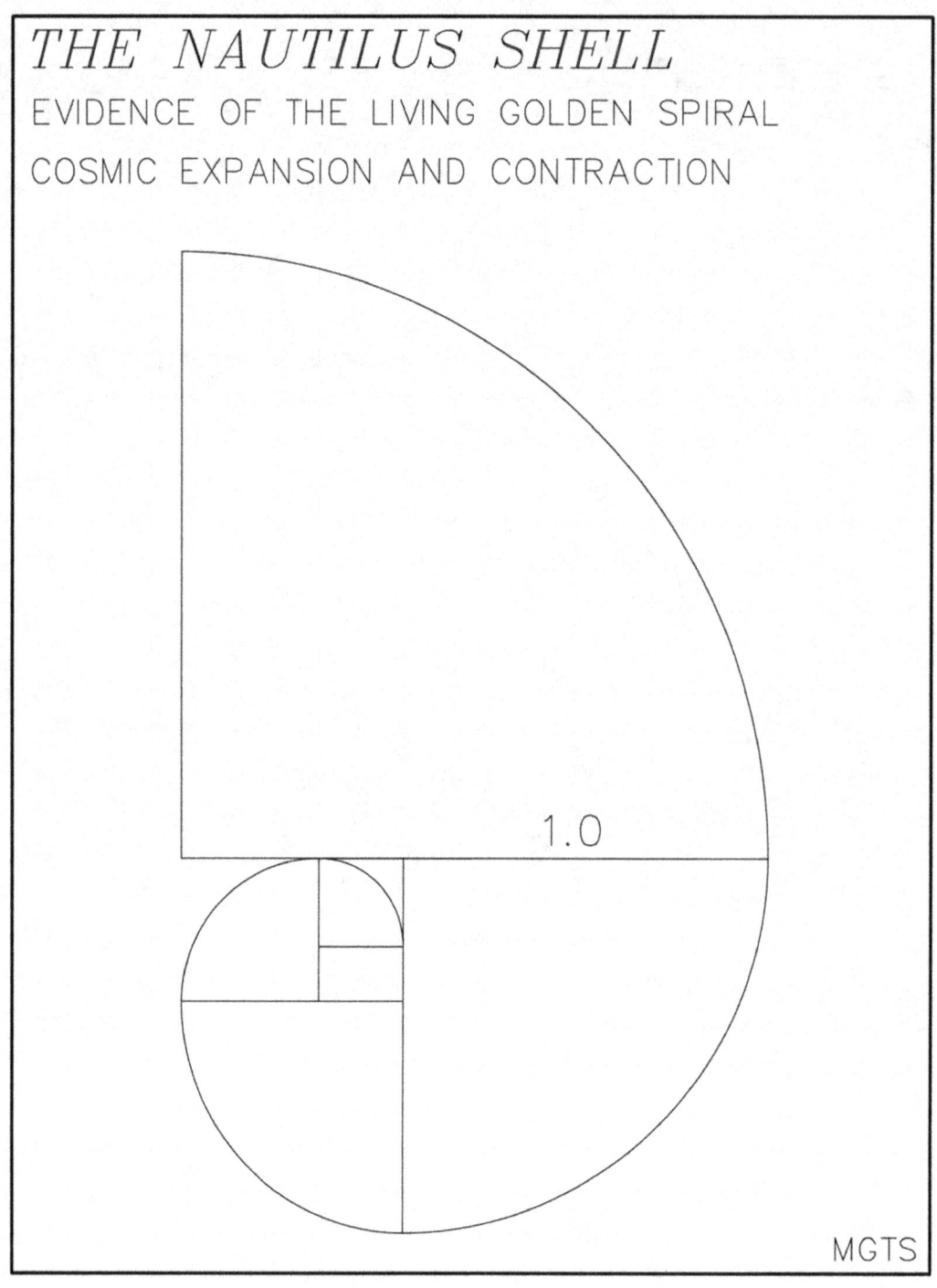

THE NAUTILUS SHELL
EVIDENCE OF THE LIVING GOLDEN SPIRAL
COSMIC EXPANSION AND CONTRACTION

1.0

MGTS

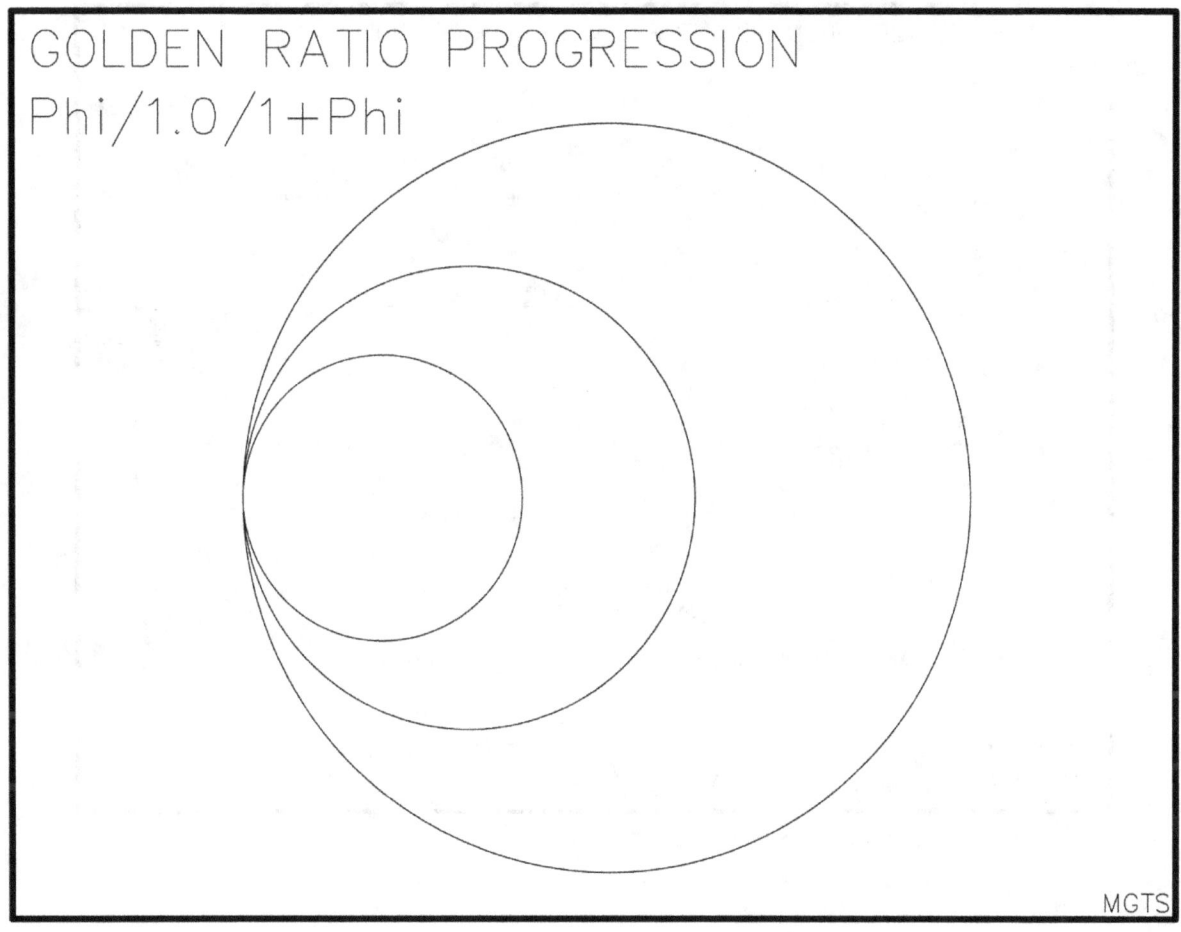

GOLDEN RATIO PROGRESSION
Phi/1.0/1+Phi

MGTS

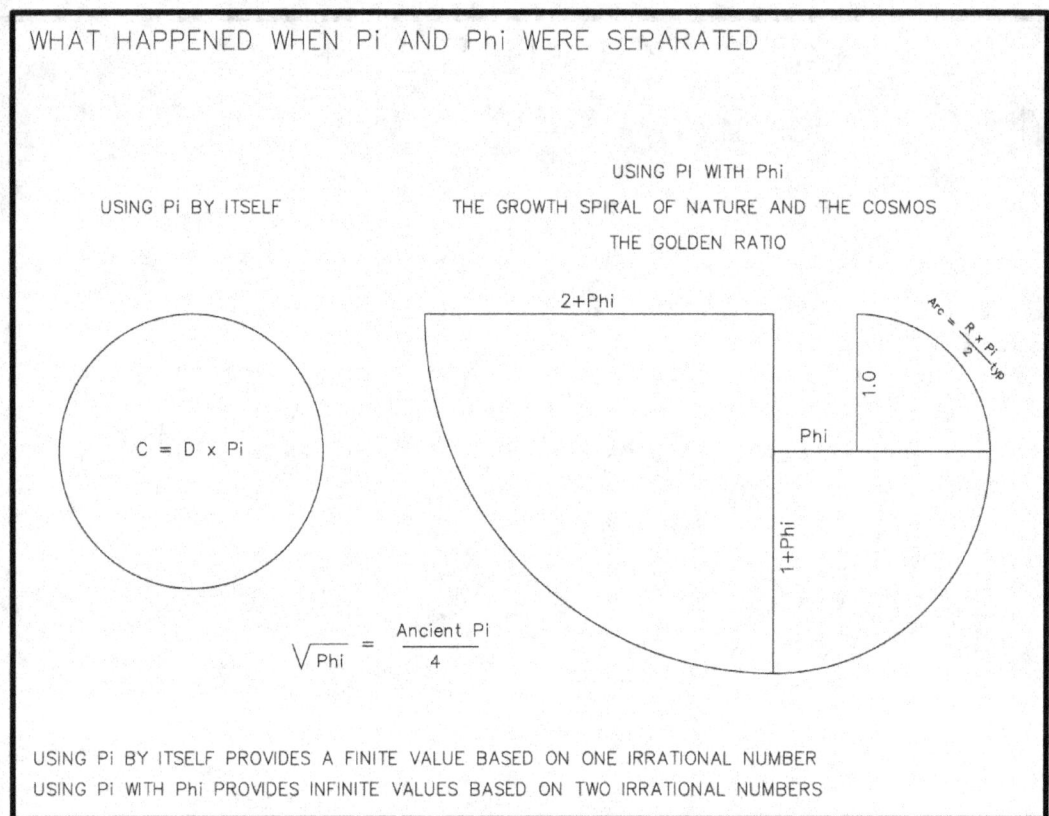

WHAT HAPPENED WHEN Pi AND Phi WERE SEPARATED

USING PI BY ITSELF

USING PI WITH Phi
THE GROWTH SPIRAL OF NATURE AND THE COSMOS
THE GOLDEN RATIO

2+Phi

C = D x Pi

1.0

Phi

Arc = R x Pi typ / 2

1+Phi

$$\sqrt{Phi} = \frac{Ancient\ Pi}{4}$$

USING Pi BY ITSELF PROVIDES A FINITE VALUE BASED ON ONE IRRATIONAL NUMBER
USING Pi WITH Phi PROVIDES INFINITE VALUES BASED ON TWO IRRATIONAL NUMBERS

PHI AND A CIRCLE / D = 1.0

SYMMETRY IN THE GOLDEN RATIO

FOR EVERY FORCE OR VIBRATION IN THE UNIVERSE,
THERE IS AN EQUAL AND OPPOSITE ONE.

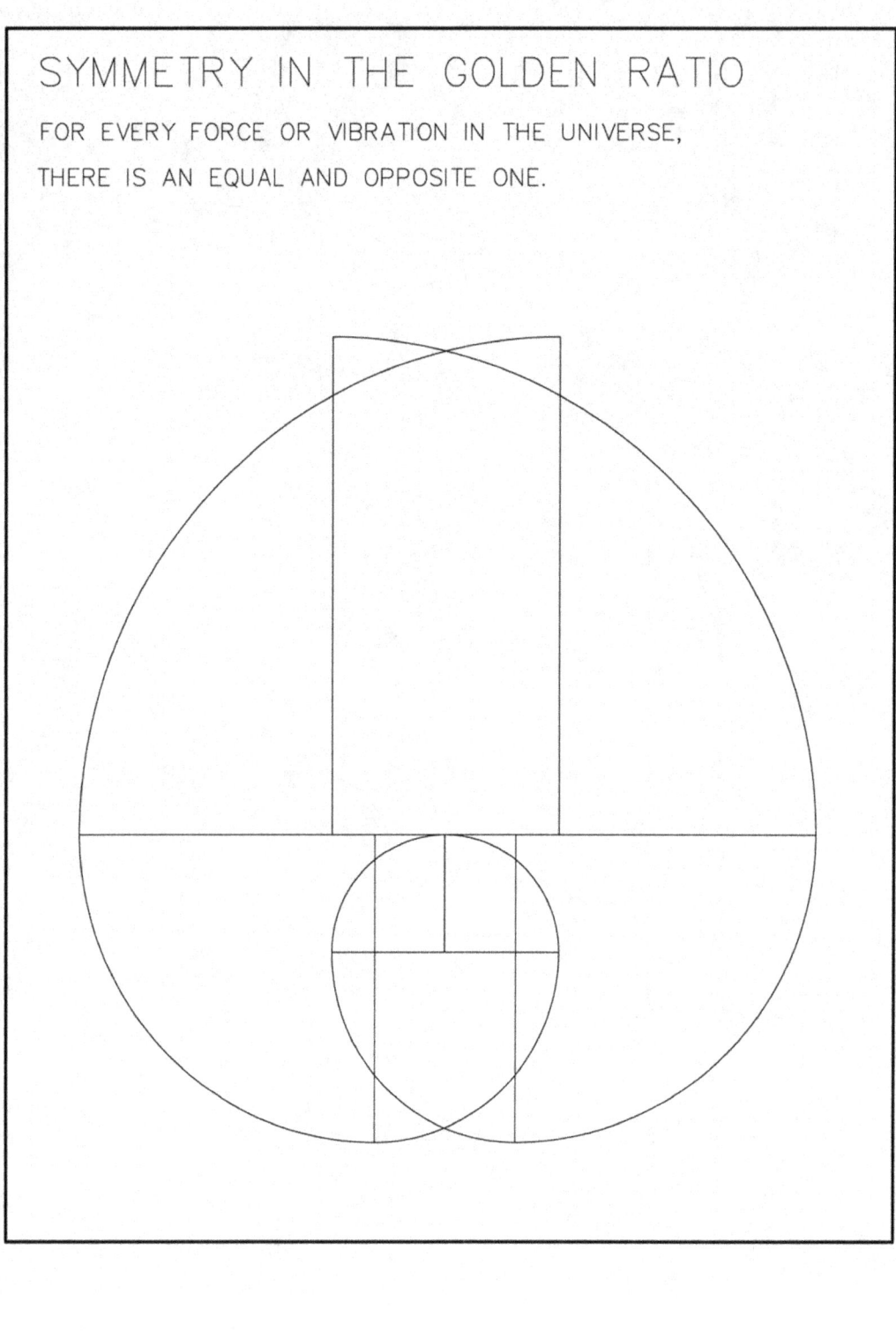

PROPORTIONS OF LIFE:

IN THE KIDNEY...FEATURING THE GOLDEN NUMBERS

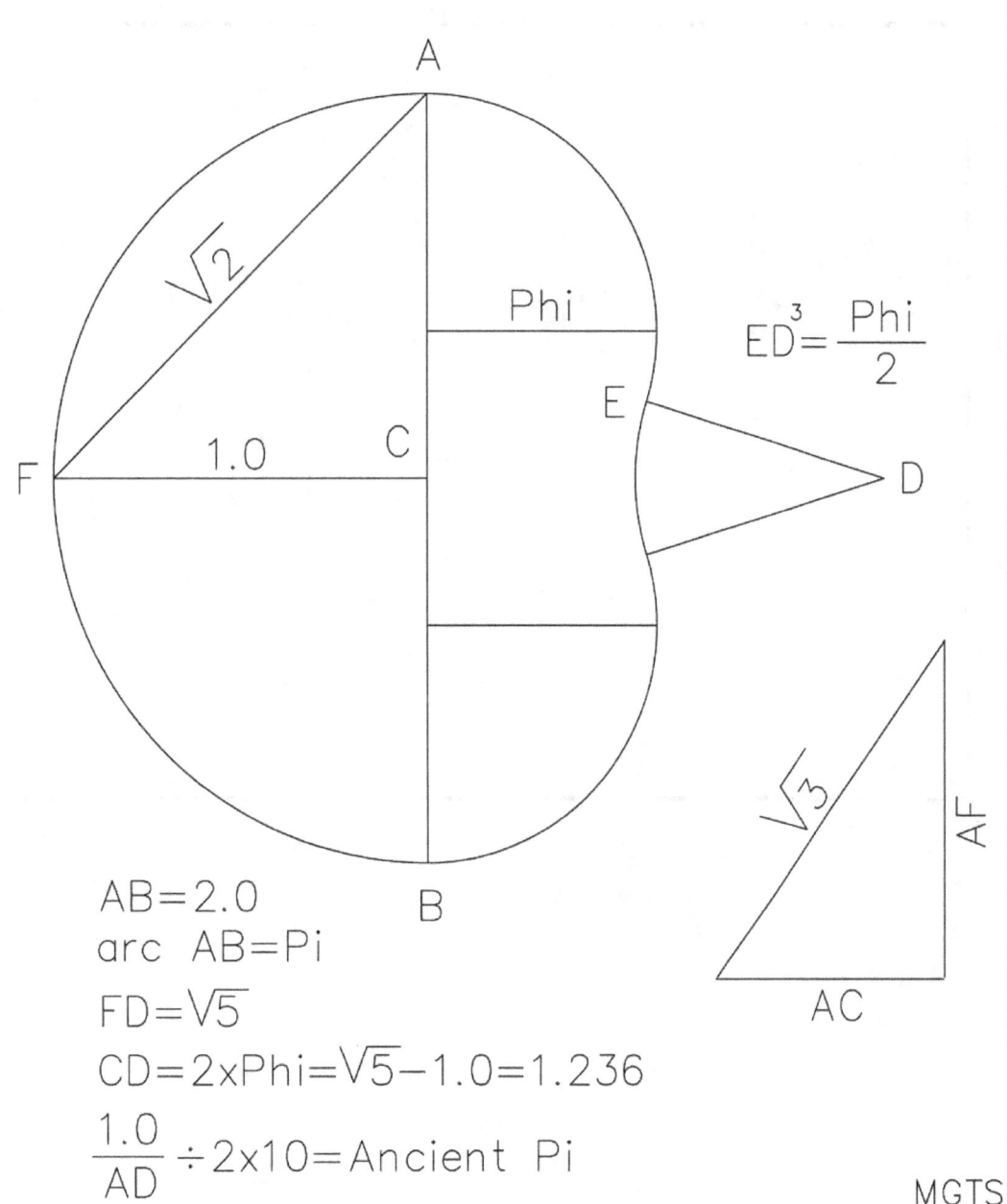

$$ED^3 = \frac{Phi}{2}$$

$$AB = 2.0$$
$$arc\ AB = Pi$$
$$FD = \sqrt{5}$$
$$CD = 2 \times Phi = \sqrt{5} - 1.0 = 1.236$$
$$\frac{1.0}{AD} \div 2 \times 10 = Ancient\ Pi$$

MGTS

WINGS

ANGEL'S WINGS

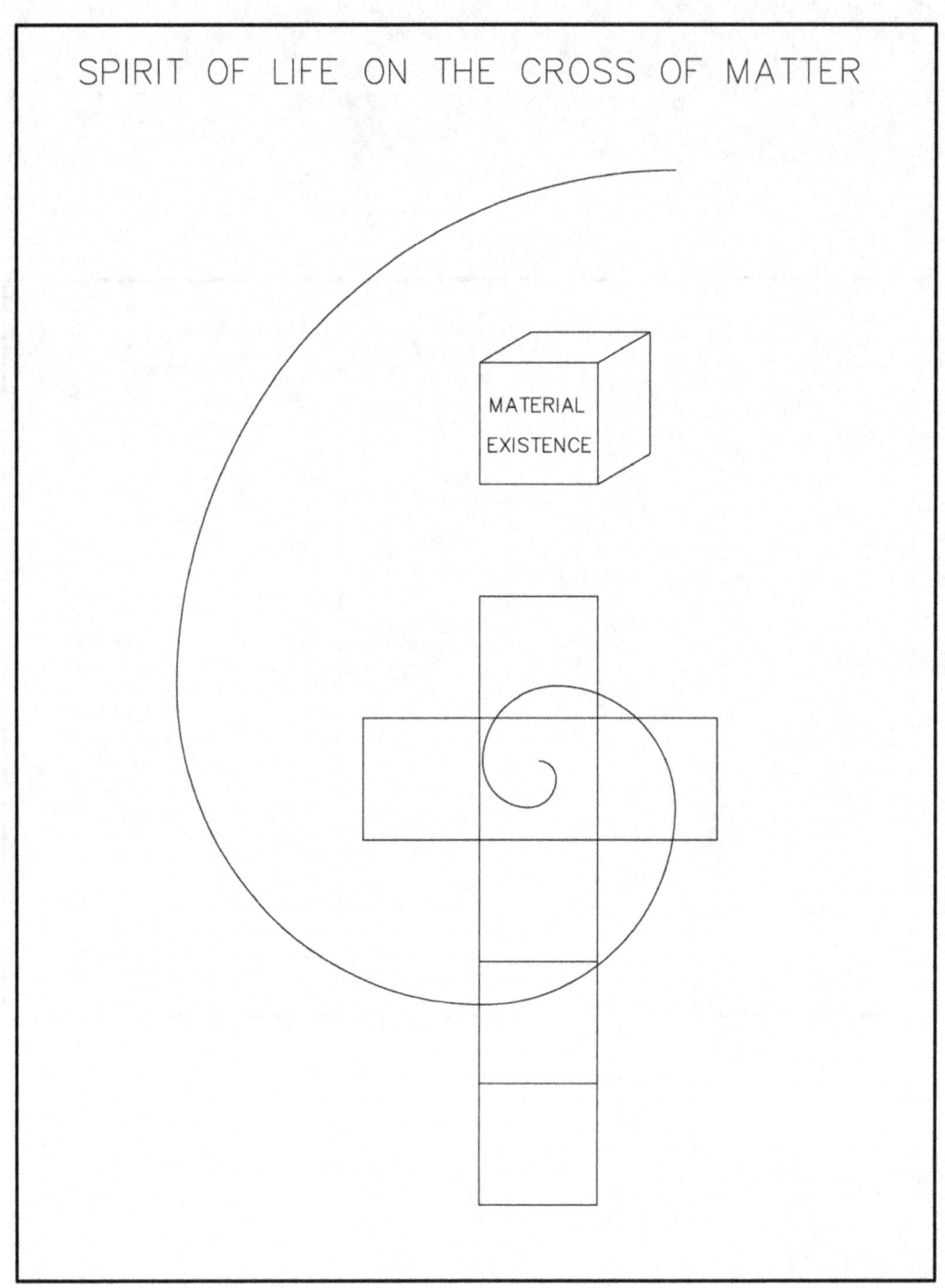

PYTHAGOREAN THEOREM

In any right-angled triangle, the square which is described
upon the side subtending the right angle, is equal to the
squares described upon the sides which contain the right
angle.

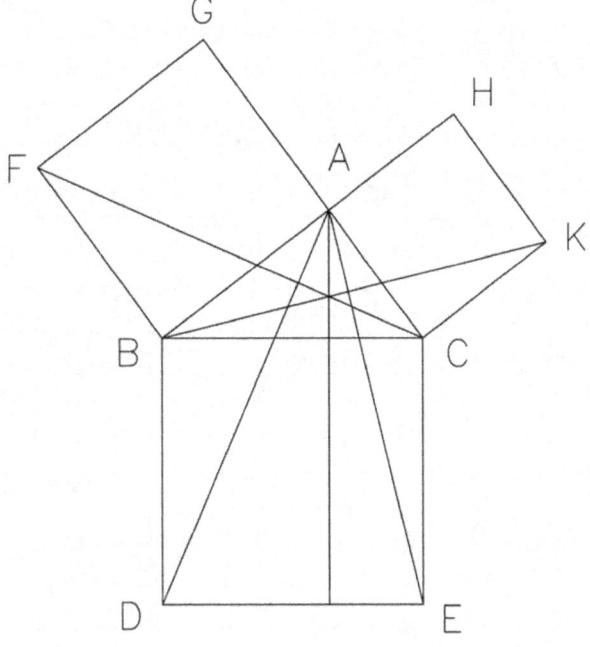

The hypotenuse side of a right angled triangle equals
the square root of the sum of the other two sides
squared.

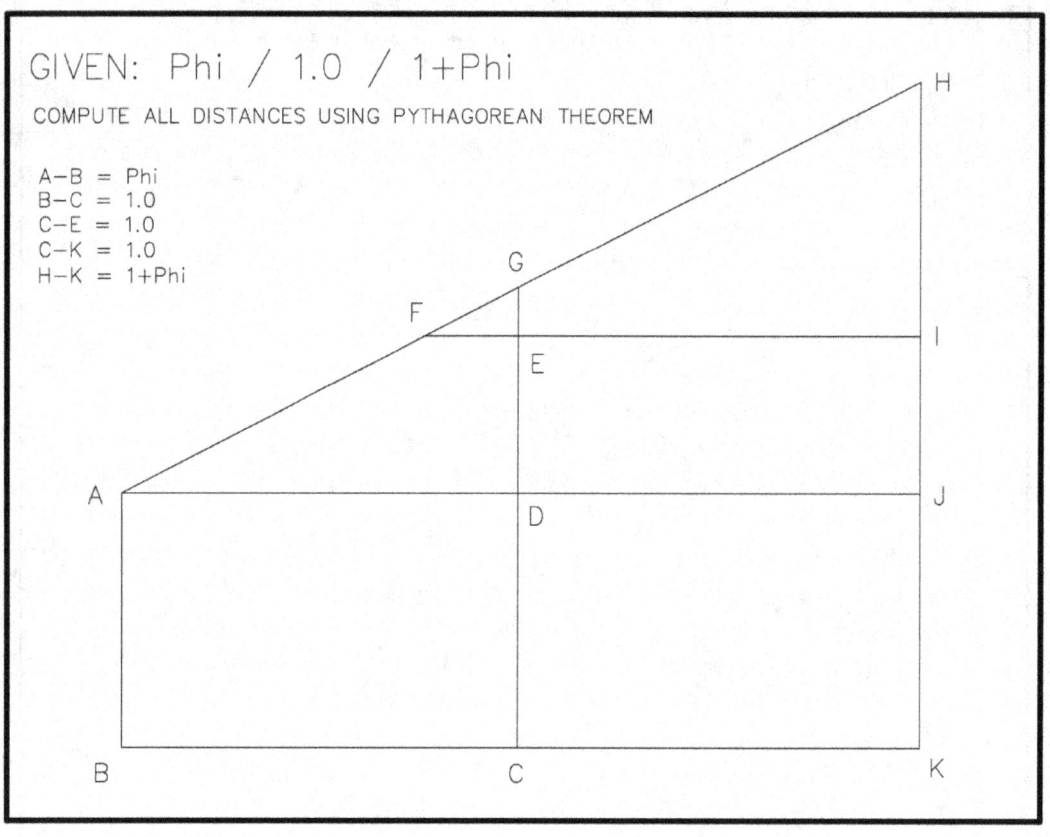

GIVEN: Phi / 1.0 / 1+Phi

COMPUTE ALL DISTANCES USING PYTHAGOREAN THEOREM

A–B = Phi
B–C = 1.0
C–E = 1.0
C–K = 1.0
H–K = 1+Phi

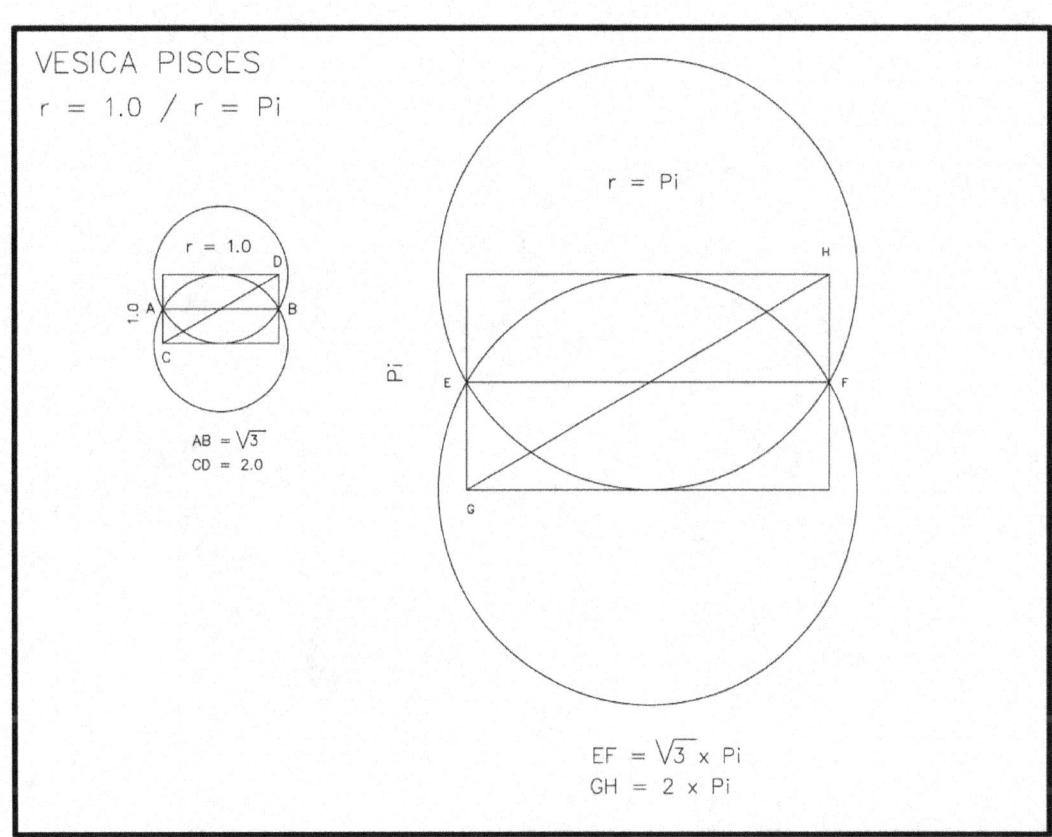

VESICA PISCES
r = 1.0 / r = Pi

r = 1.0

r = Pi

D

1.0 A   B

C

Pi

E   F

H

G

AB = √3
CD = 2.0

EF = √3 x Pi
GH = 2 x Pi

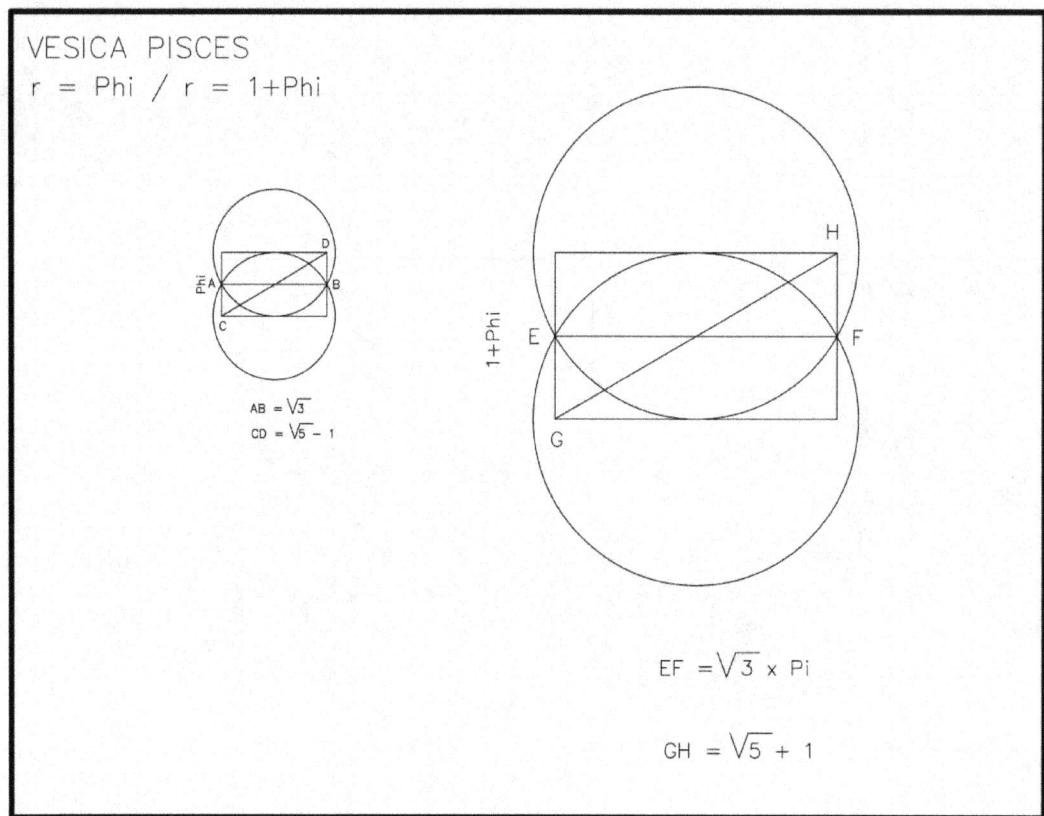

VESICA PISCES
r = Phi / r = 1+Phi

AB = $\sqrt{3}$
CD = $\sqrt{5} - 1$

EF = $\sqrt{3}$ x Pi

GH = $\sqrt{5} + 1$

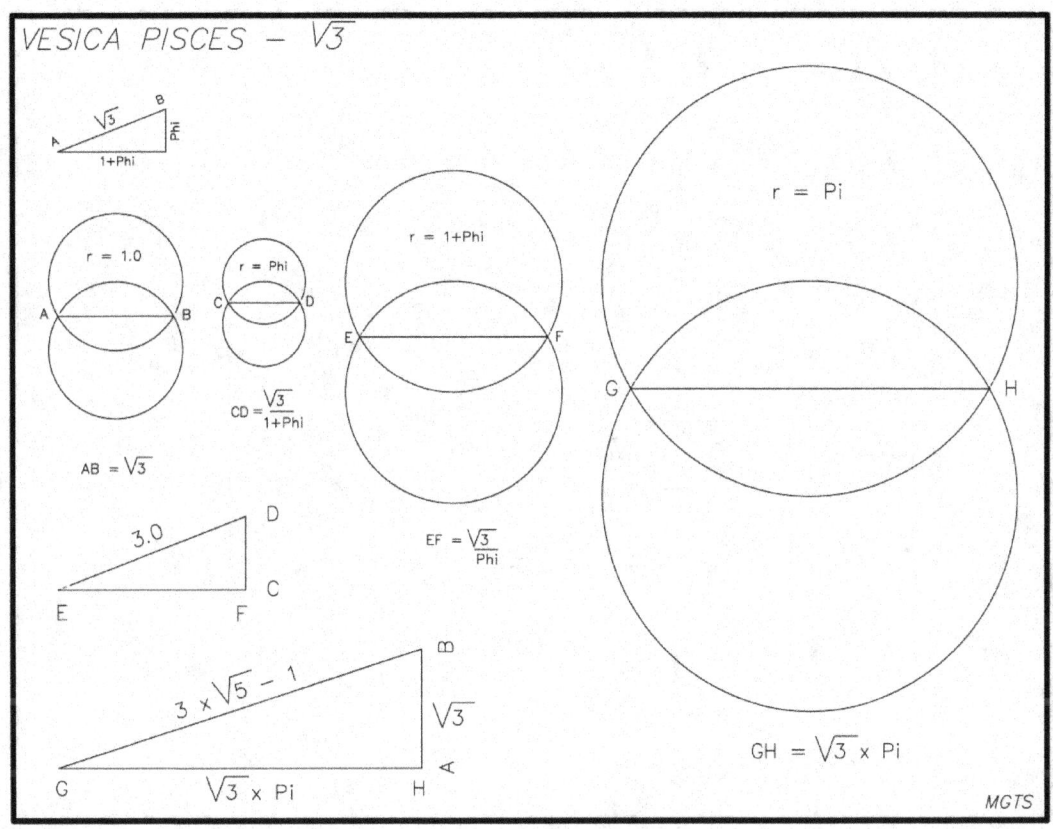

VESICA PISCES – $\sqrt{3}$

$\sqrt{3}$ — B — Phi
A — 1+Phi

r = 1.0

A — B

AB = $\sqrt{3}$

r = Phi

C — D

$CD = \dfrac{\sqrt{3}}{1+Phi}$

r = 1+Phi

E — F

$EF = \dfrac{\sqrt{3}}{Phi}$

r = Pi

G — H

GH = $\sqrt{3}$ x Pi

3.0 — D
E — F — C

$3 \times \sqrt{5} - 1$ — B — $\sqrt{3}$ — A
G — $\sqrt{3}$ x Pi — H

MGTS

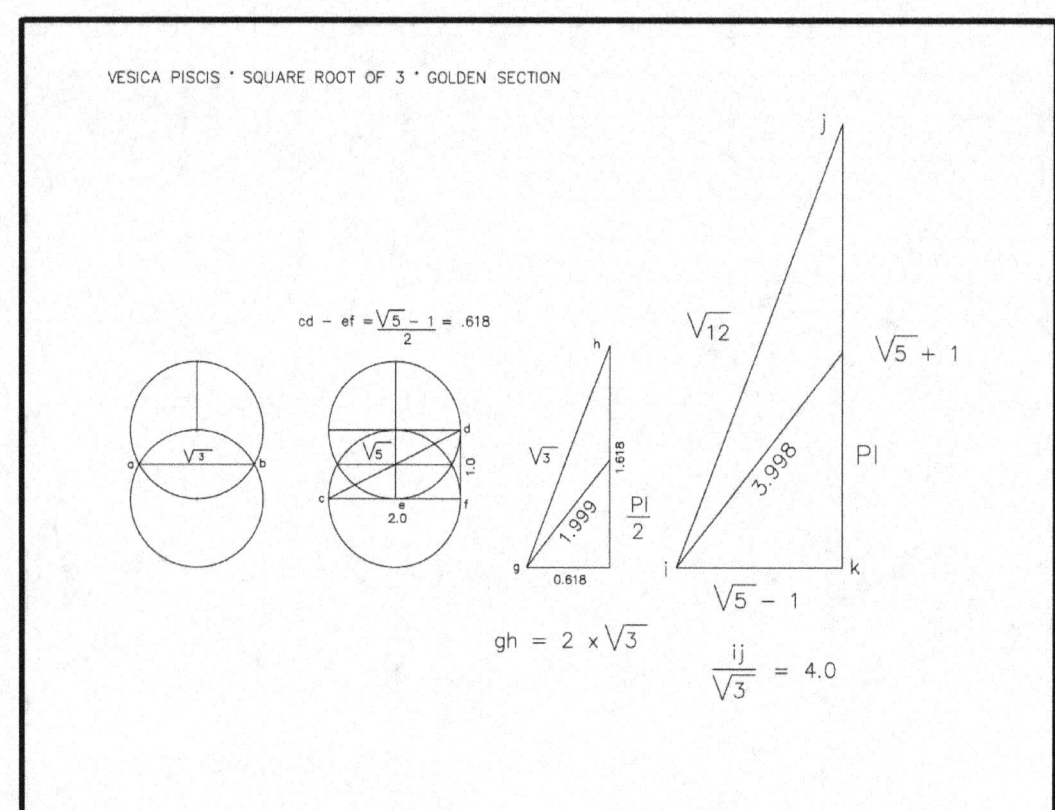

VESICA PISCIS · SQUARE ROOT OF 3 · GOLDEN SECTION

$cd - ef = \dfrac{\sqrt{5} - 1}{2} = .618$

$\sqrt{3}$

$\sqrt{5}$

2.0

1.0

$\sqrt{3}$

1.999

1.618

$\dfrac{PI}{2}$

0.618

$gh = 2 \times \sqrt{3}$

$\sqrt{12}$

$\sqrt{5} + 1$

3.998

PI

$\sqrt{5} - 1$

$\dfrac{ij}{\sqrt{3}} = 4.0$

GOLDEN SECTION

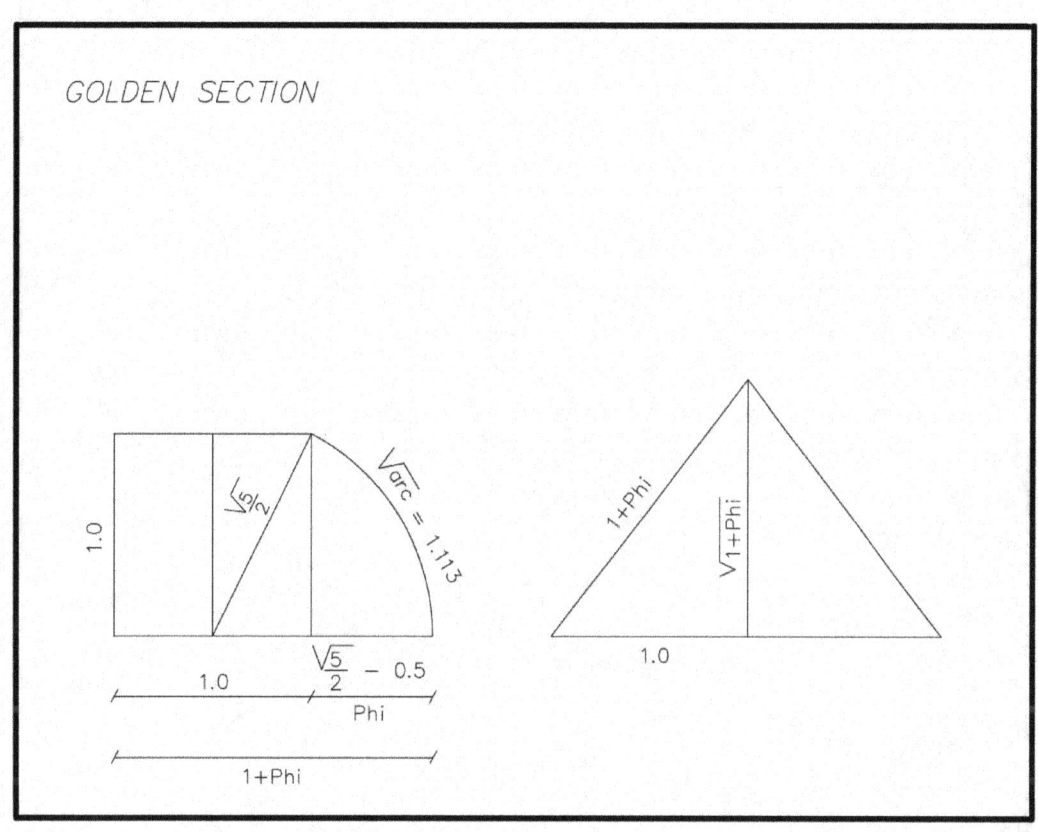

VESICA PISCES INQUIRY

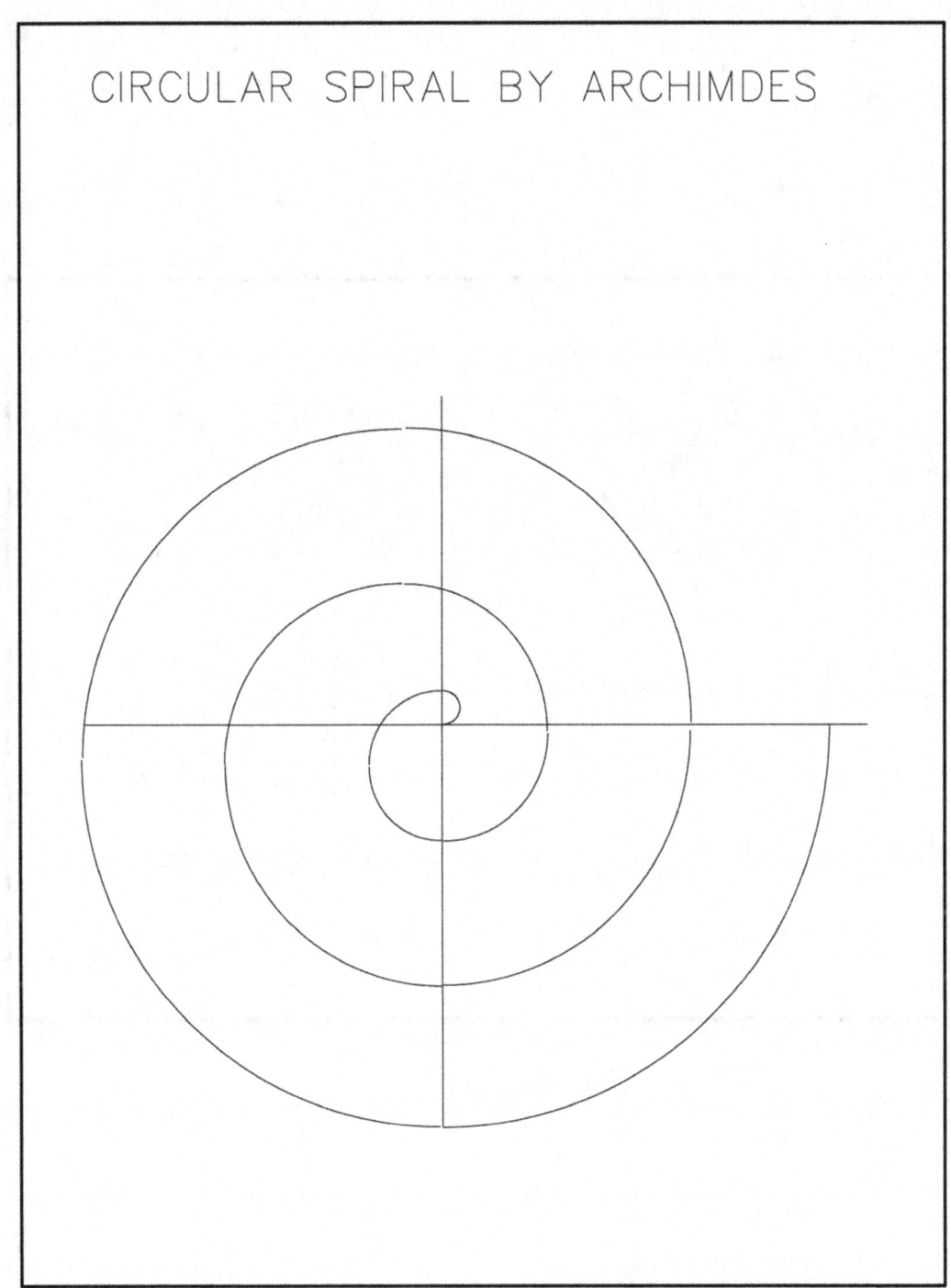

CIRCULAR SPIRAL BY ARCHIMDES

SQUARING THE CIRCLE
A SQUARE EQUAL IN AREA TO A CIRCLE
USING r = Phi

b-a/2 = .618

AREA CIRCLE = 1.1992413

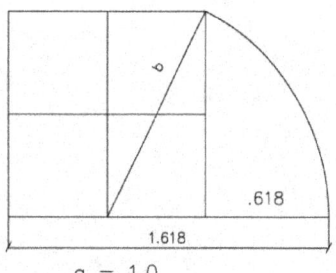

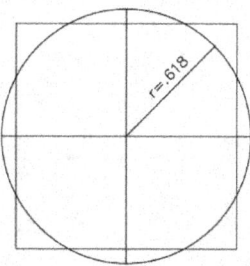

.618

1.618

a = 1.0

AREA SQUARE = 1.199025

DOUBLING THE CUBE

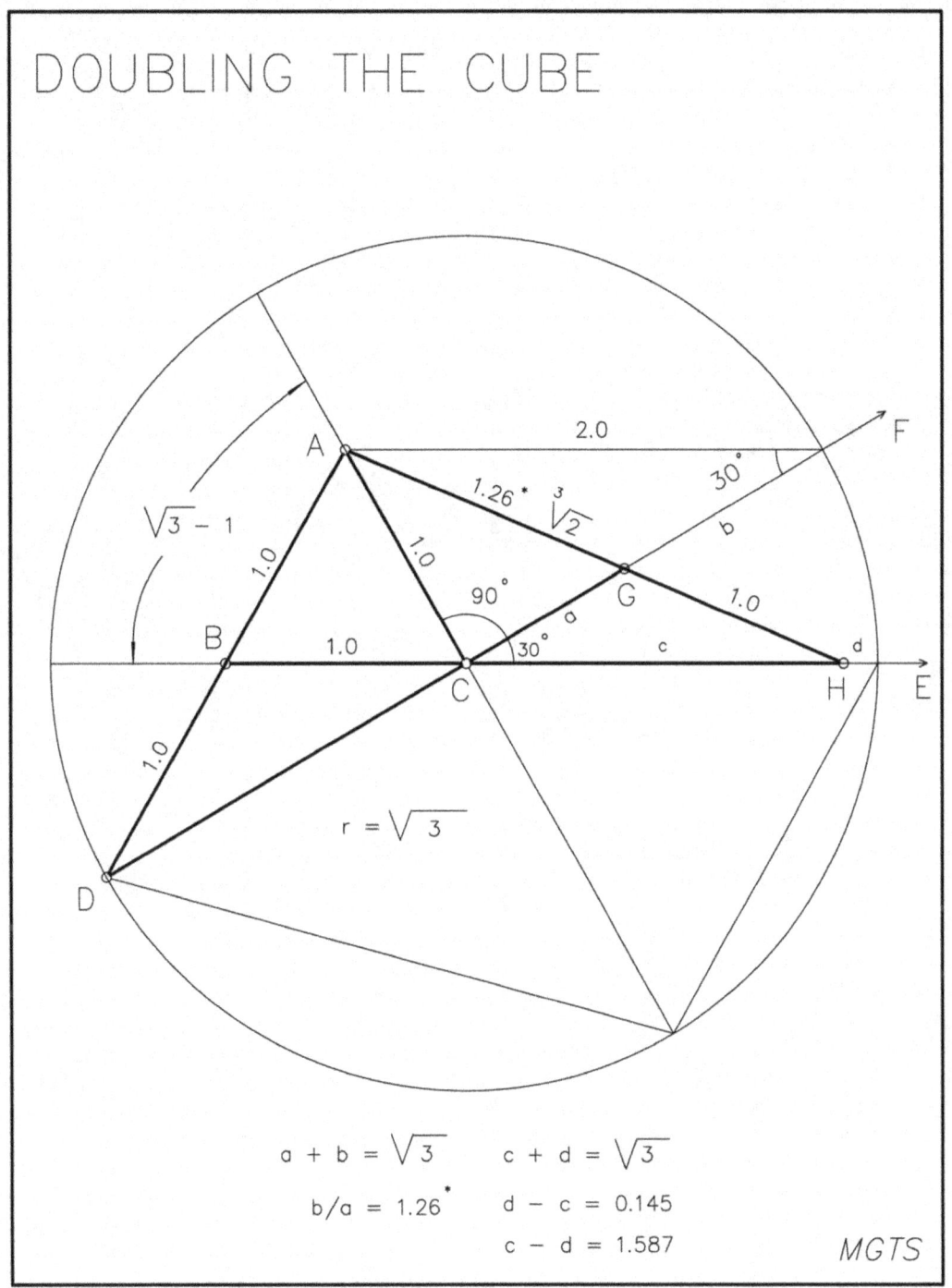

$a + b = \sqrt{3}$ $c + d = \sqrt{3}$

$b/a = 1.26\,^{*}$ $d - c = 0.145$

$c - d = 1.587$

MGTS

ARCHIMEDES TRISECTION THEOREM

... USING ONLY COMPASSES AND STRAIGHT EDGE

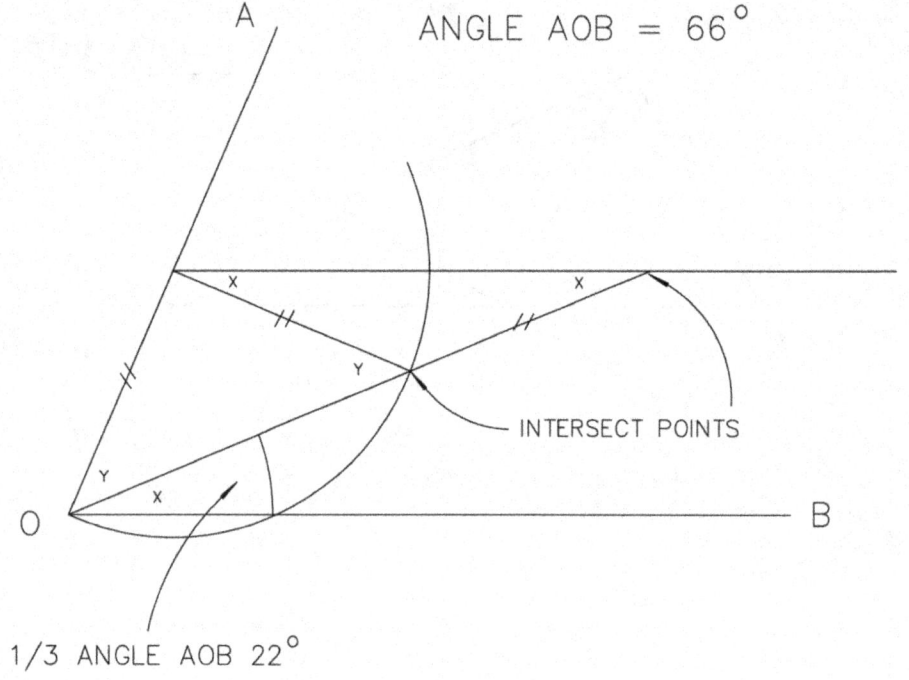

ANGLE AOB = 66°

INTERSECT POINTS

1/3 ANGLE AOB 22°

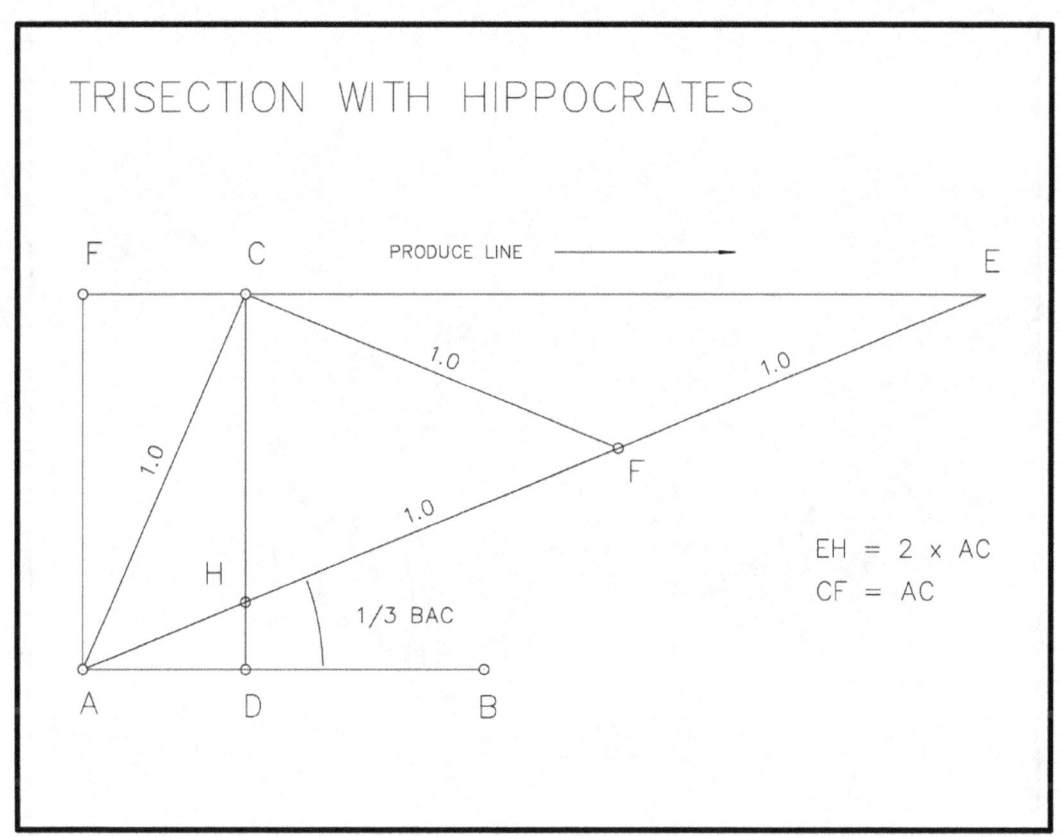

TRISECTION WITH HIPPOCRATES

PRODUCE LINE

EH = 2 x AC
CF = AC

1/3 BAC

ENNEAGRAM — THE NUMBER 9

ORIGINS UNKNOWN

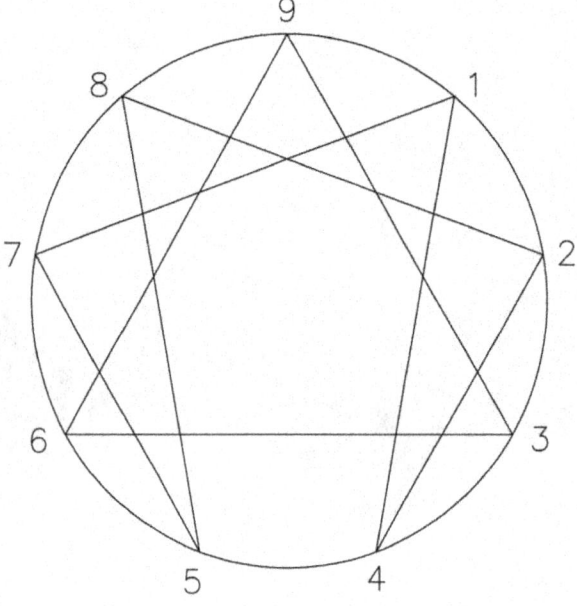

ENNEAGRAM STELLATED

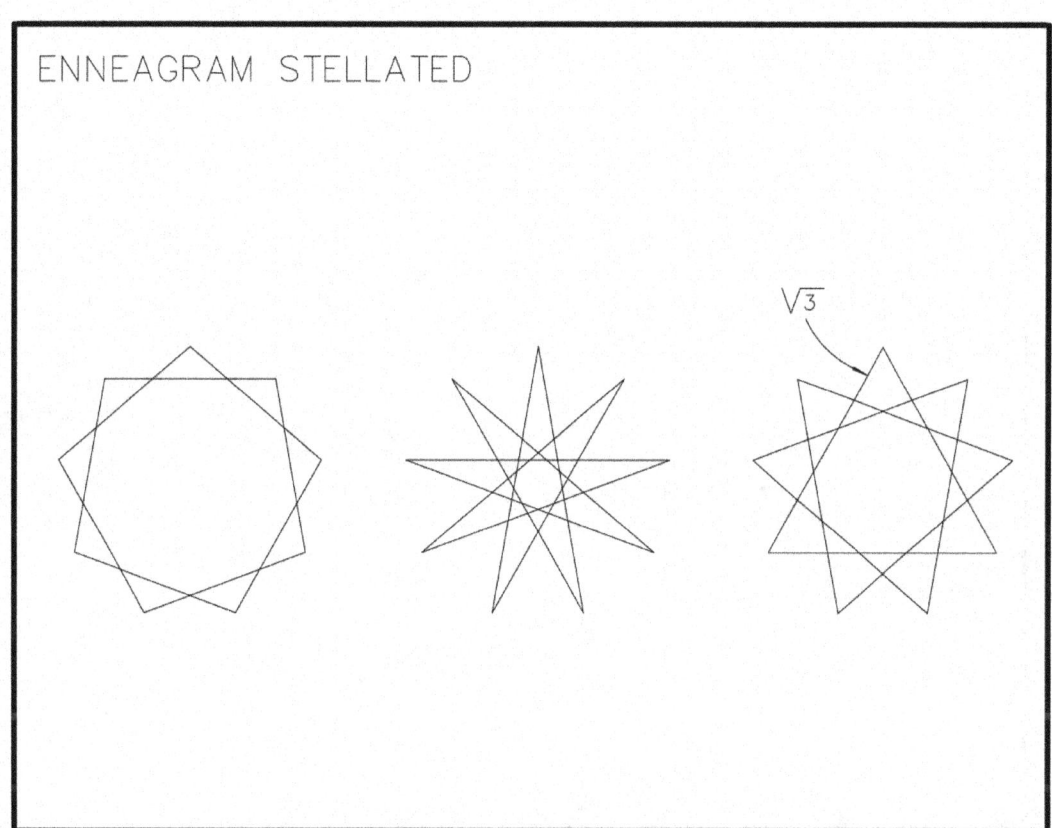

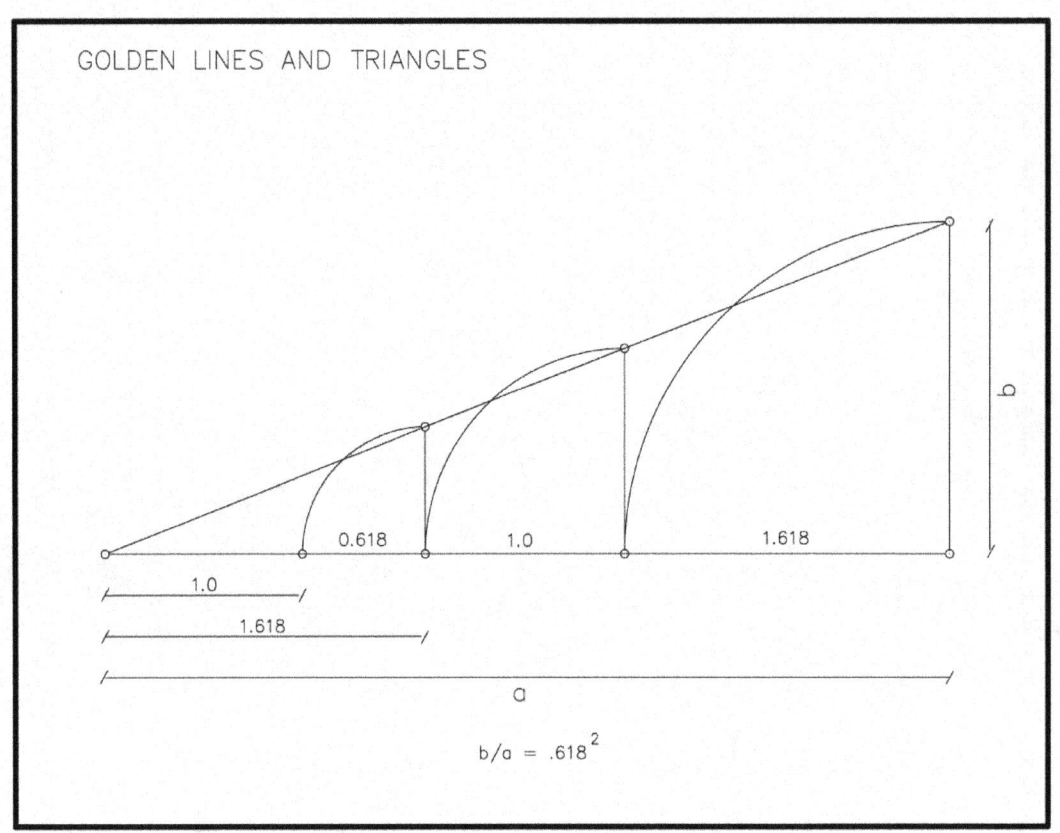

GOLDEN LINES AND TRIANGLES

0.618   1.0   1.618

1.0

1.618

a

$b/a = .618^2$

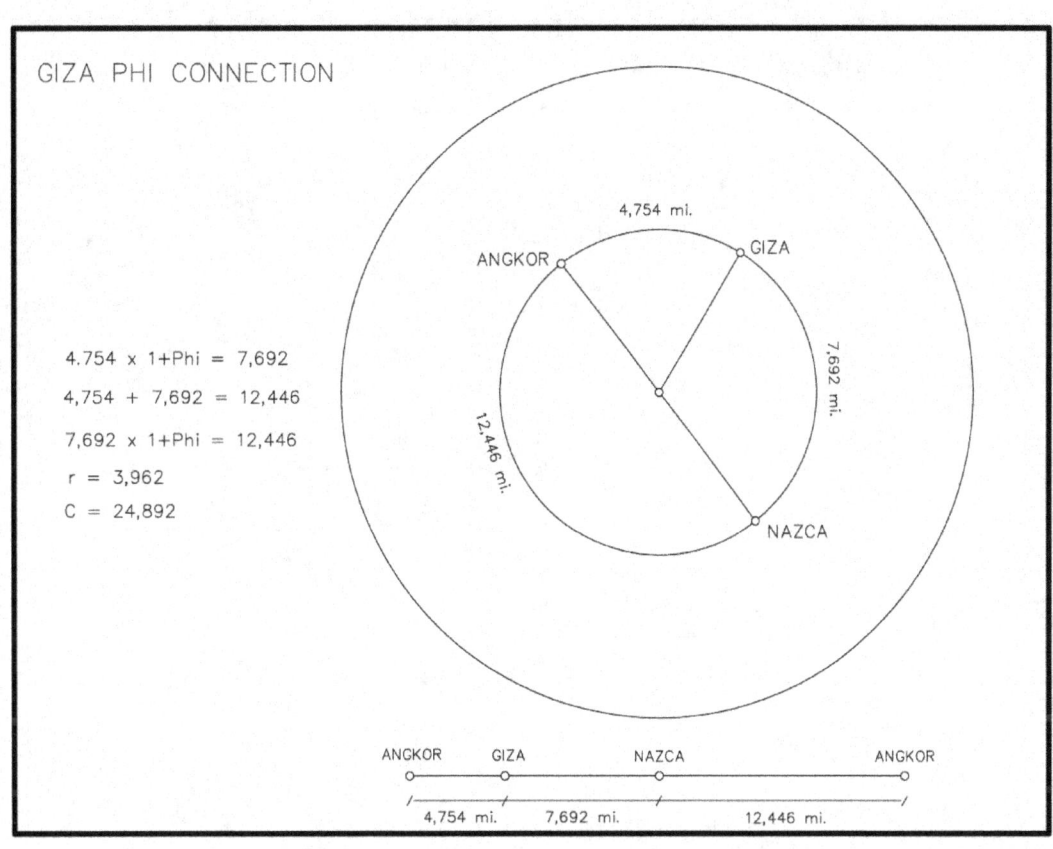

GIZA PHI CONNECTION

4.754 x 1+Phi = 7,692

4,754 + 7,692 = 12,446

7,692 x 1+Phi = 12,446

r = 3,962

C = 24,892

4,754 mi.

ANGKOR   GIZA

7,692 mi.

12,446 mi.

NAZCA

ANGKOR   GIZA   NAZCA   ANGKOR

4,754 mi.   7,692 mi.   12,446 mi.

PHI BASED PYRAMID — GREEN HOUSE AND SITTING CHAMBER

RAMSES IV

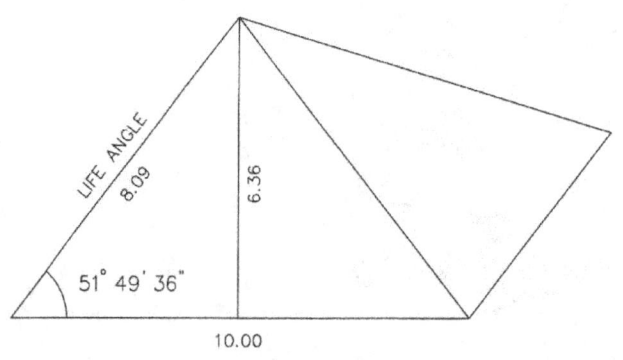

SIDE VIEW

ALIGNED WITH MAGNETIC NORTH

MGTS

PYRAMID SLOPE CONNECTION

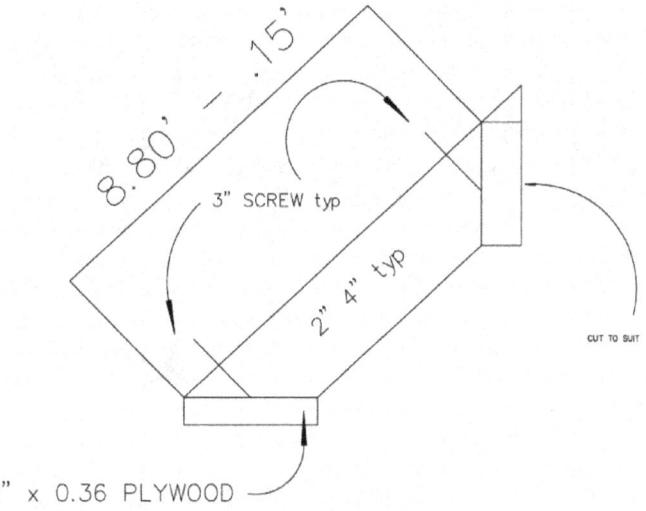

8.80' — .15'

3" SCREW typ

2" 4" typ

CUT TO SUIT

1/2" x 0.36 PLYWOOD
SCREWED IN PLACE AT BASE CORNERS

CONSTRUCT A TEEPEE

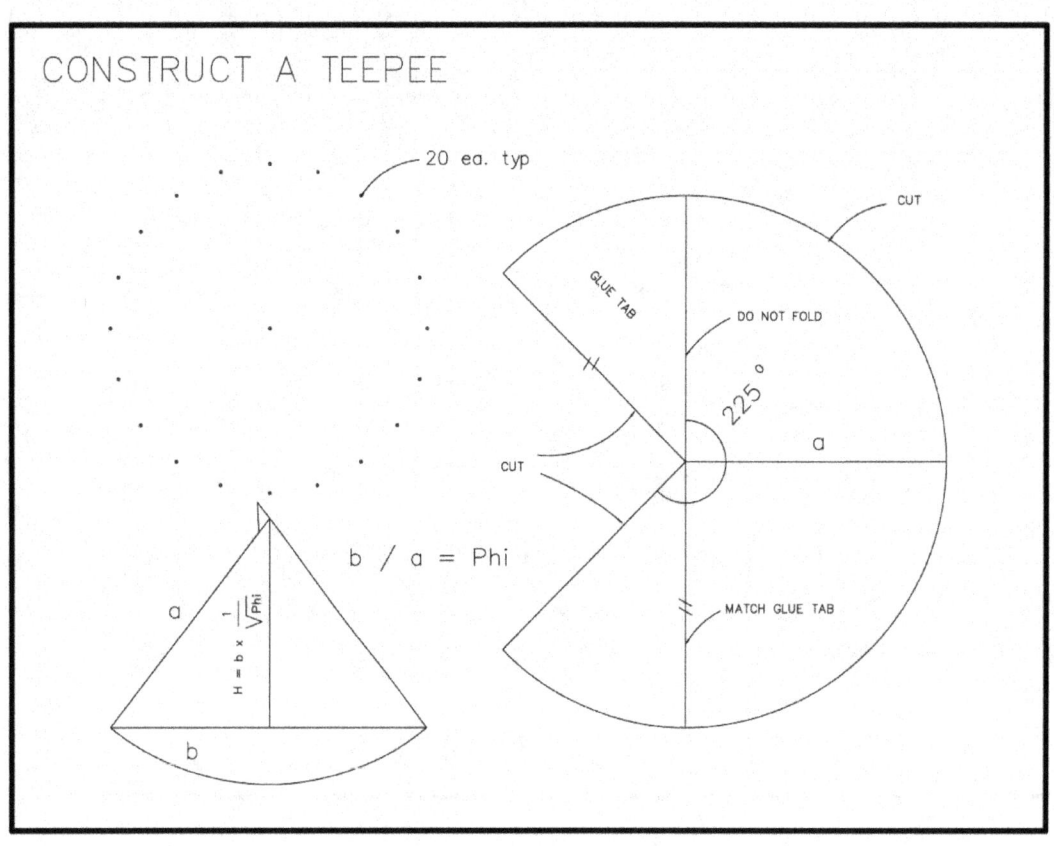

20 ea. typ

b / a = Phi

$H = b \times \dfrac{1}{\sqrt{Phi}}$

a

b

GLUE TAB

CUT

DO NOT FOLD

225°

a

CUT

MATCH GLUE TAB

GOLDEN SPIRAL

COSMIC EXPANSION AND CONTRACTION

COSMIC EXPANSION AND CONTRACTION
THE GOLDEN SPIRAL — THREAD OF LIFE

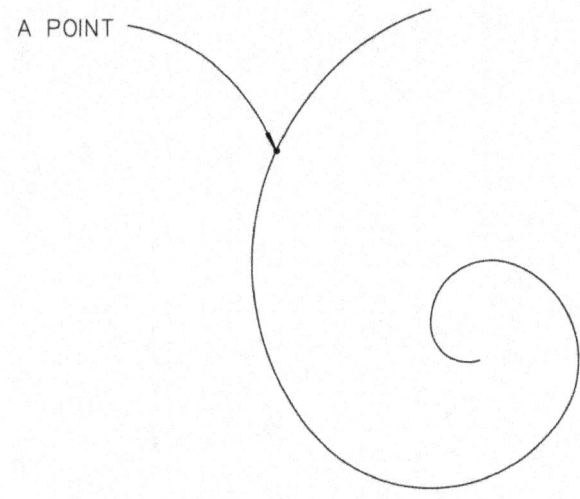

A POINT

*
THE BEGINNING AND END OF COSMIC EXPANSION AND CONTRACTION
OCCUR SIMALTANEOUSLY AT INFINITE LIMITS. A FINITE POINT OF REF-
ERENCE ON THE GOLDEN SPIRAL IS RELATIVE TO TIME, SPACE AND
CONSCIOUNESS WITHIN THE UNLIMITED EXPANSE OF THE UNIVERSE
AND AT THE SAME TIME IT IS THE CLOSEST WE EVER GET TO
INFINITY ITSELF.

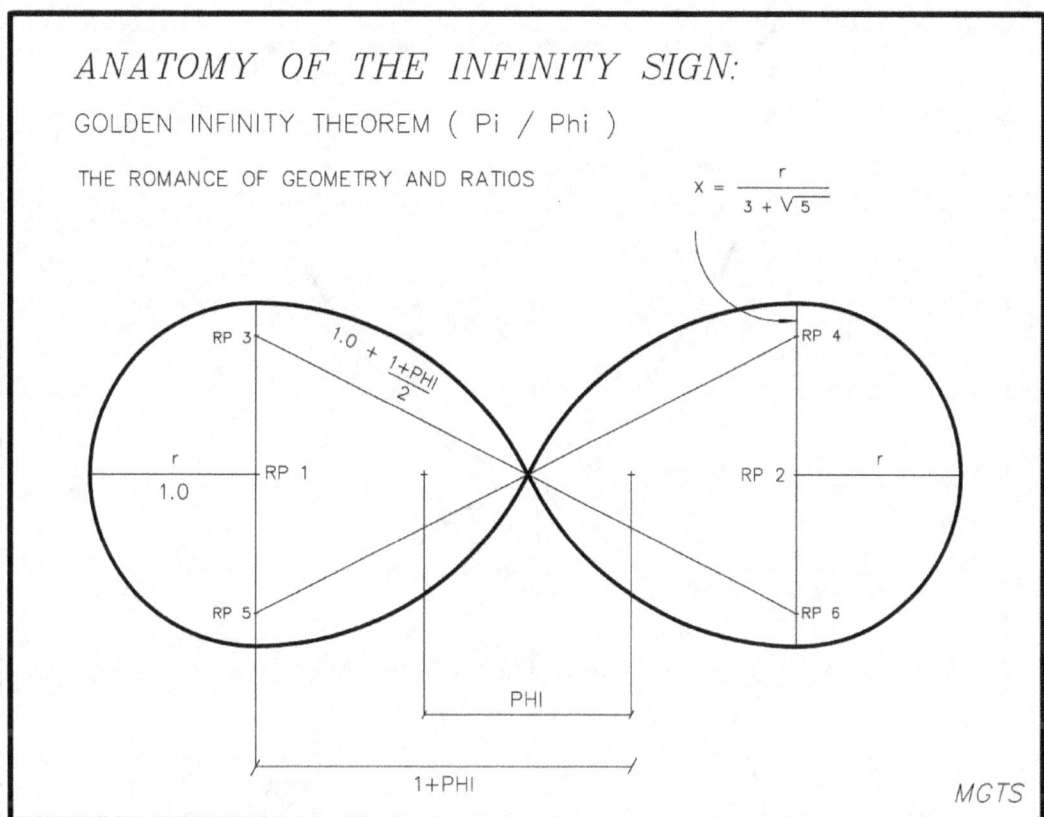

ANATOMY OF THE INFINITY SIGN:

GOLDEN INFINITY THEOREM (Pi / Phi)

THE ROMANCE OF GEOMETRY AND RATIOS

$$x = \frac{r}{3 + \sqrt{5}}$$

RP 3

$$1.0 + \frac{1+PHI}{2}$$

r

1.0

RP 1

RP 4

RP 5

RP 2

r

RP 6

PHI

1+PHI

MGTS

GOLDEN HOUR GLASS

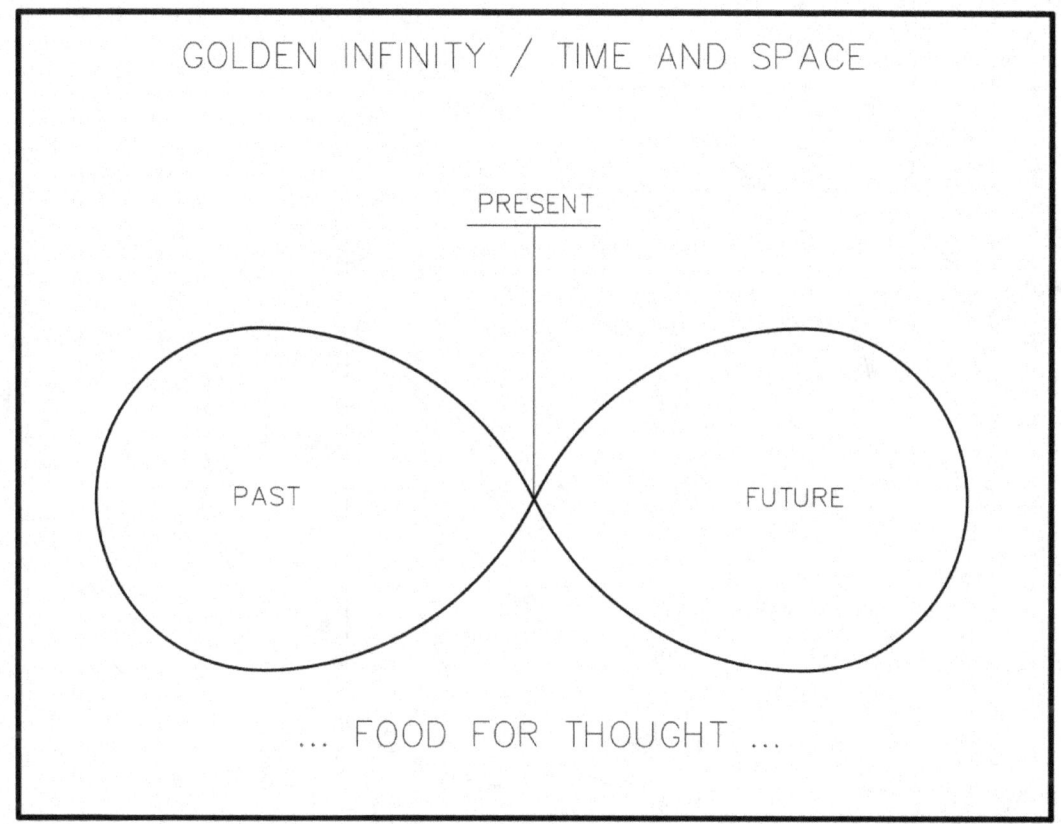

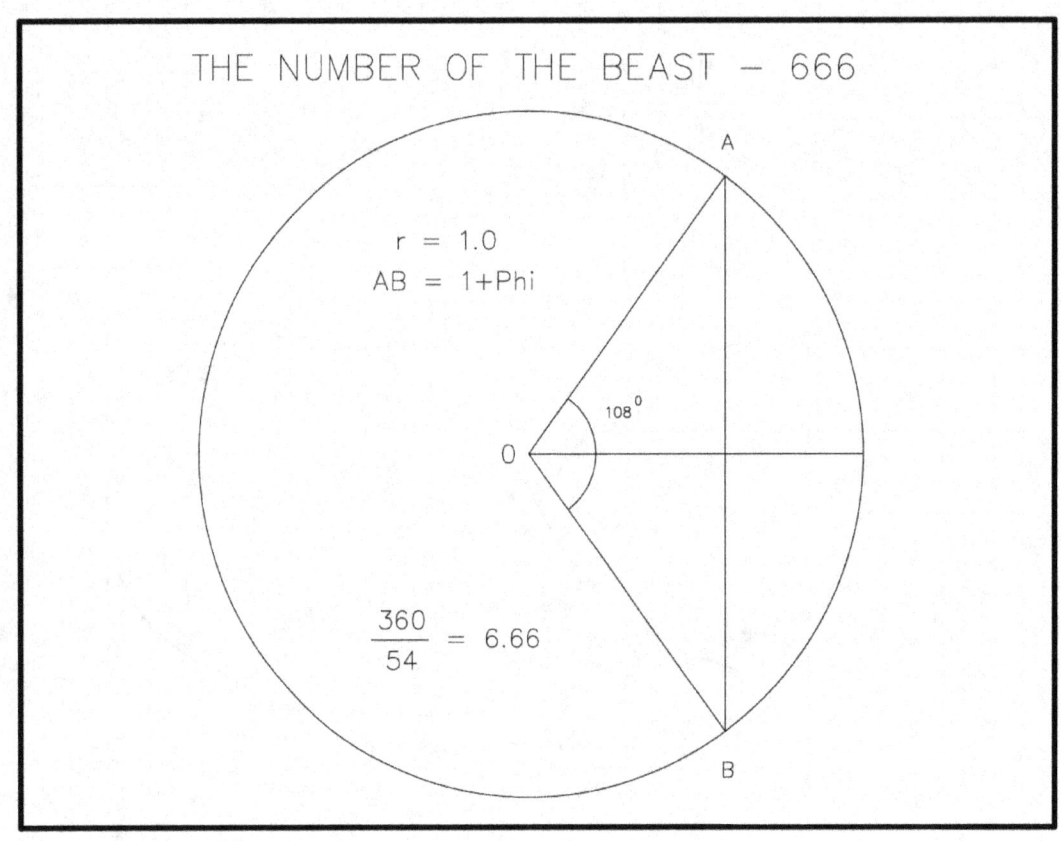

TEMPLE FASHIONED AFTER THE LOTUS BLOSSOM

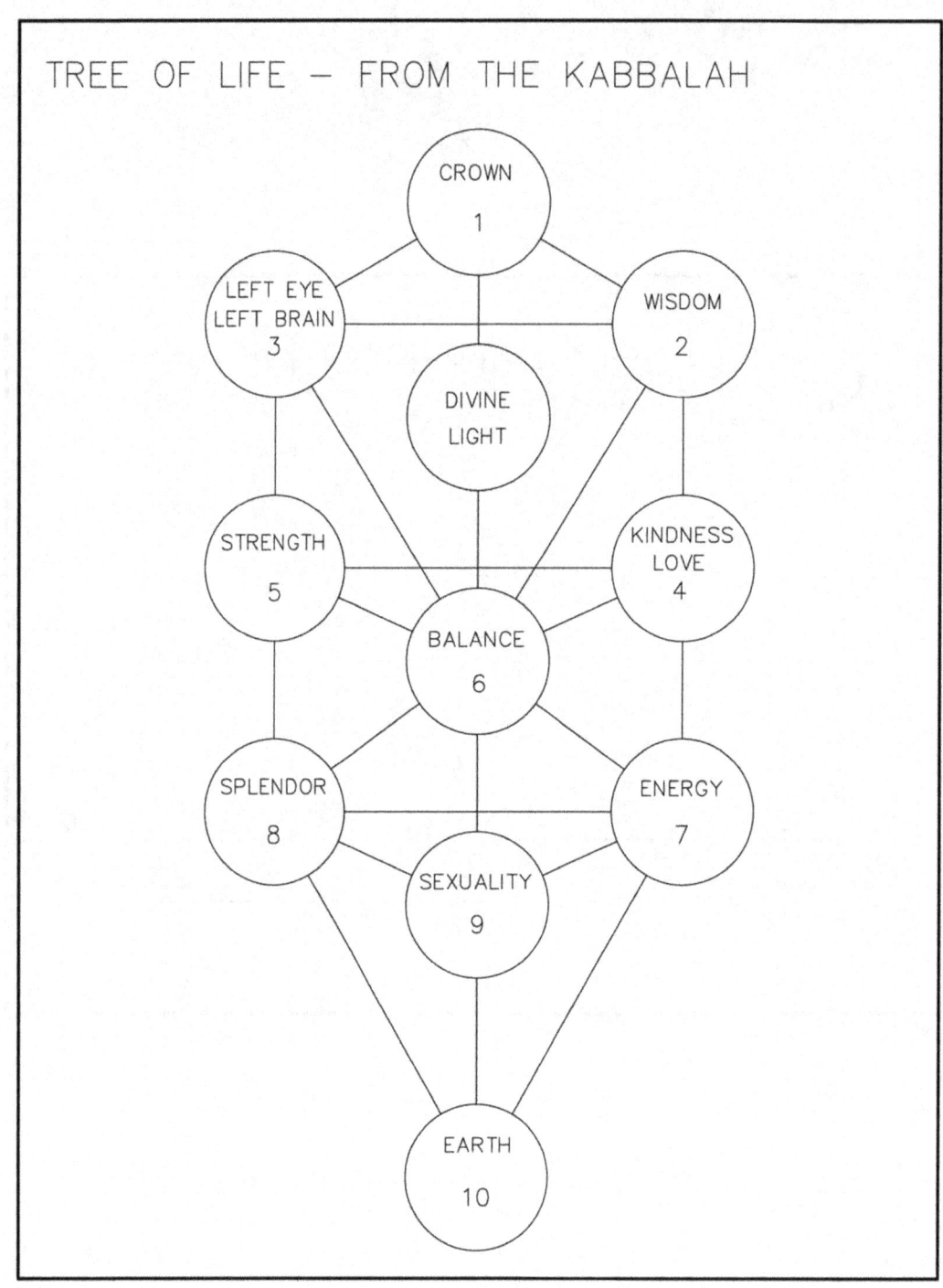

TREE OF LIFE — FROM THE KABBALAH

CROWN
1

LEFT EYE
LEFT BRAIN
3

WISDOM
2

DIVINE
LIGHT

STRENGTH
5

KINDNESS
LOVE
4

BALANCE
6

SPLENDOR
8

ENERGY
7

SEXUALITY
9

EARTH
10

THE FLOWER OF LIFE

TREE OF KNOWLEDGE

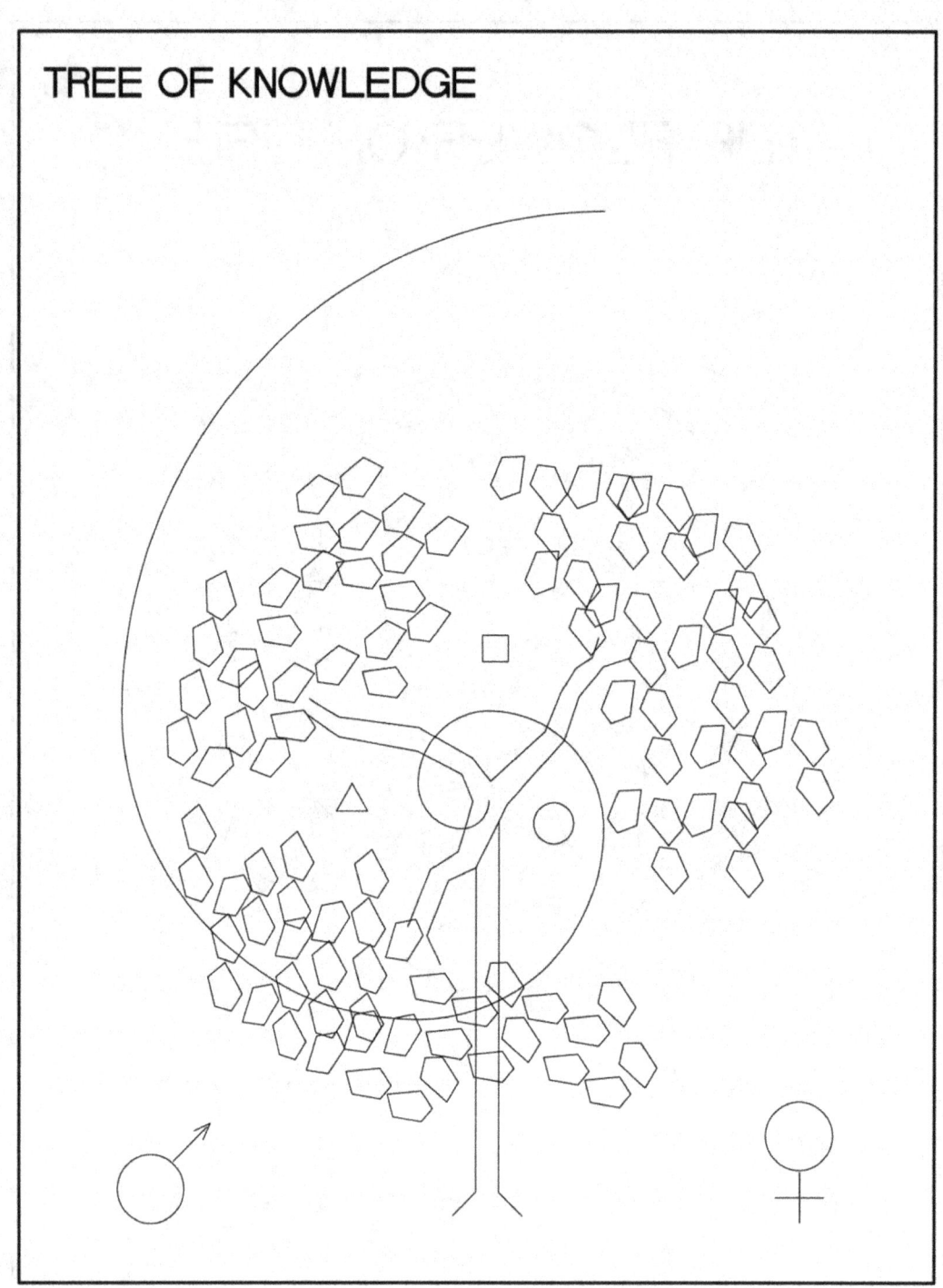

I FOUND MY GLASSES

MGTS

MGTS Technologies Inc.

Survey / Engineering & Technical Services
Michael Green, C.E.T. ~ Senior Survey / Engineering Technician

Career Practices Commencing in the Mid 1960's, now spanning a 40 year + period

Support Services & Expertise:

- Engineering Surveys   - Layout & Inspection

- Design Drawing   - Project Coordination & Management

- Cost Estimating   - Select Construction Administration

- Community Planning & Development

Areas of Application:

- Municipal	- Structural	- Transmission	- Petrochemical
- Highways	- Bridges	- Buildings	- Mining
- Railways	- Marine	- Airports	- Environmental
- Geotechnical	- Athletic Facilities	- Landscape Architecture	

- Archaeology Sites etc.

* Current Drawings: via Field Genius Data Collection, Micro Survey Cad

Featuring: A proven record with a hands on, practical approach as a team player amongst other professionals, from the physical study phase to successful project completion.

- Agent to the Engineer, Land Surveyor, Architect and Planner
- Asset to the Owner / Developer / Support for the Contractor

Michael Green, C.E.T.

Senior Survey / Engineering Technician

Phone / Fax / Message: (250) 957-2977 / e-mail: mgts_surveys@yahoo.com
www.outskirtspress.com/livingwithgeometry

The Great Pyramid Speaks...
the Solar System and Milky Way Theorems etc.

The Thing About Royal Cubits, Yards, Feet, Inches and Metres

Measuring has been such a large part of my process in life this section commences by providing a discourse on the subject to help set the record straight on the units we are using today and where they come from and what they are based on. In the process important discoveries were made that the scientific community and the world populace in general should be aware of. Like a story within a story that involves coincidence or a divine form of fate that has a lot to do with the type of work I have been practicing for many years, which has been interrelated with my interest and sideline studies into the ancient sciences. To begin with, what appears to be a coincidence is the number of this chapter, 3, and its square root. This becomes an integral point of interest in the study as a rather remarkable journey and a new beginning of understanding emerges that evolve into a number of profound conclusions by the end of this edition. There will be other coincidences with assigned numbers as the story unfolds.

The French Academy of Science, in 1793 had determined the length of a metre as being a one ten millionth part of the arc distance of the globe between the equator to the north pole through Paris at the transit of Venus, not knowing the earth is made up of a patch work of five major ellipsoids each having different radii, therefore the metre was falsely derived as a true earth measure to begin with and in my opinion it was ridiculous to use an arc section of the globe to base a world measurement standard on and that has always made me boil. Traveling back to an ancient time to get a handle on the history of measurement we are told by one source, in early Egypt the Royal Cubit was decreed to be equal to the length of the forearm from the bend in the elbow to the end of the middle finger plus the width of the palm of the hand of the Pharaoh or ruling king at the time. This measure was approximately 18 inches plus 3 inches, or 7 palms equal to one cubit. Quite frankly I disagree with this because the measuring system of a scientifically sophisticated people such as these might be considered to be based on something more intelligent than the use of body parts that would vary from one individual to the other in order to determine the length of the Royal Cubit because, after all, the early Egyptians were masters of measurement and geometry on a scale we can barely grasp. Therefore, the object is to determine from the dimensions of earth a measuring unit as it might have been derived in ancient times. An estimate of the Royal Cubit we are given is between 20.61 and 20.63 inches or 1.718 to 1.719 feet, however it could be determined the Royal Cubit was originally determined by knowing the polar and equatorial radii of earth and it has been proven the early Egyptians, or those before them, indeed had a good handle on these dimensions due to their advanced practices in astronomy. These dimensions would be reliable constants and this proposal makes sense because the Royal Cubit would then be based on an earth measure just as the French had incorrectly attempted to do. At least, the basis of this being so is there and quite possibly the original value of the Royal Cubit has been lost over an enormous time span. There have been so many other types of cubit versions throughout history it can get quite confusing.

The investigation proceeded with what I thought to begin with might be an elusive goal and perhaps an exercise in futility, after all, who was I to question the authority, not that I have ever been particularly pleased with their performance. Irregardless, I took it on as a challenge to see where it would go. After all number crunching was nothing new to me. After some trial and error with the given earth dimensions to see what might have been done, a few simple calculations provided a value amazingly close to the Royal Cubit value 1.719 feet we are given by the establishment. Following was the process: When the value for earth's polar radius, 3949.9 mi. is subtracted from the equatorial radius 3963.5 the difference amounts to 13.60 miles. Subtract this from the polar radius 3949.9 - 13.60 and as follows 3936.3 x 5280 ft. = 20783664, divided by 12,000,000 = 1.7321972 or 20.78 inches. I could hardly believe my eyes. I rechecked the calculation and not by any coincidence it seemed, this just happened to be the square root of 3, accurate to within three places of decimal based on a logical calculation involving ancient earth measurements. What could this mean, I muttered to myself. When the 1.719 ft. given as the Royal Cubit is subtracted from 1.732 the difference is 1.3 hundredths of a foot, or 0.013, a little less than one quarter of an inch. Then I questioned myself as to why I had used 12,000,000 and not 10,000,000 like every other party has done in the past. First of all 10,000,000 didn't do anything for me and the number 12 is commonly used. I got into some number jargon as follows: Hmmm…there are 12 months in a year and the globe is usually subdivided into 12 equal meridians of 30 degrees each and 12 = 2 x 6 the double trinity number. There are 12 inches in a foot and it is likely those before our time used the 12 factor in one way or the other in their measuring systems and coincidentally the square root of 12, divided by 2 = 1.732 so I went for the 12,000,000, based more or less on intuitive thought as much as anything and I got a "bingo" in the form of the square root of 3 and was intrigued by determining it this way. * See pages 114 to 118 further along in this chapter when it is proven there are 12 planets in the solar system and it is this among other influences that validates the use of the 12 factor. Then I got busy in the number crunching department using a basic electronic calculator, one of my best friends at my working station. In an attempt to explain the ancient use of 12,000,000 please view the following calculation using the given radii: 3936.6 x 5280 = 20785248, divided by 1.732 = 12,000,720.55, virtually 12,000,000. The exact earth dimensions to three places of decimal have been unknown for eons but the following answers this question, assuming it is allowed to use the square root of 3 to three places of decimal to begin with: Therefore 1.732 x 12,000,000 = 20784000, divided by 5280 = 3936.363636 miles and this gorgeous looking number appeared. When 13.6 miles is added to this number the total = 3949.963636. Then it occurred to me that the true polar radius of earth might be closer to 3950 miles so a subtraction was carried out, 3950.000 - 3936.363636 = 13.636364, then I added this value to 3950 and determined the equatorial radius must be 3963.63636 and another great looking number value emerged. These numbers looked good but I needed to find a logical way to check, or verify them so they would make sense and the following table was generated based on the distances in feet divided by 3.0 and 5.0 because these numbers and their square roots play such a significant role in the inquiry:

First using 3.0:

3950 x 5280 = 2085600, divided by 3 = 695,200

3936.363636 x 5280 = 20,784,000, divided by 3 = 6,928,000

3963.63636 x 5280 = 20,928,000, divided by 3 = 6,976,000

Total = 14599200, divided by 0.618 = 23,623,300...

Then using 5.0:

3950 x 5280 = 20,856,000, divided by 5 = 4,171,200

3936.363636 x 5280 = 20,784,000, divided by 5 = 4,156,800

3963.63636 x 5280 =20,928,000, divided by 5 = 4,185,600

Total = 12,513,600, divided by 0.618 = 20,248,544

Because the quotients work out to nice round, full numbers I believe the true earth dimensions have been reestablished in the above process:

Polar Radius = 3950.000 miles x 2 = 7900 = polar diameter
Equatorial Radius = 3963.636 x 2 = 7927.272 = equatorial diameter

Therefore, with no equivocations, these dimensions will be used from this point on.

This will challenge the Royal Cubit dimension and pyramid inches of the Great Pyramid given by reigning authority on the subject. For example their 440 Royal Cubits for the base length of the Great Pyramid works out to 440 x 1.719 = 756.36 ft. but this becomes 762.08 feet when multiplied by 1.732. The difference is approximately 1.00756% and this difference will be will discussed shortly. The original surveys of the Great Pyramid were carried out in feet and many of the more recent ones in metres then the distances were converted using the 1.719 foot value. It was never absolutely clear to the archaeologists and surveyors who were not well versed in Golden Section mathematics to begin with, where the foundation sockets of the Great Pyramid exactly were under heaps of sand and shattered limestone so differences here and there were likely. Each socket was found at different elevations and the limestone casing was 100 inches thick, they say, and this equates to around 8 feet. How then would they know just how these sockets related to the original exterior dimensions? And furthermore, ground shifts and settlement plus erosion and shrinkage factors cannot be overlooked. These foundation sockets might originally have been at the same elevation until the ground settled. If these were brought to the same level based on the survey data, what effect would this have on the outer dimensions? Another question might be, could these foundation stones have moved out of their horizontal positions due to ground settlement over time? Is it possible the inside corners of the foundation were surveyed after the outer ones had long eroded away? Could there have been a devious plan to remove the outer foundation stones in order to throw off future visitors who would go there to study the site? All the possibilities should be examined and no stone should be left unturned. That sounds almost hilarious. What we have here is a detective story involving this grand piece of work that has led from one insoluble mystery to another for centuries and forthcoming it will be determined, the reason for this is because the wrong dimensions were being used all along and the investigators were not well grounded in the workings of Golden Section Mathematics.

It would only make sense that the Royal Cubit would be based on a sacred value such as the square root of 3. For years the experts have been telling us a Royal Cubit is 1.719 feet and this is so close to the square root of 3 it really isn't all that difficult to miss. This should have been noticed a long time ago. The story is; A copper alloy bar was found in a dig dating back to 2650 B.C, at Nippur, Egypt and its length was "assumed" to be one Royal Cubit. Its length measured to be the 1.719 ft. and we have been stuck with this value ever since. The problem is this bar might have been fashioned after a Pharaohs arm length and hand width which would vary from a 7 th. part of the square root of 3, hence this bar would not be a true royal cubit, or the square root of 3. It is either that or the material in the bar oxidized over this long period of time and its length became shortened. Another story about a red granite piece that was assumed to be a royal cubit could have been based on the shortened oxidized bar or due to air and moisture loss over such an extensive period. No one will ever know for sure. When Herodotus, the well known ancient Greek scholar visited Egypt he commented on their meticulous ways of keeping records and said it was superior to the capabilities of all other groups of people he knew. Everything the ancients constructed was sacred to them and the theme was always, application of the Golden Section - Phi Ratio, therefore the Royal Cubit could also be called the Sacred or Golden Cubit. When it comes to mind the square root of 3 is the most sensible unit of measure to use because it is a universal value. One proof of this proposal is to draw intersecting circles with radii values of the square root of 3 (1.732 ft.) and find the Vesica Pisces value to be 3.0 ft., which is the square root of 9, that full number embodying the triple trinities and it is seen this doesn't work using 1.719. In this is there another theory on how the yard came into being before its time? Another interesting feature we get when 5,280 ft. is divided by 1.732 = 3048.4988, approximately ten thousand times the metric conversion factor. This could be presented as 0.5280, divided by the square root of 3 = 0.3048. Then we have 1.732 or 20.784 inches converted to millimeters = 528 m.m. and there are 5280 feet in a mile. What is interesting about this is 5280, divided by 3 = 1760 yards, divided by 4 = 440 yards, the old track and field quarter mile foot race. It gets more interesting when 5280, divided by 1.732 = 3048.50, divided by 4 = 762.12 and this is very close to the 762.08 foot base I have derived for the Great Pyramid which is based on the earth dimension to 3 places of decimal with regards to the square root of 3. Here are some coincidences and food for thought but before we get ahead of ourselves, please refer to the drawing titled Golden Section Elements and it is seen how 1+Phi, Phi and the Square Root of 3 relate in a right angled triangle. Then please see the drawing dedicated to the "Royal Cubit Theorem". Save this drawing alongside the one for the Elements of the Golden Section because it is important. The foundations of the perpetrators of a bogus Royal Cubit value will be shaken and it is back to the drawing board for them.

Furthermore and quite amazingly, when the so called Egyptian polar radius value converted to feet is divided by 7,000,000 a number within 0.7% of a yard comes up while the metre is within 0.914% of a yard. When the above mentioned altered polar radius, 3950.00 is used in this arrangement the number, 2.979 comes up, within 0.993% of a yard. It would seem logical to use the value of 7 in the calculation as it is perceived by ancient philosophy to be not only the solar number but also the number of completion. Whatever the case might be it works out this way. Oh, if only the French had worked out the value for the polar radius of the earth, a trustworthy constant, and used the 7 factor to derive the so called metre a lot of difficulties would have been avoided. It is as if the measurements of the earth, or earth itself has been trying to tell us something all along. This is possible, at least to those with a metaphysical or occult bent or disposition because indeed, if Earth is perceived as a living organism, then it would also have a mind of its own and we, its occupants might be functioning under the direction of its influence and will. It is possible some of the rational scientists of today might be in agreement with this notion, however they would likely view it with detached reservation. Soon enough though, their position in the world of science will be known.

Interestingly, in mentioning the 7 factor, when the square root of 3 is divided by 7 a value of 0.247 ft. comes up and this is very close to 3 inches, a palm width, and we are told there are 7 palms in a Royal Cubit. When it is seen 1.732 - 0.247 = 1.485...rounded off this would become 1.5 ft. or 18 imperial inches. Another observation is made: To the ancients the polar radius of 3949.9 miles less 13.6 miles, divided by the square root of 3, divided by itself provides the quotient 1.0 unit which happens to be 1.0 foot because the square root of 3 was derived from it and the oxidized 1.719 foot shortened copper alloy bar referred to earlier proves this. To the ancient geometers this value would not be a foot but an earth dimension used for the derivation of the square root of 3.0 (1.732), through the Vesica Pisces application, which is proposed to be the true length of the Royal Cubit. Here, coincidence or not, is where the foot might have come from before its time. The numbers seem to be telling the truth and for now the proposal and possibility is put forward: the Royal Cubit = the square root of 3. It is an entertaining concept at this stage to say the least, and so far some important information has come to light from the investigation process.

This is an area the researchers should have worked on more because the debate on the Royal Cubit and the Pyramid Inch have been hotly disputed since Isaac Newton's time when he was searching for a measuring unit that made sense. He sent his entourage to the Great Pyramid, following up on a hunch that what he was looking for was there. How right he was with this notion, but it will be found they missed the mark and this input proposes to help resolve the issue. I believe some confusion developed over the pyramid inch in Newton's day. The researchers had determined there were close to 25 imperial inches in a Royal Cubit using the following formula: 3949.9 x 5280, divided by 10,000,000 = 2.0855472 ft., divided by 0.08333 = 25.02666 inches. Yet if the above formula is applied using the 12,000,000 factor and the square root of 3 is derived from it the following is put forward: 1.0 divided by 12 = 0.08333, one imperial inch, while 1.0, divided by 1.7320508 = 0.5773502 = 1/3 square root of 3. Then 1.732, divided by 20 = 0.0866, a twentieth part of the square root of 3, a proposal for the pyramid inch and this will make total sense in time. Furthermore it is noted that the square root of 3, divided by 40 = 0.0433 the half pyramid inch emerges.

It is interesting to find that the equatorial circumference, 24,903.525 x 5280 = 131490612, divided by 1.732 = 75918367.21, divided by 0.0866 = 876655510.5, divided by 100,000 = 8766.555 which proves out to be very close to the number of hours in a terrestrial year. The value for this is 365.24 x 24 = 8765.76 hours and a mental note is made. With reference to the base length of the Great Pyramid telling us the number of days in a year using the square root of 3 and the proposed 0.0866 pyramid inch, we need to know exactly what its length is. The problem is, we are dealing with an erosion factor that has been in progress for over 5,000 years and there are disagreements on the dimensions between various parties who have actually surveyed the sites, therefore the precise measurements of the Great Pyramid are unavailable to us at this time. There are other parties who also believe the original height of the Great Pyramid was close to 485 feet and that erosion has been a factor. One group says the base dimension is anywhere from 755 to 756 feet while another claims it is between 756 to 757, then another says it is 760 feet and so on. Amazingly Piazza Smyth, the Astronomer Royal for Scotland said its height was 484 feet and this equates to a base of 762.20 in terms of the Golden Section but the influence of Isaac Newton at the time proclaimed the 25 inch Royal Cubit based on the earth's polar radius divided by 10,000,000 was the correct way to go. Neither of these parties were particularly knowledgeable in the Golden Section department and it is this, as I have claimed, that has hurt us. It wasn't part of the education system then and it isn't today either. My intuitive feeling on this matter is, it had to be somewhat wider and higher originally but by just how much it is very difficult to say because to date there have not been any real references available to go on except the input of other parties on the subject. No one involved with the study at the time had any conception on how to deal with a structure as ancient as this. We have ascertained the Great Pyramid was designed and constructed in proportion to the earth's dimensions for reasons not quite clear to us and we are told the base length tells us the number of days in a year, therefore at this preliminary stage I have estimated the base length to have originally been an outrageous 762.08 feet as opposed to 756.36 feet after putting some of the facts together and following up on a hunch I have had about this issue for many years, therefore I am sharing the process I went through that has brought me to some conclusions on the matter. I kept it to myself thinking no one would believe it but the more I thought about it, and the more I played around with the numbers, the more possible it became because the reference I had on the subject just didn't add up. At an earlier time I was so busy managing survey / engineering projects the window of time opportunity to explore my notions in depth were just not there. The difference is a whopping 5.72 feet and this sounds scary but when this is divided by 2 = 2.86 feet, this is how much longer the base would have been at the sides along its perimeter and the eroded depth at right angles to the slope is very close to 2.25 feet. Please bear with me, I realize there is some explaining to do on this. Using this proposed base length the height would then be 484.68 feet since the slope angle is 51 degrees, 49 minutes, 36 seconds. This is based on what is given, the base length is 440 Royal Cubits and though I may be out on a limb on this I will use this number times the square root of 3 and see where it goes. Therefore, as indicated, 440 x 1.732 = 762.08 feet. To keep it simple I'll use 3 places of decimal unless otherwise indicated in an attempt to prove this hypothesis.

The erosion factor and other influences on the physical state of the Great Pyramid need to be dealt with realistically. Wind driven sand and the acidic levels of the air have contributed to the erosion factor and at a time when the climate in this area was more temperate rainfall would also have been a contributing erosion factor as well. Aside from this an awesome amount of time has gone by since its construction. By the time the surveyors and archaeologists showed up on site some 5,000 years or more after its initial installation tons of the limestone facing had crumbled away and tumbled down the slope and was used for other structures in nearby Cairo and scavengers had broken into the structure and it is very likely the lower outside stone blocks around the perimeter base were carted off after they had disintegrated into smaller pieces due to erosion of the 2.25 foot or so mentioned depth and this left inner wall corners which the surveyors tied into. Though there wouldn't be any record of it, it is possible there was a decree by a Pharaoh a long time ago for the workers to chisel and square off the rock face around the perimeter to tidy up its appearance out of respect for this treasured monument. Then after Egypt went into decline the sand blew in from the Sahara Desert and the archaeologists and their crews had much excavation to deal with when they arrived at the site. A reference to the general interior stone block size in the Great Pyramid is 2.5 ' x 10' x 10'. Further along input will be provided on the dimensions of the sloped limestone exterior perimeter blocks. At this stage it is unknown just how these enormous stones were cut to such high precision or how they were placed in position at such great heights but they are most definitely in place, what is left of them. The perimeter blocks were triangular, or wedge shaped to suit and it was a portion of these and the upper face that eventually vanished long before the surveyors and archaeologists showed up on site. These blocks would have been cut in such a way so that they could be keyed in and interlocked into the body of the structure and footed on a well compacted load bearing surface. The observation is the outer perimeter stones consisted of high grade limestone and these were cut to suit and stacked from the bottom to the top of the structure with no tolerance for error. I believe the above helps to describe the sequence of events that led to the shortening of the base length because there is no other more plausible explanation except the rock shrinkage factor that was accelerated due to the geometrical design of the structure. The survey measurements taken were relative to the erosion and shrinkage factor, therefore the slopes and proportions of the structure would seem about right but at the same time these would have been deceptive because the erosion and shrinkage factors were proportional as well. For now this can be added to the mystery list that surrounds the Great Pyramid but the erosion and rock shrinkage factor theory does help to answer this question. Along with this topic is a simple graphic entitled: Great Pyramid - Erosion Factor. It basically explains the situation and sheds some light on the study. It will be proven the original dimensions of the Great Pyramid were altered in one way or the other, mostly due to erosion and today's measurements have mislead many researchers throughout the past 300 years or so. It will become very clear shortly why I am staying with the base dimension of 440 x 1.732 = 762.08 feet and the reasoning behind this. In chapter 2 it was mentioned the square root of 3 and its relationship to Phi and 1+Phi is important, well, here we are, this is it, while on the thresh hold of great discoveries. I have never visited the Great Pyramid in person but if I did go there I would have a close look at the foundation corners that were included in the surveys and I would wonder what might be found below the ground surface 2.86 feet outside of these walls. If I saw tell tale parallel rock cube lines atop the corner stones and these were limestone and not granite my theory would gain support. Many others have carried out surveys of the Great Pyramid and the surrounding areas so there would be no call for me to set up my instruments.

The crystalline capstone on top of the structure which must have been in place after construction would have eventually become dislodged from its moorings and broken into thousands of small pieces on the way down the 617 foot long bumpy slope, or perhaps it was retrieved by some party in time before it was destroyed and safely stored away. I have given some thought to the concave nature of the pyramid faces and wondered if this was a result of erosion or did it have something to do with its design, I believe this is the case, but no one seems to know about this for sure. One look at the Great Pyramid, even in a photo and the state of its poor condition is apparent. Around its base only a hazy notion of its original limits is deductible. If the proposed height was 484.68 and the height we know of is close to 481 feet according to today's estimate the difference is 3.64 vertical feet, divided by some 5000 years = 0.000728, perhaps a little less than one ten thousandths of a foot, close to a finger nail thickness of erosion per year and this seems reasonable enough. It isn't as if the structure sunk 3.6 feet into the ground, if it was only that easy. Another way of looking at progressive erosion is to divide 5000 into 5 equal periods of one thousand years each. In the vertical plane there would be 3.64, divided by 5 = 0.728 feet of erosion during this period. In the vertical plane on the slope it would be 2.25, divided by 5 = 0.45 feet. This doesn't seem unreasonable. Imagine for a moment the amount of erosion there would be over this huge span of time and it becomes more believable. Limestone is a sedimentary material, softer than granite, therefore the erosion factor is more so supported. In a time period such as this lakes and rivers disappear and new ones form, mountains erode or become higher, entire continents move substantially, the earth's crust shifts, sinks and heaves sometimes dramatically, wooden structures exposed to the elements turn to pulp and disappear, and weather patterns change. Some quarters believe the Great Pyramid is much older than this, and if this is truly the case the erosion theory has even more support. For a drawing project at a workable scale show the smaller dimensions of today's estimate of its size inside the larger one based on the square root of 3 and it will be found that a pencil dot is about equivalent to the distance difference. This exercise not only shows some perspective on the matter, it also provides an appreciation for how truly enormous the structure is.

In this attempt to establish the original base length as 762.08 ft. using the Royal Cubit Theorem the following data is provided. The height is computed at 484.68 ft., divided by 1.732 = 279.84, therefore the slope would be 616.53 ft., divided by 1.732 = 355.96 and these numbers rounded off are very similar to the Royal Cubit dimensions provided by others that show the number of cubits as Base = 440, Height = 280 and Slope = 356 in the order as listed above. Below is a proof run on the computations showing the arrangement for the proposed square root of 3 cubit and pyramid inch propositions to explain how they work.

Base 440 Royal Cubits x 1.732 = 762.08 feet, divided by 0.0866 = 8800, divided by 100 = *88, the number of Constellations.

Height 279.84 x 1.732 = *484.68288 feet.

Side slope 355.96 Royal Cubits x 1.732 = 616.52 feet.

Slope from corner to apex = 418.461 Royal Cubits x 1.732 = 724.774 feet.

Diagonal from corner to corner 622.25 Royal Cubits x 1.732 = *1077.74 feet and another mental note is made of this valuable number.

****

The Following is Pertinent to the Study ~ basic mathematics of the Great Pyramid

****

To begin with, solve the dimension of a right angled triangle having sides of 8800, divided by 100 = 88.00. Hence 88.00squared x 2, the square root of the sum = 124.4508, divided by 2 = 62.2254 becomes a radius. This is the hypotenuse length in pyramid inches, divided by 2, between the foundation corners of the Great Pyramid at the time it was constructed. Multiply this radius by 2 x modern Pi = circumference 390.962. Then divide by 1.07044, the dimension shown in the Royal Cubit Theorem Drawing which is the hypotenuse dimension of a right angled triangle having an adjacent side 1.0 and opposite side Phi squared. Therefore, 390.962, divided by 1.07044 = 365.24, the number of days in a year. It makes proper sense the solution to this question would be derived from a Phi Ratio related arrangement. To be clear on this, the right triangle above is seen in the drawing titled Elements of the Golden Section, center page, where the relationship between Phi, 1+Phi and the square root of 3 are shown. This unique triangle is shown on the drawing titled Royal Cubit Theorem. From the apex of this triangle use 1.0 on the vertical line where 1+Phi is indicated and the distance left on the horizontal, intersecting on the hypotenuse is Phi "squared", and the distance from there back to the apex is the 1.07045. This is the equivalent of 1.732 divided by 1+Phi = 1.07045 Note: It is interesting to see that 88.00 divided by 62.225395 = 1.4142146, very close to the square root of 2 and that when the hypotenuse length of a right angled triangle with these dimensions, square root of (88.00 squared + 62.225395 squared = 107.77731, divided by 62.225395 the result is 1.732047, very close to the square root of 3. In both cases the values are the same to within four places of decimal. Compute the square roots of 1.99999 and 2.99999 and it will be seen that numbers don't necessarily need to be whole in order to acquire correct results. When 4 x 8800 = 35200 is divided by 1.414216 the result is 24,890.14. This is of interest because it works out to within 11.86 miles of the 24,902 circumference value for some reason. The question is, what can be done with this? A mental note is made for now. Some other areas of interest will be covered.

We know that half the perimeter over the height = ancient Pi. The circumference of a circle based on half the width times ancient Pi, divided by the perimeter = the square root of Phi and these details add to the knowledge of structure's geometry. They also relate to time. For example the above base value, 762.08 feet (squared) = 580765. 9264 x Ancient Pi to three places of decimal, 6.289 = 3652436.911, divided by 10,000 = 365.24 days in a year rounded off to two places of decimal. It will ultimately be proven beyond doubt that 762.08 feet was the original base dimension of the Great Pyramid.

Then there is that mysterious dimension from the base corner to the apex which equals 724.774. Using this value the number of days in a year are also computed as follows :

Square root of Phi = 0.786, divided by 100 + 1.0 = 1.00786 x 724.774 = 730.472, divided by 2 = 365.24.

The slope distance, 616.52751 cannot be overlooked. One way to compute this dimension is by using the Pythagorean Theorem:

Step 1.) Compute the diagonal from the base corners using 762.08, then divide by 2. The square root of the sum of the base lengths squared.

Step 2.) Compute the square root of the sum of the squares of its height and half the diagonal. The result is 616.53.

* A simpler way to compute this dimension is:
762.08 x 1+Phi = 1233.0454, divided by 2 = 616.53. This is interesting because no squaring theorem or trigonometry is required to solve the problem. Other simple ways to compute the height and base lengths are: 762.08 x 0.636 = 484.68, 484.68, divided by 0.636 = 762.08.

The number of days in a year can be computed using the slope dimension, the square root of 3 + 1.0 and Phi:

616.53, divided by (the square root of 3 + 1.0 x Phi) = 1.688

616.53, divided by (2.732 x .618) = 1.688

616.53, divided by 1.688 = 365.24

Another rather awkward yet workable way to compute the number of days in a year is as follows:

1.6180339 - 1.61653761 = 0.0015064 x 0.61652751 = 0.0009287,
divided by 2 = 0.0004643 + 1.618 = 1.6184643
Therefore:

1.6184643 x 616.52751 = 997.82776 divided by square root 3 + 1.0, in other words,

997.82776, divided by 2.732 = 365.24...

The solution is based on a combination of interpolation and extrapolation and appears somewhat unconventional yet the correct answer is provided and since this is the case the methodology must be legitimate. The side slope of the Great Pyramid is that particular dimension and a year will always have 365.24 days in it. I am delighted with the result yet it is a surprise at the same time, therefore another mental note is made and I am left wondering where else this unique application might be applied.
Note: the function 0.6165275 is the side slope dimension divided by 1000, used to evaluate astronomical distances further along in this chapter. This function is given the notation A ~ Phi,

meaning astronomical evaluation in progress. It is nice to see that it is applicable in this case and it is rather unique the number of days in a year can be computed using the outer dimensions of the Great Pyramid. Another application of the side slope dimension is, the height, 484.68, divided by 616.5275 = 0.786 = the square root of Phi. Also, out of interest, 484.68, divided by 100 x 2.618 = 2.236, the square root of 5.

Following is yet another rather unconventional way to compute the number of days in a year using the height, and it may draw some skepticism.
Height x Ancient Pi, divided by(6.75 x Phi) = 365.24
484.68 x 3.1439 = 1523.785,
6.75 x Phi = 4.172

1523.785, divided by 4.172 = 365.24...

It is seen that Ancient Pi was adjusted somewhat but 1.9999, divided by 0.636 = 3.144 and 3.1439 is a reasonable approximation of Ancient Pi if the theory that any number is not truly whole is applied and the value 6.75 is easy enough to construct therefore the solution has some merit since it provides the correct answer. There may be some hoots and jeers focused in my direction over this one but the intent is to determine the wondrous capacities of the true dimensions of the Great Pyramid now that they are available. I feel like a kid that got a new toy to play with, or else its the mad scientist in me that can't resist experimenting.

The following alternate method of computing the numbers of days in a year based on the height, the number of days in a week and the number of weeks in a year is interesting but stage trash up to my eyebrows could be the reward for this one:

Opening Statement: 7 is the Solar Number
365.24, divided by 7 (days in a week) = 52.177 (weeks in a year), true...

a.) Phi, 0.6180, divided by 0.6165, the slope divided by 1000 = 1.002... O.K.

b.) 1.5 x 0.618 = 0.927 x 1.002 = 0.9289 x 10 = 9.289... Sounds all right...

c.) 484.68, divided by 9.289 = 52.1778 x 7 = 365.24... Interesting...oh, oh...

Seeing is believing and the numbers aren't lying. The method appears to be legitimate and it beats using those so called pyramid inches and royal cubits others have been using for the past 300 years. This topic will be addressed further along, and by the time I get finished that theory will be smothered in camel dung.

Following is yet another interesting way to compute the number of days in a year using the base length and height first divided by 100, then multiplying by 10:

Given: The height of a Phi based pyramid with a base of one unit = 0.636, the one unit, or 1.0 is added to 0.636 = 1.636 and some decimal point manipulation.
  Length of base = 762.08, Height = 484.68

7.62 x 4.847 = 36.93, divided by
0.636 + 1 = 1.636 x Phi (0.6180339) = 1.0111034
36.93 divided by 1.0111034 = 36.524 x 10 = 365.24…sounds reasonable, hmmm…

When 762.08 is multiplied by 484.68 the quotient is 369364.93 and if Phi to 3 places only is multiplied by 1.636 = 1.011048 then divided into that = 365328.77, divided by 1000 = 365.33, the answer is incorrect. As shown, the above procedure provides the correct solution.

Such is the way with these special numbers, we know that 1 + 1 = 2,
2 + 2 = 4, 3 x 2 = 6 and 2 x 4 = 8 etc., but for example:

1.0, divided by 0.636 = 1.572325 x 2.0 = 3.144654, Ancient Pi
or 2.0, divided by 0.636 = 3.144654
3.144654, divided by 2 = 1.572327 x 0.636 = 0.99999999…
3.144654 x 0.636 = 1.9999999…
3.144654 x 1.5 = 4.716981 x 0.636 = 2.9999999…
0.636 x 6.289 = 3.9999998
Sq. rt. Phi x 10 x 0.636 = 4.9999222
3.144654 x 3 = 9.433952 x 0.636 = 5.999999
3.144654 x 4 = 12.578616 x 0.636 = 7.999999.
3.144654 x 5 = 15.72327 x 0.636 = 9.9999997

Some interesting ones:

3.144654 x 0.618 = 1.9433961 x 0.636 = 1.2359999 = sq. rt. 5 - 1
3.144654 x 1.618 = 5.0880501 x 0.636 = 3.2359998 = sq. rt. 5 + 1
2 x 0.636 = 1.272 = 1.0, divided by sq. rt. Phi
3.144654, divided by 0.636 = 4.9444245, divided by 0.618 = 8.0006868

It is possible this is the basis of an ancient astronomic mathematical system modern science has been unaware of.

It works out that the number 3.0 x Phi, the addition of 0.001 to modern Pi and some decimal point manipulation also provides the number of days in a year.
The square root of the sum of the squares,
of 7.6208 squared x 2 = 10.777 x 6.284 = 67.72

3.0 x 0.6180339 = 1.8541… 67.72, divided by 1.8541 = 36.524 x 10 = 365.24…

Then there is the challenging case where: 1.732, divided by 4 = 0.433 x 6.282 = 2.72,

the square root of 3, divided by 4, then by 100 = 0.0043
x 1.732 = 0.007447, divided into 2.72 = 365.24 days in a year…
Using modern Pi less 0.001…
It does no harm to experiment and play around with the numbers.
Using the numbers 3 and 5 adjusted in a unique yet sensible manner also yields positive results with regards to the number of days in a year:

for 3.0, or 1.732 x 1.732 = 3.0:

2/3 = 0.666, divided by 100 = 0.00666 + 1.0 = 1.00666

3.0 x 100 x 1.00666 = 301.998 x 6.289 = 1899.27,

divided by 365.24 = 5.2, divided by 1.732 = 3.0

1.732 x 3.0 = 5.20,

therefore, 1899.27, divided by 5.2 = 365.24 days in a year.

* 1.732 squared = 5.2 x 365.24 = 1899.27,
divided by 6.289 = 301.998,
divided by 1.00666 = 3.0

for 5.0:

2/3 = 0.666, divided by 100 = 0.00666 + 1.0 = 1.00666

5.0 x 1.00666 = 5.0333 x 6.289 = 31.654,

divided by 365.24 = * 0.086666, divided by 1.732 = 0.05

1.732 x 0.05 = 0.086666

therefore: 31.654, divided by 0.086666 = 365.24 days in a year.

* 1.732 x 0.05 = 0.086666 x 365.24 = 31.654,
divided by 6.289 = 5.033,
divided by 1.00666 = 5.0

Further along another experiment of this type is carried out using Phi...

Days in a year using Square Root of 5 x 1+Phi x 2 x Ancient Pi:

(2.236 x 1.618 = 3.618 x 6.289) = 22.754,
divided by sq. rt. Phi + 2 x 2.236 = 3.6524 x 100 = 365.24 days in a year

It gets even more interesting when the number of days in a year are determined using 2 x Modern Pi again, knowing the diagonal distance from corner to corner = 1077.744 feet, divided by 2 = 538.872.
Hence:

538.872 x 6.283 = 3385.733,
divided by 365.24 = 9.27,
divided by 0.618 = 15.000 exactly.

In other words half the hypotenuse distance from corner to corner x 2 x Modern Pi, divided by 15 x Phi = 365.24 days in a year. It appears then, without speculating on who, or what body designed and constructed the Great Pyramid, the functions of both Modern and Ancient Pi were, beyond doubt, well known to its creators and this challenges historical information on the topic. The reason for this of course is, because the historians didn't know the correct dimensions in the first place, and further more, the above serves as another proof of the Royal Cubit Theorem and the dimensions of the Great Pyramid provided are most reliable.

The simple statement: 15 x 0.618 = 9.27 opens up a treasure chest of values seen in the intriguing relationships between the numbers 15, 15 x 0.618 = 9.27, Phi and the square roots of 3 and 5 and how they relate in terms of the Royal Cubit Theorem.

Please see the following list:

15 x 0.618 = 9.27 known

2.236 x 2.236 x 3 = 15.00 simple enough

1.732 x 1.732 = 3 x 5 = 15.00 yes

15, divided by 3 = 5 O.K.

15, divided by 5 = 3 Got it…get to the point

15, divided by 1.618 = 9.27 interesting

15, divided by 9.27 = 1.618 " "

15 + 9.27 = 24.27, divided by 9.27 = 2.618 " "

15, divided by 1.732 = 8.66 " "
15 x 9.27 = 139.05, divided by 0.618 = 225, divided by 5 = 45, divided by 3 = 75

15, divided by 0.0866 = 173.2, divided by 100 = 1.732 * Royal Cubit Theorem

5, divided by 0.0866 = 57.734 x 3 = 173.2, divided by 100 = 1.732

3, divided by 0.0866 = 34.64, divided by 2 = 17.32, divided by 10 = 1.732
9.27, divided by 0.0866 = 107 = 1 + 7 = 8 = Sign of Infinity,
or, 5 + 3 = 8 = Sign of Infinity
3 + 3 + 1 = 7 = Solar Number…would seem admissible, 1.0 = the all number,

or 5 + 3 = 8 - 1 = 7 = Solar Number, there is no question that 1.0 exists

7 + 8 = 15 " "

9.27, divided by 3 = 3.09 = 0.618, divided by 2 " "

9.27 + 15 = 24.27, divided by 3 = 8.09 = 1.618, divided by 2 " "

9.27, divided by 2.236 = 4.145 = 1.0004 % of Ancient Pi + 1.0 " interesting "

…none of the above are coincidences…mental notes have been made…there is a long way to go yet…the story is just warming up …

It will be seen the value of 365.24 days in an earth year reoccurs like a theme throughout because the true dimensions of the Great Pyramid are available and it serves as a proof whenever it is the result of certain computations and of course the functions of modern Pi and Ancient Pi are invaluable to the inquiry because of their inseparable relationship with Phi.

It is interesting to note that 8800 squared = 77440000 the base area in the revised pyramid inches divided by 4 x 8800 = 35,200 = 2200, divided by 10 = 220, half the length of the 440 Royal Cubit base and there doesn't seem to be any coincidence in this. The numerological significance of 8800 is 8+8=16 = 1+6=7, the solar number. When 440 is added up this way the number 8 is yielded and this is the number of infinity. It appears as though the geometric design of the Great Pyramid is so oriented in these capacities.

The Capstone

Please see the drawing titled: Ancient Pi and the Royal Cubit Theorem. It shows how Ancient Pi is derived from a Phi based pyramid that has a one unit base and how the same is derived from a Phi based pyramid that has a height which is the square root of 3. This is the dimension of the Capstone that was once atop the Great Pyramid and note is taken that when its perimeter is divided by 2 x Ancient Pi the quotient, is the square root of 3. When the Royal Cubit Theorem is applied, 1.732, divided by 0.0866 = 20. To compute the base length, 20 divided by 1.0 divided by Phi squared x 2, divided by 10 = Ancient Pi. For example 20, divided by 1.272 = 15.72327 x 2 = 31.44654, divided by 10 = 3.144654, Ancient Pi. When the perimeter of the capstone in feet is divided into the newly derived base of the Great Pyramid the result is its height. ie. 3048.32, divided by 10.893081 = 279.84 x the square root of 3 = 484.68, the height. When half the height is multiplied by 2 x Ancient Pi then divided by 2 the base length 762.08 is computed. For example 484.68, divided by 2 = 242.34 x 6.289308 = 1524.1509, divided by 2 = 762.08. When 1.732 is divided by 484.68 the quotient is 0.00357 and when this is multiplied by 1.732 the quotient is 0.00618, multiplied by 100 = Phi. And we know that when 484.68, divided by 1.732 = 279.84, the height in Royal Cubits. What appears remarkable is, the height of the pyramid with a base of one unit divided into the square root of 3 equals the base length of the capstone, 1.732, divided by 0.636 = 2.723. And this value 0.636 is interesting to work with as seen above. I was so impressed by the dimensions of the capstone I got some 1" x 2" smooth red cedar and constructed one using those dimensions. At first I set is up facing magnetic north nearby me in my studio then I set it up at the north east corner of the property because it intrigues me so. After a few weeks I tired of it being there so I placed it atop my 10 foot pyramid knowing there would be a mismatch of proportions but I just felt better about it. When I visited there for one of my think tank, chill out sessions I distinctly sensed a richer blend of energies as if they were better connected in some way, therefore that is where the capstone will remain in place for a long while to come. It will be discovered further along when this ever so unique value, the square root of 3 is divided into that number of all as in 1.0, divided by 1.732 x 's 2.0, multiplied by 100,000 a great revelation will be at hand. Perform this simple computation on your calculator and jot the numbers down, then give it some thought.

…save this information for future reference…

****

Sound the Trumpets…" The Great Pyramid Speaks"

Using its newly acquired dimensions it will be seen how they perfectly correlate with the true distance from the sun to earth and the correct speed of light (sunlight) when travelling from its source (the sun) to earth. This important section requires strict attention by the reader as was the case for the drawing titled, Elements of the Golden Section / Phi Ratio. The distance between the sun and earth is known as the Astronomic Unit = AU. Modern science, the others claim its value is 91,848,817 miles x 5,280 ft. / mi. = 1,000,581,319 feet yet the simple conclusion is the height of the Great Pyramid could only be a one billionth part of the distance between the sun and earth. With more than a leap of faith in the Royal Cubit Theorem knowing its 'actual height' to be *484.68288 feet x 1,000,000,000 = 4.8468288 E11, divided by 5,280 = *91,796,000 miles, the correct AU. Note: The 484.68 foot height to two places of decimal is used for more basic computations.

Proof: 484.68288 E11, divided by 1.732 = 2.7984 E11, divided by 1,000,000,000 = 279.84 Royal Cubits. Sound the Trumpets indeed because the "Greatest Teacher and Profit of all Times" has spoken, providing the first very important lesson with many more to follow that will come to the attention of the misinformed scientific community and the world populace. The true AU provided is the distance value between the actual gravity centers of our earth and sun making it possible to compute accurate relative distances to all other heavenly bodies in deep space, and that is just how unique and trustworthy the Ancient Astronomical System is. There is a way to go before it is understood more completely, therefore, patience along the road way to discovery is required.

The correct value for the AU is 52,817 miles more than the value used by modern science a great deal more than a stone's throw.. It will be seen how this has affected previous computations of our astronomers for locations of heavenly bodies further out in space. With regards to the pure quartz crystal capstone which is believed by many to have been in place originally, its vertical dimension could only have been the square root of 3, 1.732…in terms of the Royal Cubit Theorem. The reference on this by the Great Pyramid buffs claim it was 1.719 feet but the true story on that issue is known by now. The original base dimension was 2.723 feet but due to 5000 years worth of erosion the flat area atop the Great Pyramid is much wider by now of course.

The question on the speed of light on a chord in the vacuum of space is easily solved by the use of the Royal Cubit Theorem and the Golden Numbers which is what the science behind the Great Pyramid is all about. The square root of 12 = 3.464 = 2 x 1.732 or, two Royal Cubits. Therefore, SOL on a chord in the vacuum of space = the square root of 3.464 = 1.861182424 x 100,000 = 186,118.242 mi./sec. Because it is important the next question to know the answer to is SOL on the arc of space in the vacuum of space and again the Royal Cubit Theorem kicks in. As follows, Under the title of Royal Cubit / Golden Numbers SOL Theorem: a combination of three values are employed. The first one is the number of feet in a mile 5,280, divided by 10,000,000 = 0.000528 +

1.0 = 1.000528. The second one is the square root of 3.464 = 1.86118242 as shown above. The third one is explained as follows: The AU 91,796,000 miles = 1.56 E-5 light years x 10,000, divided by *6 = 0.26, divided by 1,000,000 = 0.00000026 + 1.0 = 1.00000026.

SOL on the arc of space in the vacuum of space:

1.000528 x 1.86118242 = 1.862165124 x *1.00000026
= 1.8621656 x 100,000 = 186,216.56 mi./sec. = SOL.

The *0.26 factor is the result of a simple computation made on a 2:1 slope = 63.435 degrees (tangent = 2.0) and (co-tangent = 0.5) from the lower left corner on a drawing of the Great Pyramid at a scale of 1:100 to the intersection point on the center vertical line. The total distance value of the line is 4.26 – 4.0 light years from earth to Alpha Centauri = 0.26.The ancient mathematical code in this tells us 0.26 x .05 = 0.13 x 100 = 13, the sun + 12 planets in the solar system and 0.26 x 2 = 0.52 x 100 = 52, the number of weeks in an earth year and the AU 91,796,000 miles = 1.56 E-5 light years x 10,000, divided by *6 = 0.26.

Run this data through on your calculator as a check and there will be no disappointments. If the reader wishes he or she might wish to sketch and compute the details for a right angled triangle with an adjacent side equaling 3.0 and an opposite side of 1.732 and find the hypotenuse length to be 3.464, the square root of 12 or, take it a step further with a right angled equilateral triangle that has sides that are the the square root of 1.732 to find that the hypotenuse length is 1.86118242. The science of this age is unaware that there is no need of a sophisticated stop watch or vacuum tube with a beam of light running through it waiting for its velocity to be incorrectly measured due to the earth's gravity factor. The answer to acquiring the true speed of light has always been available by way of the correct use of the Golden Numbers and the true dimensions of the Great Pyramid.

Another solution for SOL is available using the vertical dimension from the base of the Great Pyramid at center to the floor of the Grotto, *114.312 ft., with some help from those Golden Numbers. This detail is looked at further along but for a different reason.

As follows then:

Square root of *(114.312 x 1,000,000,000) = 338,100.58
x 1.732 = 585,590.20, divided by (2xAncientPi) = 93,108.84
x 2 = 186,217.68 x (6.0, divided by 1,000,000
= 0.000006 + 1.0 = 1.000006), reciprocal = (0.999994)
= 186,216.56 mi./sec. = SOL...For clarity on the matter, 186,217.68 x 0.999994 = 186,216.56 mi./sec.

There are other Speed of Light Theorems like these in the follow-up edition and they are all about how the Golden Numbers are employed to find the correct answer.

The fast track to answering this vital question comes with an understanding of the Golden Number code that has been locked up in the dimension ratios of the Great Pyramid these past thousands of years that has been unobtainable to the scientists and public in general of this time zone.

The following explains the process:

Height of the Great Pyramid = 484.68288 feet, a one billionth part of the AU – the height of the capstone 1.732 feet = 482.95088. Next step, 1.732, divided by *0.0866 the pyramid inch = 20, divided by 2 = 10.0. The sum of 482.95088 + 10.0 = *492.95088 seconds x SOL = the AU distance between the sun and earth. Notice the last two digits of this number, 88 = the number of Constellations and the reciprocal of (492.95088, divided by 1.732, divided by 88) = Phi, divided by 2. These details offer a more than reasonable explanation.

Therefore SOL in the vacuum of space = *484.68288 feet x 1,000,000,000, divided by 5,280 ft./mi. = AU 91,796,000 miles, divided by 492.95088 x 0.999995829 = 186,216.56 mi/sec. x 5,280 = *983,223,436.8 ft./sec. This value for SOL agrees with the ones above and is correct beyond any doubt.

A dual purpose is served when it becomes known it takes 8.215833 minutes for light (sunlight) to reach earth from its source (the sun), while the time estimate by those others is 8.333333 minutes which shows a difference of 0.1175 x 60 = 7.05 seconds.. There is a problem here because 7.05 seconds x 186,216.56 mi./sec. = 1,312,826.748 miles. The value for SOL provided by modern science is 186,282.397 mi./sec. The difference between theirs and the correct one submitted above is 65.837 mi./hr. more, and here are some serious differences that are no different than being off by that proverbial quarter inch on a survey baseline that was discussed earlier. When it comes to the number of miles in a light year their value is off the mark by some 2.1 billion miles. These differences make themselves known time and again as the study progresses, when modern science values are compared to what the trustworthy code of the Great Pyramid tells us, wherein it will be proven that the incorrect values which are being used by today's astronomers have been the root cause of false information resulting from their guess work attempts to accurately determine the dimension details, distances to and locations of celestial bodies in deep space relative to earth's position in our galaxy. The 492.95 second question is solved by applying the number of infinity, 8, divided by 10 = 0.8 and another unique aspect of the Royal Cubit Theorem. As follows: (1.732, divided by 2 = 0.866 + 0.0005359 + 1.0 = 1.0005359, reciprocal = 0.99946 x 616.5275, the slope dimension of the Great Pyramid = 616.19 rounded off to two places of decimal x 0.8 = 492.95 seconds.

For the purpose it serves the true speed of light is within 0.99968% of the estimate by the others. The diameter of the sun is based on the height of the capstone of course. The square root of 3 to three places of decimal = *1.732 ft. = One Royal Cubit, divided by 20 = 0.0866 x 10,000,000,000 = *866,000 miles, the diameter of the sun. It is within 0.9954% of the meaningless guesstimate of 870,000 miles provided by today's astronomers. The difference is 4,000 miles, a shorter stone's throw this time. The root of the problem with regards to the inaccurate values submitted by so called modern science is, since before Newton's time until today there has been no real appreciation or applications using Golden Section Mathematics by the Establishment. It has been proven without this type of input the study of astronomy is pointless. It happens that the equatorial radius of the sun is 866,000, divided by 2 = 433,000 and its circumference is 433,000 x 2xAncient Pi = 2,723,270.44 miles. Note: Use radius x 2xPi to find the circumference of planets and always use radius x 2xAncient Pi to compute the circumferences of all sun stars and galaxies.

The sun has Golden Section qualities that make it a living organism like the Universe itself. It is far more than an enormous sphere of ignited hydrogen. The existence of life on earth is totally reliant on the output of this fiery orb. It is no wonder that ancient civilizations worshipped the sun the source of light and life. They must have known something about it we don't fully understand or appreciate in this age.

The main point of this discussion is the Universe is a living organism which is embodied by the Golden Section Ratio and this is the greatest lesson of all the Great Pyramid has provided us with once its true dimensions became known.

Getting back to the basics, when a foot becomes a mile it is hard to miss the fact the pyramid inch, 0.0866 is a 10 millionth part of the sun's diameter, 866,000 miles. Simple division or multiplication provides the details. The number 1.0 and 7 zeros appear. Shortly, more of this value, the square root of 3 divided by 4 will be seen to show how the correct dimensions of the Great Pyramid relate to earth dimensions. With this information it is possible to compute the angle of the sun's diameter from earth, 00 degrees, 33 minutes, 57.7872 seconds and the angle of the earth's diameter from the sun, 00 degrees, 00 minutes, 17.7386 seconds. The difference is 00 degrees, 40 minutes, 0.05 seconds, virtually 0.666% of one degree and another mental note is made. Which values are correct and the most reliable, and of what importance is it for us to know them to such high degrees of accuracy? This is the basis of science which is in search of absolute values and it now appears as though we have a most reliable model to work with in knowing the true dimensions of the Great Pyramid. It is no wonder why the mysteries of the Great Pyramid have driven many questioners into a state of mental distraction, even illness as is the story of a John Taylor an Englishman in the 19 nth. Century who nearly had to be committed to an asylum because of his fixation on the subject. He had determined the Pi / Phi relationship in its structure but didn't fully understand it. He opposed the metric system and that makes sense. The seekers have always known the answers were there, but just couldn't get to them. That has all changed now, welcome to a new era of understanding through the magnificent power of the Great Pyramid which speaks the language of the Cosmos through the Golden Section to educate us. No doubt, somewhere in its dimensions is a correlation with the size of the Milky Way, which is determined by others to be 80,000 to 100,000 light years in diameter. The answer to this question will be covered further along, This and much more will be discovered in time.

Miles in a Light Year…

Modern science isn't much help on this issue because it has been using an incorrect value for the speed of light all along. To set the record straight on SOL and the number of miles and feet in a light year which are provided by the long lost code that involves the Golden Number Ratios that have been locked up in the dimensions of the Great Pyramid for thousands of years simply pay attention to the units and decimal places to avoid mix-ups. Under the heading of miles in a light year, 365.24 days in a year x 24 hours in a day x 60 seconds in one minute x 60 minutes in one hour = 31,556,736 seconds in one year x 186,216.56 mi./sec. = 5,876,386,823,000 mi./light year, divided by 5,280 = 1,112,952,050 rounded off = 1,113,000,000 providing an important ratio, 0.999956918 (Note 1): 5,876,386,823,000 x 5,280 = a.) 31,027,322,430,000,000 feet in a light year. A revelation is at hand when our favorite drawing titled, Elements of the Golden Section / Phi Ratio undergoes a closer review. That arc subtended by the angle converted to decimals, 63.435 degrees, with a squared value of *1.113 (a unique astronomical value based on one unit) that equates as follows: 1.113 x 5,280 x 1,000,000,000 = 5,876,640,000,000 mi./light year x 5,280 x *0.999956918 = b.) 31,027,322,420,000,000 feet in a light year. It all started with the simple act of bisecting a square of one unit along with some deductive reasoning that made it possible for a progression to be made from one unit through 5,280 ft./mi. to light years. Note 2: a.) divided by b.) = 0.9999999997, reciprocal = 1.0, therefore, a.) = b.). Quite reasonable and simple! Explanation of the arc value: It is seen there is an angle of 63.435 degrees subtended by an arc of which the radius is sq. rt.5, divided by 2 = 1.118. In solving for the arc dimension it is a mistake to use Modern Pi, divided by 180 = 0.0174533 x 63.435 = 1.0715 x 1.118 = 1.2378 which provides a meaningless number close to sq. rt. 5-1. The true meaning of this arrangement becomes clear when it is understood Ancient Pi is used to compute astronomical values in the vertical plane. Therefore, Ancient Pi, divided by 180 = 0.0174703 x 63.435 = 1.1082285 x 1.118 = 1.239, 'square root' = *1.113 x 5,280 = 5,876.64 x 1,000,000,000 x 0.999956918 = 5,876,386,823,000 miles in a light year, that value mentioned above. To avoid confusion the full details of this drawing were not addressed in chapter 2 when acquaintances were being made with the elementary concepts of the Golden Proportion. Now that the decoding process is underway discussions of these details are very important at this stage.

The closest star to earth, Alpha Centauri A is four light years distant. With reference to the velocity of space travel in an ancient era as it is put forward in the following, to make a comparison it is proposed that an astronaut traveling at the speed of light in order to get there would have to be in an induced coma state for that period of time and get his nutrients intravenously. One of the complication would be, when this hair ball in a space suit wakes up orbiting around some unknown planet in that solar system his call to earth would take four years to be received and another four years to get a response. We wish him well on his journey into the unknown. It occurs that if the planets in the solar system revolve around the sun there must be a focal point somewhere in the Milky Way from where they originated that the other stars and their moons and planets revolve around that too but we don't know enough about it at this stage to determine where or what it is and the Milky Way after all, in relative terms, is like a speck of dust, being only a small part of the infinite Universe. It is quite possible the true dimensions of the Great Pyramid can tell us about this but where to start is the question, we haven't advanced to that stage as yet. Should we worry about it? Well yes, if we want the answers. Exponential function mode is needed for these computations but so far we have done some successful navigating in our own back yard, tapping away on a $ 10 scientific calculator, and that is a decent start at least. It is almost too fantastic to believe these ancient dimensions can be used to run checks on current information by today's standards and make corrections on them. This is now possible because the original dimensions of the Great Pyramid have been determined and the implications of this are HUGE! It is concluded , the technology of the others, being those of the rational science persuasion at this time are running in a bumbling second place to the methodologies and accuracy of ancient times. It is mind boggling to say the least that this is the case. Some of the questions we might ask are, how is it that human beings in so very ancient times had this knowledge and how or why did this knowledge become lost? The invitation is there for the popular assumption that visitors with superior capabilities from elsewhere in the Universe visited earth and developed these megalithic structures then left and / or that God was incarnated on Earth in the material plane to do this work. Then there is the other conception or notion that survivors of the Atlantis cataclysm are responsible for the Great Pyramid's design and construction and so forth. Whoever they were it isn't likely they were dressed in animal skins living in mud huts along the Nile River and went out hunting and foraging for food every day. They knew the Giza Plateau was at the epicenter of the earth and were well equipped to do what had to be done there. The problem is there are no accurate records for history to comment on that period on this planet and it is inaccurate about many events as far back as two to three thousand years ago. For all we know there could have been lighted streets, hydrogen or crystal powered air, land and water vehicles and a host of other high standard amenities available for human habitation 50,000 or more years ago. On the other hand, war, famine, disease, drought, cataclysms etc. must all be factors that would lead to losses one would think. Another way of looking at it is, the majority of the worlds human populace have degenerated since those ancient times and the root cause of this I believe is man's great capacity to stray from that which is right and workable in life. He gets a good thing going for himself then indulges in decadence and victimizes himself by becoming engulfed in the downward spiral syndrome of his own making. These factors have been seen in the rise and fall of Greece and Rome which were but brief periods compared to Egypt's long reign, but in the end, this great empire also went into decline for reasons that have been a mystery. The reason for this could be partly due to climate change but one theory speaks of an incidence of incest in the Pharaoh blood line a long way back and that was most likely the beginning of the end for the Egyptian ruling class after many, many generations of tremendous social and scientific development.

Who were the Ancients ?

Since there is no official account of who the Ancients were, where they might have come from and if they ever existed in the first place it is a mystery in itself, therefore, like others have done in the past a challenging and fictional theory is put forward as a tantalizing and possible answer to this riddle:

The Ancients were a super human race from certain planets in the solar systems that revolve around suns which appear as stars to us on earth. One we call Alnilam is the central star in the Orion Belt and the other is in the Sirius Constellation we know as 9 Alp Cm a. These planets are virtually identical or parallel in nature to Earth in many respects except they are much larger. They had an advanced technology and were so highly evolved they were like light beings or highly enlightened immortals, and lived a long, long time functioning as agents with the powers of the Universe at their aide. Their lifetime was equivalent to a Zodiac period of 5150 years in earth time. They had the ability to teleport themselves at the speed of thought = 0.0866 light years per second from their Great Pyramids to earth in the area of the Giza Plateau thousands of years ago where they had joined forces on those projects. Using simple arithmetic the time it took to travel from Anilam in Orion, 1381 light years from earth was 120 seconds or two minutes. From Alp cma in Sirius, 8.66 light years distant from earth it took only 0.75 of a second. They knew all about Earth and the solar system it is in beforehand from research using their fantastic science of astronomy and knowing that traveling vast distances in space at the speed of light was a lost cause they had developed a far superior means of space travel which involved mind power, and cosmic energy. The most abundant mineral in the Universe is silicon dioxide, quartz, and that is what their technology was based on. From their research a measuring system that suited Earth, its solar system and how it related to the Milky Way Galaxy and beyond had been developed by them and the Sphinx and Great Pyramid at Giza were constructed in order that they could return to Alnilam and 9 Alp Cm a from its King's Chamber, the Galactic Transportation Room, once they had completed what they had set out to accomplish on Earth. Using their advanced technology the tools and equipment they needed were manifested with ease from the earth's resources. Because the ancients were from larger planets they could fly and perform fantastic physical feats and operations. Because they were well versed in the nature of all forms of energy they could slip into other dimensions and become invisible at will and reappear at distant locations. They built air machines that were powered by hydrogen and cosmic crystal energy that were capable of lifting enormous weights and could swiftly travel to any location on the globe in them. They had long developed the means for cutting large stone blocks with great precision using laser beam technology and for mortar cement needs the rock was reduced to a crushed state and mixed with local limestone and certain additives such a tree resins etc. Also enormous cranes were constructed to help lift and place the stone blocks into place very accurately. Six other pyramids were constructed at Giza so that an exact image of the seven main stars of the Orion Constellation were proportionately in place. This was very important to the process because it allowed communications with their home bases. They developed relations with the nomadic humanoid types that populated earth at the time and while the projects were conducted around the globe in the planet energy zones the local inhabitants in each area became organized by them to assist with the work and were trained to some degree in the processes. Some of the tradeoffs were, the Ancients taught them about the basics of medicine, the arts and sciences, metallurgy, agriculture and animal husbandry etc. but they could never curb the humans from corruption and waging war on each other. The ancient civilizations from elsewhere had outgrown this waste of time and energy thousands of years prior to their extraterrestrial exploits but it wasn't their place to interfere directly with human affairs on earth. In time the children of the ancients who had impregnated certain earth women with their consent became leaders and within a few generations a more advanced level of earth people began to flourish. Hence, a great and powerful empire known as Egypt emerged and came into its own for thousands of years of practice in the arts and sciences. Other empires such as in Asia, India, Central and South America and Europe etc. developed around the same time as well and testimonial to this is there are remnants of ancient pyramids and other megalithic structures everywhere in the world. The greatest gift to the worlds people was the encoded messages in the structure of the Great Pyramid and many other megalithic structures that told the story about the Phi Proportion, our solar system, the Milky Way and beyond so that earth beings would eventually find their way out there. Once their mission was completed some of them departed for Orion and Sirius from Earth by way of the Sarcophagus inside the King's Chamber that was within the energy zone of the Great Pyramid, and some traveled to other habitable planets at certain locations in the galaxy and throughout the Universe where they continued with the work, while those that stayed had developed a great passion for Earth because it was teaming with life, had a healthy food supply and an abundance of minerals and the people there still needed assistance and leadership. The plan behind these works was to keep tabs on Earth's condition from their distant planets by way of remote sensing using the sacred sites they had installed as signal beacons because of it strategic location in the Milky Way. Their intention is to return to Earth some day to restore the Great Pyramid and Earth then explain everything to us and during the process we will go through a retraining program with them and we will have no choice but to listen to them.

It sounds like the beginning of a script from a science fiction novel not unlike the theme of Super Man and Planet Krypton with similarities to the show known as Star Gate or some of Isaac Asimov.s Star Trek episodes,... beam us up Scotty. The true story of who the ancient people were that are responsible for these scientific marvels is up for grabs because our history knows nothing about them or that era on this planet, yet it is more than apparent, many wondrous undertakings and applications of an ancient science were undertaken by some parties with spectacular capabilities, who possessed a supreme level of knowledge and intelligence to accomplish them. We don't fully understand what these ancient structures are all about to this day and they will never be duplicated by us, but the proof of these works actually taking place during an ancient era on this planet are there before our eyes. The only way we can learn about them is by deciphering the code, or story left behind in stone. For want of a better theory it is a good read and even makes some sense, and for those of us who appreciate thought provoking and mind expanding concepts it provides some entertainment at least. Who knows, when the possibilities and evidence are considered it might be found there are some shades of truth to it. If there was an influence on Earth from elsewhere in the Universe in ancient times I think it is a better story and more realistic than Van Daniken's Chariots of Fire, the UFO theme, because in practical terms no spaceship of any type could travel vast astronomical distances of light years unless it somehow dematerialized and traveled at the speed of thought. However, aside from the UFO theme in Van Danikens writings, I believe he has made some good points. Either the above has merit to it or, indeed, there were highly advanced civilizations developed on earth long before any recorded history was made of that era, and if this was the case, then any records of that time were lost forever. For certain, the following revelations will inspire the reader to think more than just a few times about it.

The Great Pyramid and the smaller ones in the cluster on the Giza Plateau that were constructed there relate to the Orion and Sirius Constellations for some reason and many other megalithic structures throughout the world relate to certain other constellations. Will the truth on the matter ever be known? Well, I would say, put it to the side for now, though the theme of the story is at least partially believable it cannot be proven or disproved. All we can really do at this stage is continue with the study, follow up on any leads and learn as much as possible about the ancient era even if some of the deductions need to be made on a somewhat intuitive level at this time. One way or the other Man's beginnings and destiny are somehow linked with the stars. Some details and thoughts about the potential of the Sarcophagus in the King's Chamber with regards to the above enchanting story will be discussed shortly and none of it will be totally unrealistic because some grand discoveries that relate to the workings and encoded information within our own Great Pyramid is just around the corner.

I had promised myself from the beginning of this book I wouldn't get into the paranormal or the science fiction UFO thing because a lot of it is unacceptable to my way of thinking and what I know for sure, but once the following information is revealed there might be second thoughts about the potential for space travel at the speed of thought, extraterrestrial visitors, and certain events that take place on our planet we don't understand.

The most interesting characteristic about the Royal Cubit pyramid capstone that has a vertical dimension of 1.732, the square root of 3, is when it is divided by 2 the quotient is 0.866 and by 20 the quotient is 0.0866, and it becomes the true pyramid inch as previously discussed because this value is a ten millionth part of the sun's diameter. When 0.0866 is divided by 2 = 0.0433 the ½ pyramid inch becomes available, a ten millionth part of the sun's equatorial radius. Using the Royal Cubit Theorem it will be seen that the base length perimeter of the Capstone, divided by Ancient Pi = 2 x 1.732 = 3.464 = sq. rt. 12 = two Royal Cubits. This detail will become evident further along. These values are a piece of the sun and now it all makes absolute sense how this ancient measuring standard came into being. It was these dimensions and the square root of 3 that Newton was looking for but due to his misinterpretations of the facts the wrong ones were found and have been used for over 300 years and this is what modern science is based on along with an incorrect value for the speed of light. The reason for this is, because he and his following were not wise in the ways of the Golden Ratio and I make no apology for harping on this issue.

Inside the Great Pyramid

Something interesting happens when the vertical height of the Great Pyramid is divided by 100 x 1+Phi. The quotient is 2.9954945 when 161.803 is divided into the height and 161.803 feet on the vertical from center is the level of the top of King's Chamber and its floor level matches the top of the Grand Gallery, where it is said the main energy zone exists, approximately one third the height and its access is through the Antechamber. Therefore 161.803 x 2.9954945 = 484.68. When 484.68 is divided by 3 the quotient is 161.56 and the difference 161.8 - 161.56 = 0.24 x 10 = 2.4 x 8, the number of infinity = 19.2, which is precisely the height of the King's Chamber. This dimension is a bit of a mystery. It's value also comes from this expression: Phi Squared + 4.0 = 4.3819 squared = 19.2. Nevertheless it has been measured many times and it is a reliable reference. The mystery of what the dimensions of the King's Chamber are based on will be dealt with shortly. The 1883 publication by Flanders Petrie came up with the following measurements: H = 19.2 / W = 17.19 / L = 34.38 but air and moisture loss since its construction a very long time ago were not considered. A hands on approach to answering this question is going back to that percentage derived when 762.08 divided by 756.36 = 1.00756%. Then 17.19 x 1.00756 = 17.32 = 10 x 1.732 and 34.38 x 1.00756 = 34.64 = 10 x 3.464, the sq. rt. of 12. This is reasonable because the length and width are now compatable with the sq. rt. of 3 theme that has surfaced throughout the inquiry. He also claimed the slope of the Grand Gallery is 26 degrees and that has proven to be incorrect. A mathematical Golden Ratio proof that also answers this question appears to be as follows: (Ancient Pi divided by 180 = 0.0174703 x 63.435 degrees = 1.10823) / 19.2 divided by 1.10823 = width 17.32 x 2 = length 34.64 creating a 2:1 ratio. Therefore, the question of the length and width of the King's Chamber has been answered two ways. It gets interesting when the diagonal distance is computed, 17.32 squared = 300 and 34.64 squared = 1200 and the square root of the sum of these squares = 38.73 divided by the sq. rt. 5 = 17.32. The length of the King's Chamber to its height is a right angled triangle with a hypotenuse of 39.6 and when this value is divided by 9.0 the quotient is 4.40 which, x 10 = 440, the length of the base in royal cubits. Further along an enlightening reference drawing descriptions which addresses the geometry and proportions of the Grand Gallery, ascending and descending passages is put forward.

Between the top of the Grand Gallery and the King's Chamber is the Antechamber that has a width of 9.6 feet and a height of 12.4 feet. These dimensions are converted to feet using those bogus pyramid inches that I prefer to stay away from, however, with regards to its width and perimeter they do yield some interesting numerical features worth taking note of. For example: 2 x 9.6 = 19.2 = height of the King's Chamber, a 2:1 ratio. Perimeter, 4 x 9.6 = (38.4) + 4 x 12.4 = (49.6) = 88, the number of constellations. Here is where it gets more interesting: 88, divided by Ancient Pi, 3.144654088 = 27.984 x 10 = 279.84, the height of the Great Pyramid in Royal Cubits, divided by sq. rt. 1+Phi = 220, x 2 = 440 Royal Cubits = base length at the time of its construction and 440 x 4 = 1760 Royal Cubits, its perimeter. What is known as the Great Step which is 4.8 feet in height where it meets at the Grand Gallery and it ends 9.6 feet from the north end of the step at the Antechamber entrance. It was determined that the width of the Antechamber, 9.6 feet, divided by 1.732, divided by 0.0866 = 64 pyramid inches, sq. rt. = 8, the number of infinity. Then when 64 is divided by 2 = 32 pyramid inches x (2 x Pi) multiplied by a unique number arrangement provides the number of days in a year. Some drawings and more details about this area of interest and the sq. rt. 5 x 10, divided by 2 = 11.18 foot passageway from the south end of the Great Step into the Antechamber and the sq. rt. 5 x 10 = 22.36 foot passage way to the King's Chamber from the south side of the Antechamber will be made available in the follow-up. Additional information on the dimensions of the Great Pyramid in terms of feet and Royal Cubits will be presented in the final chapter. What has been ascertained by this part of the investigation is that the Antechamber is not just an ordinary access to anywhere but is in actuality a "Star Gate" to the King's Chamber at its north east corner that provides a Golden Section Mathematics understanding of the heavens and the dimensions of the Great Pyramid in terms of Royal Cubits as well. Inside the King's Chamber is a body size rectangular granite box laying north / south along its west wall which was strategically located there by its ancient designers. It is known as the Sarcophagus. It is seen that when its volume is divided by the volume of the King's Chamber the quotient = 0.008 x 1000 = 8.0, the number for infinity. The researchers to this day marvel that the Sarcophagus was carved out of a solid block of red granite. It has been determined by the experts it wasn't used for entombment but that an initiate would lay within it for a length of time, unknown to us, and somehow become metabolically, mentally and spiritually aligned with the earth and cosmic energies. A thought on this is, its use was part of a ritual ceremony for an individual of high office and/or for only the Pharaoh himself during certain cosmic events. One would think the effects of such an activity in the Great Pyramid being where it is located therein, and when the capstone was in place when it was fully functional, could only have been beneficial to whoever was placed in it. The effects must be at least a thousand times more potent than what I experience in my own 10 foot meditating pyramid and yet I do experience positive effects from it because it was designed and built in agreement with the Phi Ratio Proportions. It is possible the Sarcophagus was used to cure disease in upper echelon members of that society or a pregnant noble woman was placed in it to ensure success with the outcome. Atop the chamber laying horizontal are 5 huge 70 ton stone blocks which are spaced by smaller stone blocks and above these are even larger blocks that form a sloping roof cap affair. I believe this rather serious looking affair is or was a condenser of some sort that focused and amplified energies such as solar, piezoelectric, electromagnetic and other cosmic types into the King's Chamber below. From this location and its setup who knows, perhaps long distance communications with distant sacred sites or to other peoples and locations elsewhere in the Universe took place. With such powerful forces at hand the recipient of this type of energy exposure might be capable of teleporting themselves elsewhere or be enabled to explore other times zones and dimensions, or gain a larger degree of longevity or immortality itself. As discussed, from elsewhere in the Universe others might have teleported themselves to earth. There might have been holograms available in the King's Chamber of different places or objects to study. A Pharaoh in his last hours of life on earth might have been placed in the Sarcophagus in order to teleport his soul to a planet in the Orion or Sirius constellations where it would live a new life forever in form. It is conceivable there was a control panel of some sort perhaps in the Antechamber adjacent to the King's Chamber at one time but no evidence of one was ever found. It is possible some treasured knowledge of how these rooms worked is hidden in their floors or elsewhere in the structure. I believe the answers lie within its the geometric makeup and this is the code that needs to be deciphered. There are what is called air shafts from both the Queen and King's Chamber that line up with the Sirius, Orion and Vega Constellations to name some of them and we are mystified by it. This is beyond science fiction as we know it and the full potential of its capacities stagger the imagination of the human mind. The possibilities with this study are immense and the concept is absolutely intriguing but it is doubtful we will ever understand exactly how it worked. There is certainly no harm in trying though.

It can only be guessed at how long the Great Pyramid was fully functional. Would it have been operational for 100, 500 a 1000, perhaps 2000 years or so? At what point in time did it become dysfunctional due to the erosion factor and what was the true purpose it served in those ancient times? A good question is, why was it made inaccessible? From what we have learned it appears as though its interior was made inaccessible for a very long time for some reason. The Queen's Chamber and Grand Gallery etc. will be addressed further along in this chapter. This was an eye opener once the dimensions in feet were available. In using those bogus pyramid inches and royal cubit one needs to be aware because there seems to be different values for these in the references and it can get confusing and misleading. I believe the terms King and Queen's Chamber are incorrect and misleading. The so called King's chamber was a space for transformation, transmutation or galactic transportation of some sort and the Queen's Chamber could have been the preparation area where its west air shafts line up directly with Sirius and I believe once these stars were observed to be in alignment with the shafts during the process it was time to visit the Sarcophagus where the constellation starlight from Orion into the King's Chamber had the desired effect. There is speculation in some quarters that there was a Sarcophagus in the Queen's Chamber but there is no proof of such. If there was the looters must have broken it up and carted it away thinking that the stone had valuable gems in it. Experiment with the Royal Cubit Theorem and a true adventure will be at hand, much progress has been made with its use to date and there is much more to come.

Great Pyramid ~ Sun to the Planets

For opening statements, it has been determined the Sun and Earth are in the Phi Proportion and are therefore living organisms. Actually, the Universe is a living organism because the Phi Proportion matrix is the basis of the energy it operates on. In short, all is life, or life energy and that is a wondrous thing to come to terms with. The following is an investigation and proof that the distances from the Sun to all the planets in the solar system are uniquely arranged in the Phi Proportion and the correct dimensions of the Great Pyramid provides these. Like an observatory without a telescope its dimensions and ratios truly do contain encoded messages that tell the story about the aspects of Earth, its moon, the Solar System and beyond. My honest opinion on the science behind the Great Pyramid and other sacred sites of this vintage is, no rocket ships or space capsules were ever needed to explore the solar system and beyond. The encoded messages tell where everything is, how far, what size and probably much more. All we need is time to fully decipher the code based on the Royal Cubit Theorem. As far as can be determine those who traveled from within the sarcophagus on earth to other locations in the Universe did so at the "Speed of Thought". A most worthy project will be to reconstruct the Great Pyramid to make it functional again so that we can begin to understand how it works and though the task will be daunting it isn't beyond our means to do so now that its true dimensions are known. These challenges lay before us and we have an opportunity to take the right approach now. We need to retrain our thinking and perspectives then carry out more research before moving ahead, and the following will help with the process.

The Great Pyramid and the Solar System

The correct Astronomic Unit is 91,796,000 miles beyond doubt. As stated before, in applying the Royal Cubit Theorem to this value x 5,280 ft. / mi. = 4.8468288 E11, divided by 1,000,000,000 = *484.68288 ft., the original height of the Great Pyramid including its 1.732 capstone. In the table below are guesstimates of the mean radius distances between the sun and the 9 planets by today's rational science astronomy group or, the Newton following it might be said.

Sun to:

Mercury - 36,000,000 miles
Venus - 67,252,952 miles
Earth - 92,977,004 or, 93,000,000 or, 91,848,817 miles
          *Great Pyramid Input AU = 91,796,000 miles
Mars - 141,653,200 miles
Jupiter - 483,884,400 miles
Saturn - 890,944,680 miles
Uranus - 1,782,454,000 miles
Neptune - 2,793,697,900 miles
Pluto – 3,648,017,400 miles

Following is a table showing the mean radius distances from the sun to the planets starting with the distance to our moon and its equatorial and polar diameter to serve as an example based on simple computations using the Astronomical Unit derived by employing the true dimensions of the Great Pyramid and the Golden Ratio:

Earth to Moon: The AU 91,796,000 miles, divided by Phi squared x 1000 = 381.924. The answer comes in as follows: 91,796,000miles, divided by 381.924 = 240,351.48 miles.

Moon diameter, divided by (the square root of 5, divided by 2 = 1.118 x 100) = 111.8. The answer comes in as follows: 240,351.48, divided by 111.8 = 2,149.83 miles.

Moon polar diameter 2,149.83 x 0.997 = 2,143.38 miles.

Moon diameter 2,149.83, divided by earth diameter 7,927.272 = 0.271 + 0.001 = 0.272 = the square root of 1+Phi − 1.0.

Input on the topic by modern science: the moon is 240,000 miles distant from earth and nothing is offered by them on its polar diameter.

Note: The value 1.118, the square root of 5, divided by 2 is Phi + 0.5 and there are special qualities attached to this value with regards to computing planet diameters. It will be seen how this works a little further along..

Below is a table provided by modern science based on its version of the AU value and mean radius distances from the sun to each planet in an attempt to show how Phi is the basis of the solar system geometric arrangement.

Sun to:

Mercury ~ 35,991,298 miles ~ 1.0
Venus ~ 99,275,229 miles ~ 1.86859
Earth ~ 92,977,004 miles ~ 1.38250
Mars ~ 141,653,200 miles ~ 1.52353
Ceres ~ 257,172,150 miles ~ 1.81552
Jupiter ~ 483,884,400 miles ~ 1.88154
Saturn ~ 890,944,680 miles ~ 1.84123
Uranus ~ 1,785,245,400 miles ~ 2.00377
Neptune ~ 2,793,697,900 miles ~ 1.56488
Pluto ~ 3,648,017,400 miles ~ 1.30580

Total…….. 16.18736
Average……. 1.61874
Phi……. 1.61803
Degree of Variance……. (0.00043)

The problem with the above is none of the multiples between the planets are true phi values, though they are somewhat close, but they are meaningless as presented and the hokey values that are used could be easily fudged to make the total to add up to 1+Phi. I find this unacceptable and have taken steps in the following to set the record straight on this making use of what the Great Pyramid tells us. Actually the above list of multiples provided on the right hand side were of assistance while dealing in an intuitive process to ascertain the true Phi values and the planet mean radii. For example, in terms of the Phi Ratio the number 1.86859 could only be ½ Phi + 1+Phi, and the number 1.52353 could only be 2.5 x Phi, and the number 1.30580 could only be 1+ .5 Phi and so on but there is a catch to this which will be clarified shortly. After I determined the mean radius from the sun to Mercury which could only be based on Phi Squared times the AU the rest of the puzzle started to fall into place. The proof of the multiples used below will prove out because a great surprise is at hand when the planet orbit distance values are computed. It will shake the foundations of what modern science thinks it knows about astronomy and the solar system. It has the right ideas but misses the mark just as Newton did.

It needs to be explained again that Phi, like Pi is a number combination that repeats to infinity...0.618033908749894848304586834365638117720309180... The following computations for distances from the sun to the planets are carried out using a minimal number of decimals with the exception of the value for *Phi Squared which will be theoretically determined at its *Infinite Limit. In the process it became apparent an adjusted Phi value for these computations was needed from the sun to Mars and beyond to Pluto. The explanation of the value used is as follows: The slope dimension of the Great Pyramid is 762.08, divided by 2 = 381.04 x 1+Phi = 616.52, divided by 1000 = 0.61652 and becomes known as Astronomical Phi or, A ~ Phi for short. Since the sun to earth distance is known it will be seen how the AU 91,796,000 miles x *Phi Squared equals the distance from the sun to Mercury. Some interesting ratios appear when A ~ Phi is compared with regular Phi, for example:

A ~ Phi, divided by Phi = 0.9976 rounded off to four places of decimal, reciprocal = 1.0024. When 0.9976 is divided into *0.997 the quotient is 0.999 rounded off to three places of decimal. This is the case for all heavenly bodies, a valuable constant to be aware of. Note: 616.52 divided by *1.732 = 355.96 Royal Cubits and a simple conversion from feet to the ancient measuring system is made.

All aboard, no seatbelts are required, we will be traveling in the footsteps of the Ancients to the far reaches of the solar system at the speed of thought, so to speak. The flight will be a Great one because the Pyramid we have so long admired is at the controls. Relax and enjoy the scenery.

Sun to:

Mercury: AU 91,796,000 x *Phi Squared approaching the infinite limit (0.3819709702) = 35,063,407.18 miles.

Venus: 35,063,407.18 x ½ Phi + Phi (1.927) = 67,567,185.64 miles.

Earth: 35,063,407.18 x 2+Phi (2.618) = 91,796,000 miles. O.K.!

Mars: 91,796,000 x 2.5 A ~ Phi (1.5413) = 141,485,174.8 miles.

Jupiter: 141,485,174.8 x 5.5 x A ~ Phi (3.391) = 479,776,227.7 miles.

Saturn: 479,776,227.7 x 3.0 x A ~ Phi (1.850) = 887,586,021.3 miles.

Uranus: 887,586,021.3 x 3.25 x A ~ Phi (2.0037) = 1,778,456,111 miles.

Neptune: 1,778,456,111 x 2.5 x A ~ Phi (1.5413) = 2,741,134,404 miles.

Pluto: 2,741,134,404 x 1.0 + 0.5 x Phi x A ~ Phi (1.309) = 3,588,144,935 miles.

Total: 16.1813 / Average 1.61813 / Degree of Variance = 0.0001.

A great trip, very informative it could be said. There were no telescopes or sophisticated electronic gadgetry involved. A space craft traveling at some mach speed wasn't needed and there were no problems with a nerve wracking blast off, bulky space suits or the problems with weightlessness, oxygen supply and re-entry into the atmosphere etc. We got there and back safe and sound along with a taste of performing this miracle at the speed of thought. In the way the branches grow out from the trunk of a tree in Phi Proportioned gaps similar to those between the planets is seen the structure of the living which is apparent in the workings of the natural domain, our solar system, on Earth, in us and our fellow creatures on this planet and all elsewhere in the Universe. I had taken note of the spacing between the branches of an alder tree that shows a certain lengthy gap spacing on its trunk which is a common constant of their species and with a plot of the planet mean radii from the sun in hand a visual comparison was made and this gap in an alder tree is amazingly similar to the one between Mars and Jupiter. Of course I have done the machete thing through alder brush for clearing survey lines many times but this time it was a special situation. I got curious about it and played out a hunch by uprooting a 6 foot alder tree, roots and all, then I measured the distance between the ground line above the root cluster to the first branch growing out of its trunk. Based on the assumption this distance represented the sun to mercury distance in proportion I performed the computations same as above based on this distance to see if the distances to the upper branches would compare to the planet distance ratios starting at Mercury. To my delight I found corresponding measurements to branch shoots on the alder trunk for Venus, Earth, Mars and Jupiter but the tree was too short to find the branch that matched with Saturn. Had the tree been more mature, 40 - 50 feet the other planets would have been found represented by certain branches I am quite sure. The tentative conclusion with this observation is the position of earth's location is relative to the branches as the sun is to the planets. Between these branches that appear to represent the planets there are other smaller ones that must be moons and large asteroids etc. I will have to find a more mature alder tree and a long ladder and then I might have a problem with gravity, so I'll wait for a calm day and find another party to help me. Actually I think I'll pass on this venture unless I find a large alder tree ripped out of the ground by a backhoe on one of the construction sites I frequent. Never the less, this is very interesting, trees grow from earth toward the heavens and tell us about them. As Above, So Below is a ring of truth that echoes from the wisdom of the Ancient World. This leads me to wonder if certain other plants tell us about other arrangements in the heavens. As it is the patterns of the zodiac constellations are identified with earth creatures and beings such as Aquarius the water bearer, Sagittarius the Archer and Virgo the virgin. A curious reference to the five pointed star fish is made further along with regards to the geometry of the Milky Way. To show what Phi has to do with the planets in terms of the music of the spheres and color are two drawings entitled Planet Color Wheel and The Planets and Rainbows. In these is shown that aside from the regular seven colors there is white which is all colors and black, the absence of color and this makes sense. Both color and sound phenomenon are related through Phi Ratio frequencies or vibrations under the influence of the electromagnetic energy spectrum. In both of these drawings the ratio of green, blue, indigo and violet, the colors of vegetation and water, to the whole is close to 60% which is about right with regards to earth colors as seen from space. It might also be seen that the red, orange and yellow colorations represent the mineral domain which is seen above water from space. The Golden Section based gap relationship between the planets relative to their mean radius distances from the sun can be portrayed accurately now and these show an intriguing relationship especially with color as we perceive it on earth and this has been a mystery for thousands of years. What Newton and Kepler would give to see this. Their concepts were sound in theory but lacked the full and accurate details and the same can be said for Newton's theory of gravity. With regards to astrophysics the true numbers can now be entered into the equation.

The total Phi interval value from Venus to Pluto is: Average 1.61813 – 1.6180339 = 0.0001, very decent. We can live with this result and another proof is at hand. As discussed the way to adjust the Phi multiples past earth is uniquely provided by the slope distance of the Great Pyramid divided by 1000 for astronomical purposes. This must be so because the total is virtually 1+Phi. What is at hand here is a good start and all looks to be decently on track and in order for the most part to carry on. When it comes to astronomical values, precision is of great importance but it doesn't appear as though the ancients were using the nth value of Phi in their astronomical calculations whereas seven places of decimal seems to get the job done quite well. It is noted that the Phi squared value used to compute the Sun to Mercury distance times the AU value added to the 2+Phi value used to compute the distance from Mercury to Earth times the AU value as shown: 0.3819660 + 2.618034 = 3.0 exactly, and the theme of the square root of 3 is well known by now. This is of great interest because it lends further support to the Royal Cubit Theorem and of course the square root of 3, squared = 3, and none of this seems to be a coincidence. Here again it is mentioned, the yard and the foot must have had earlier beginnings than is thought.

In the process the astronomical distance from the Sun to Mercury becomes one unit. In other words it is seen the distance from the Sun to Mercury is based on the multiplication of the AU 91,796,000 miles times the *Phi Squared factor then multiplied by the ½ Phi + 1+Phi factor arrangement from Venus to Earth and the same is carried out from Mars to Pluto using the A ~ Phi values. The Great Pyramid is just like a book of encoded information and the great challenge is to learn how to read it and we are getting there. The Phi multiples are those which occur everywhere in the natural world and the above listed computations for mean radius distances from the sun to the known planets is a proof showing that this Proportion exists everywhere and at all levels in the workings of the living Universe. Now that the distances between the sun and planets are known, the distances from earth to the other planets can be determined with relative ease in terms of a straight line and this serves as a good reference.

Planet Orbit Details

The planets orbit around the sun counter clockwise on an elliptical pathway and the simple law of physics at work here is, a lighter object such as a planet in space orbits a heavier object such as the sun and the same principle is at work in the molecular theory as electrons orbit the nucleus of an atom. The baseline alignment of the planet orbits around the sun are on an azimuth of 63 degrees, 26 minutes, 06 seconds. Please see the accompanying graphic proof of this statement titled, Planet Ellipse Orbit. Further along this critical azimuth becomes well known after a more thorough acquaintance is made with the geometry of the Great Pyramid when the Grand Gallery, ascending and descending passage ways are discussed in detail. In this drawing the basic unit 1.0, superimposed on the proportions of the Great Pyramid are used to help explain the details of the way in which the short and long axis's of an ellipse are computed. Of course the square roots of 2 and 5 makes their golden presences known in the dimensions of the triangle on the left side in the drawing. It is a good time to refer to the drawing titled, Elements of the Golden Section for a moment. The mean radii are given above in the sun to planets list in miles. The dimensions of the long axis are determined by multiplying the planet mean radius from the sun by 1.159869983 or, 1.16 rounded off to two places of decimal, reciprocal = 0.8621656 which is the speed of light divided by 100,000 – 1.0. In order to explain the planet orbits more clearly the following information is put forward: To begin with the sun to earth distance AU 91,796,000 x 1.16 = 106,483,360 miles = one half the major axis of its elliptical orbit which is on a north east / south west alignment and 91,796,000 miles on a north west / south east alignment = one half the minor axis of the elliptical orbit. Hence 2 x 91,796,000 = 183,592,000 miles is the minor axis of the orbit ellipse and 2 x 106,483,360 = 212,966,720 miles its major axis. The distances from the sun to the other planets in the solar system are shown above. The arithmetic is straight forward to equate their elliptical perimeter values and the velocities of each of their orbits. An ellipse perimeter calculator program is available on the net and this will be discussed further along. A drawing titled, The 12 Orbits of the Solar System is made available at the end of the chapter which shows the planet orbit pathways around the sun. Following this drawing is another titled, Inner Solar System - Golden Spiral. There are an infinite number of Golden Spirals emitting gravitational and all other energy forms and forces counter clockwise from the sun, in this case just one is shown to serve as an example and it suits very well. Several more Golden Spirals could have been drawn but they would serve to crowd the details in the drawing. Now that there is clarity on this issue another rather astounding revelation discovered is our sun is dead on line with the Great Pyramid side slope, the Life Angle of 51 degrees, 49 minutes, 36 seconds from the center of the Milky Way which equates to an azimuth of 38 degrees, 10 minutes, 24 seconds. This should stir up a few of today's astronomers. To avoid confusion the above mentioned drawing shows earth on its orbit around the sun at the AU distance to the north west and south east. An azimuth is the angular value clockwise of astronomical north. In the example of the latter, 90 degrees, 00 minutes, 00 seconds = 89 degrees, 59 minutes, 60 seconds - 51 degrees, 49 minutes, 36 seconds = azimuth 38 degrees, 10 minutes, 24 seconds. At this time please see the above mentioned drawings along with the one titled, Geometry of the Milky Way. The angle formed by the intersection of the life angle azimuth and orbit baseline is 25 degrees, 15 minutes, 42 seconds, Tangent = 0.4718, divided by 1.732 = 0.272 = 1.0, divided by the square root of 0.618 - 1, therefore, Gold is present in this value. In terms of the Royal Cubit Theorem other sources of gold become known from the discussion, Tangent, 63 degrees, 26 minutes, 06 seconds = 2.0, divided by 1.732 = 1154734411 x 100,000 = 115,473.44 = diameter of the Milky Way / Tangent, 45 degrees = 1.0, divided by 1.732 = 0.577367206 x 100,000 = 57736.72 x 2 = radius of the Milky Way etc. / Tangent, 51 degrees, 49 minutes, 36 seconds = 1.272 = 1.0, divided by square root 0.618.

With no exceptions the processes of the Universe and life operate on the principle of the Golden Ratio. One is to the whole, as the whole is to one and the circular ellipses of the planets around the sun show the unbreakable relationship between Pi and Phi. Life is in all, all is motion, number, form and the Phi Proportion. A drawing of this arrangement at a suitable scale for what has been derived for the above sun to planet mean radii is found to be highly entertaining. In this is seen the music of the spheres. The planets orbit in tune with the Golden Spiral which emanates from the center of the sun. The distances from the sun to all planets are based on the Pi and Phi proportions which orbit around a revolving sun which spins on its axis and itself orbits a nucleus which is the center of the Milky Way that also revolves around an infinitely distant center. All is living motion in the unlimited expanse of the Universe.

The variable Phi intervals between the planets are like a geometric sequence or cosmic fingerprint like waves that are caused by the influences of gravity, electromagnetism, cosmic radiation and sunlight and are unique to any solar system and there would be similar types of patterns in other solar systems based on the Phi Proportion, depending on the size of their suns one would think. What is determined by this observation is, the actions of all energies in the Universe are Phi and Pi based. It is assumed the ancients must have been very knowledgeable in this entire subject area. It has become apparent that in their grand science they could study the solar system and beyond without the use of a telescope or space vehicles and the design features of the Great Pyramid are all about this and we are in the process of learning how this works. It appears as though the complete record of this information is locked up within the Great Pyramid dimensions and angles. With further research we will be able to go beyond the basics to reach out beyond the solar system and more than scratch the surface of this intriguing topic. My phone is still ringing off the wall for 911 survey emergency needs but I am making time for this extraordinary opportunity to progress with the study. Sometimes I don't answer the phone and just record the messages because I am just too busy with this undertaking. When I come up for air and get a chance to, I get back to the party who called and there I am, out there somewhere in the back forty with the three legged what's it over my shoulder doing the thing I have such a great passion for.

It has been seen how the sun and moon dimensions are derived and shortly the same will be shown for the planets in our solar system and this will be found to be very informative. I am intrigued to see where this inquiry will end up. At this time there are some other proofs to be made which will show beyond doubt that the Royal Cubit Theorem is valid and at this stage all I can say is, whoever wants to embrace this theorem, be my guest. They can proceed with the study and get their own bingo's, then we can discuss it some time if ever our paths should cross. The promise is, there will be no dull moments. So far in this presentation I have merely scratched the surface of the theorem's full potential, yet more progress will be made and the above and following will serve as more than an introduction to get us started and in the end a decent working relationship with the concept will have been established.

In our solar system there are 9 major planets modern science tells us of plus the sun = 10 and that, it might be considered this is the main influence on our mathematical processes, besides that we have ten fingers and ten toes, however in ancient times both the 10 and a hexagonal system were used and there are theories that other planets exist in our solar system and perhaps the above and following will help today's astronomers find them without the use of a telescope. Some years ago there was an intriguing edition on the market titled The 12 th. Planet and at this time it might be worth looking into. On that note perhaps, there are more than 9 planets in the solar system and the Ancients knew about them. What has been determined is the planets are on the alignment of a Living Golden Spiral emanating from that great power house the sun through Mercury which is shown at an azimuth of 333 degrees, 26 minutes, 06 seconds from the sun. This is done to freeze the picture of the planets in place so the geometry of the solar system configuration can be appreciated and understood more clearly. One only needs to think of the geometry of a nautilus or common snail shell to see that " Above is so Below ". Briefly a proof will be shown that there are 12 planets in the solar system making this a ground breaking discovery that has a mathematical basis to back it up, therefore, it could be called a special theorem within a theorem. Please view the drawing entitled: The Great Pyramid Speaks / Showing the Phi Intervals Between the first nine planets we are told of which are shown in position to the Phi Proportion distances between them. The distance from the sun to Pluto is 3.6 billion miles indicating that the diameter of the solar system is 7.2 billion miles but that is rather an odd total. Of course too, there isn't much wrong with the 7 factor, the solar number. On the other hand it could be speculated the solar system actually has a diameter of around 10 billion miles since ten is such an agreeable multiple to work with, and there is a 10 nth. planet within the limits of the solar system. No one seems to know and the question tends to leave us in the dark. However, since Pluto is the ninth station from the sun a reset at ten appears to be applicable. The distances from the sun to this planet could be determined without using a telescope as shown above.

Very Important: Introducing Planets10, 11 and 12 of the Solar System

*** The following information might be called speculative astronomy however the findings present a number of surprises for the reader and astronomers of today. It started as an entertaining concept there might be a Planet 10 nicknamed XB10, short for a planet that has a mean radius of 10 billion miles making a degree more sense than 9 planets + the Sun = 10. Then it took off when it came to the ancient duodecimal system we have heard so much about and the distance from the Sun to the 12 th. Planet became the object of the inquiry. It was discovered the distance from Pluto to Planet 10 is based on the cyclical number Pi + 3.0 = 6.14, divided by 10 = 0.614 x 2.5 = 1.535 x 3,588,144,935 miles (the Sun to Pluto distance) = *5,507,802,475 miles (Sun to Planet 10) x 2 = *11,015,604,950 miles (Sun to Planet 11) x 2Pi = 69,211,253,750 miles (Sun to Planet 12) x 5,280 feet in a mile = 3.654 E14, divided by 1.0 E12 = 365.4 – 365.0 = 0.4 x 60 = 0.24 + 365.0 = 365.24 days in a year. Bingo!

Distance from the Sun to Planet 12 in proportion to the diameter of the Sun:

69,211,253,750 miles, divided by 866,000 miles = 79,920 x 1.001001001 = 80,000, divided by 10,000 = 8.0, the number of Infinity. Proofs of there being 12 Planets in all solar systems have become apparent. Please review the following:

The time it takes sunlight to reach Planet 12:
69,211,253,750 miles, divided by the correct SOL = 371,670.7781 seconds, divided by 3600 seconds in an hour = 103.2418828 hours, divided by 24 hours in a day = 4.30 days, divided by 2 = *2.15. We know 186,216.56, divided by 0.0866, the Pyramid Inch = *2,150,000 and the diameter of the Sun 866,000 miles x 2 = 1,732,000, divided by 1,000,000 = *1.732, the Royal Cubit x 2 = 3.464, the square root of 12 from which the speed of light is derived and again the Royal Cubit Theorem kicks in. A proof that indeed there are 12 planets in the solar system is at hand and this explains why a duodecimal, or dozen system was used in ancient times. The code that has been locked up in the Great Pyramid for thousands of years informs us of this and it is now understandable why the 12 factor can be used to derive the square root of 3 using the earth dimensions and this is where the value, 12, one dozen comes from and the 12 inch foot and 5280 foot mile live on, while the metric system is left behind in the junk yard where it belongs. To make it brief, in the New Testament with reference to the Son and 12 disciples, an uncanny parallel can be drawn, because, in our solar system there exists, the Sun and 12 planets. It might make one wonder if it was ancient knowledge again, that somehow filtered into the minds of the scribes and became a basis for the storyline in the scriptures. What ever the case might be, at this stage, the above sun to planets list can now be completed:

Sun to Planets 10, 11 and 12 review:

Planet 10: 3,588,144,935 x 1.535 = 5,507,802,475 miles.

Planet 11: 5,507,802,475 x 2 11,015,604,950 miles.

Planet 12: 11,015,604,950 x 2Pi = 69,211,253,750 miles.

Since the distance from earth to Planet 12 is 69,211,253,750 – 91,796,000 = 69,119,457,750 miles Voyager I is far from the outside of the solar system. If we are to believe it has been travelling at a rate of 3.6 x the AU per year since 1977 the arithmetic is simple using a basic calculator. As of 2008 it will take another 22 billion years.. I trust there are no disappointments in this revelation, after all outer space is a big place. The counter clockwise planet orbits are appreciated in the drawing titled

Planet Orbits - Solar System. An attempt to show all 12 planet orbits is shown in the accompanying drawing titled, The 12 Orbits of the Solar System but it falls short of expectations because only the orbits from Jupiter to Planet 12 are visible. One remedy to this problem is to expand the drawing by 200% and all orbits will be seen. Beyond that, a little imaginative visualization is a help. This sort of detail could be portrayed on a live screen showing the planets revolving around the sun at their various velocities and intervals between each other. The positions of the planets relative to the sun as if they were motionless to see where they are in the cosmic plan are determined by multiples of 90 degrees from the azimuth of the solar system 63 degrees, 26 minutes, 06 seconds, or 2:1 slope. The logic in this is 12 x 360 = 1080, divided by 360 = 3, divided by 1.732 = square root of 3... See Royal Cubit Theorem.

Comparisons of the dimensions derived by using the Great Pyramid AU value, divided by Modern Science Values can be computed easily enough. To put it bluntly, their information is wrong. The story is, today's science is stumbling around in second place and those stone throw approximations are getting a lot more serious out there at Pluto, some 3.6 billion miles from earth. Here there is a difference of 62.7 million miles. This equates to 0.983 %, a rather a huge difference. The problem is, modern science has been using the wrong values for the Astronomical Unit and the speed of light all along. It will be seen what this means when the divisions for percentage differences are carried out. I am left wondering if these true values have ever been seen before in this time era? I think not, therefore, life is full of surprises and there is a first time for everything. Let the history and text books handle it, there is much more work to do before this inquiry is concluded.

The straight forward way to determine the time it takes sunlight to reach Pluto or any of the other planets in our solar system is by knowing the correct distance from the sun to them then divide by the speed of light and these have been accurately determined in this presentation. However another way to compute the time it takes sunlight to reach Pluto and the other planets is as follows: Example: Pluto *3,588,144,935 miles x *492.95088, divided by *91,796,000, divided by 3600 = 5.35 hours. As discussed previously the alternate solution mentioned above works out as well.. By traveling at a velocity of Mach 3.6 to Pluto the space craft would likely be destroyed by asteroid strikes and aside from that the crew would surely die from cosmic radiation exposure. In a mental sense we have gotten there much faster than the speed of light because the mind was traveling at the speed of thought and there is something of great importance to be said for this. The question is, just how does the Sarcophagus work? Some day in future we might figure it out when more is known about mind power and cosmic energy. Today's physicists and egg heads are tearing their hair out trying to figure out schemes for space travel at rates beyond the speed of light and to date they are having zero success with their concepts, yet it would seem the answers lie in the Ancient technologies. This is the way it stands as the above computed dimensions will check out and are to be relied upon. It might be agreed the science behind this is absolutely staggering, perhaps it could be called Universal Geometry, after all, it is about the Geometry of the Universe, either title might do. Hmmm, any one for Celestial Geodesy covering astronomical distances at the Speed of Thought ? My survey instruments were not required for this operation, it looks as though I am out of a job for the time being, but promising results are coming in just the same. We have certainly gone a long way since 440 Royal Cubits was multiplied by the square root of 3 to establish the true base, height and slope dimensions of the Great Pyramid and now we are on a roll and there is a ways to go yet. The following should be very informative as well.

Planet Dimensions ~ Compliments of the Great Pyramid

Note (a): Multiply the equatorial radii by 0.997 for the polar diameters. All planet diameters are rounded off to the nearest mile unless otherwise indicated. Aside these values are modern science dimensions in brackets. Note (b): The multiples used in the following list are to 3 places of decimal, such as Phi = 0.618, the square root of 5 = 2.236 and the square root of 5, divided by 2 = 1.118 etc.

Mercury Diameter: 35,063,407.18, divided by 10,000 x 1.118 = (11,180.0) = 3,136 miles. (3,033)

Venus Diameter: 67,567,185.64, divided by 10,000 x 1.118 x 1.618 = (18,809.24) = 3,592 miles. (3,761)

Earth Diameter: Earlier in this chapter the Royal Cubit 1.732 ft. to three places of decimal was derived from the Earth dimensions when the equatorial diameter and polar radius were determined. These values are 3,963.636 x 2 = 7,927.272 miles and 3,950 miles.

Out of interest the alternate way of determining the diameter of Earth is as follows:

(2.236 + 1.0 x 10 x 0.618) = 19.99848 squared = 399.9392 + 10,000 x 1.118 = 11,579.939 91,796,000, divided by (11,579.939) = 7,927.157 miles, very close to the diameter value we have been using. (7,899.8)

Mars Diameter: 141,485,174.8 – 91,796,000 = 49,689,174.8, divided by 10,000 x 1.118 = (11,180) = 4,444 miles. (4, 217)

Jupiter Diameter: 479,776,227.7 – 91,796,000 = 387,980,227.7, divided by 1,000 x 1.118 = 1,118 x 4 = (4,472) = 86,758 miles. (88,846))

Saturn Diameter: 887,586,021.3 – 91,796,000 = 795,790,021.3, divided by 10,000 x 1.118 = (11,180) = 71,180 miles. (74,145)

Uranus Diameter: 1,778,456,111 – 91,796,000 = 1,686,660, divided by 50,000 x 1.118 = (55,900) = 30,173 miles. (31,771)

Neptune Diameter: 2,741,134,404 – 91,796,000 = 2,649,338,404, divided by 100 x 1.118 = 111.8, 111.8 squared = 12,499.24 x 7 = (87,494.68) = 30,280 miles. (30,782)

Pluto Diameter: 3,588,144,935 – 91,796,000 = 3,496,348,935, divided by 100 x 1.118 = 111.8 squared x Pi = 39,266.362 x 1,000 = (39,266,362.46) = 89.0 miles. (55 to 124)

Planets 10, 11 and 12 Diameters:

To begin with, modern science has no idea what the following diameters are because, in fact it has only a vague inkling that there are 12 planets in the solar system, yet an undisputable proof of this has been put forward. The following values in this uncharted area are based on a review of what takes place when the distances from the sun to Mercury, Venus and Earth are dealt with to acquire their diameters. Above, the total of the Phi and A ~ Phi multiples from Mercury to Pluto was 16.18 and this worked out very well. From Pluto though Planets 10 and 11 through to Planet 12 the multiples are 2.5 + 2.0 + Pi = 7.64, divided by Pi = sq. rt. 3, divided by 4 + 2.0 = 2.433 and the Royal Cubit theorem is in evidence. In the case of finding fixes for these planets the multiple values are augmented proportionately to suit in order to handle these large astronomical distances and the computations are the same because the solar system ends the way it began in this respect to complete the cycle. For example the 1000 multiple used for the Mercury diameter computation is increased by 100,000 times to handle the distance value to planet 10 and as shown for planets 11 and 12. It makes sense to handle it this way and though the proposal might be disputed, at this stage it seems reasonable enough, therefore it is worth the effort which will at least provide a reference. There is no harm in trying, the planets won't stop rotating around the sun if this proposal happens to be wrong, however, even though it will be a very long while before these dimensions can be physically proven, a mathematical proof has made itself known. For the record the following presentation on Planets 10, 11 and 12 and the others is not merely a mind expanding exercise that ends up nowhere. Because the long lost code from the true dimensions and ratios of the Great Pyramid have been deciphered making it possible to determine all aspects of the solar system details.

Planet 10 Diameter: 5,507,802,475, divided by 100,000,000 x 1.118 = (111,800,000) = 49 miles.

Planet 11 Diameter: 11,015,604,950, divided by 1,000,000,000 x 1.118 x 1.618 = (1,808,924,000) = 6 miles.

Planet 12 Diameter: There are two ways to determine its diameter. Please note: 1.732, divided by 2 = 0.866, divided by 1,000 = 0.000866 + 1.0 = *1.000866.

Solution (a): 69,211,253,750, divided by (2.236 + 1.0 x 100,000,000 x 0.618) = 199,984,800 squared = (39,993,920,000,000,000) = 1.73054 x *1.000866 = 1.732 miles.

Solution (b): The diameter of Planet 11, divided by the square root of 12 = 1.732 miles or, to express it in more simplified terms 6 miles, divided by 3.464 = 1.732 miles. The impact of this revelation will be seen as the following information unfolds.

Another detail lest we forget: 1.732 ft., divided by 2 = 0.866 x 1,000,000 = 866,000 miles, the diameter of the sun.

The diameters of Planet 10, 11 and 12 are diminished in size compared to the others because they are at the far end of the Phi Energy Zone in the solar system. The fact that the diameter of Planet 12 is 3 miles appears to be of significance because the square root of this number is a factor that is one of the major parts of the mathematical theme in the inquiry. The following might shed some light on the discussion when it is seen that 3 x 5280 = 9145, divided by 1000 = 9.145 - 6.0 = 3.145, Pi, the cyclical number. Also, the total of the diameters of Mercury, Venus, Earth = 14,695, divided by those of 10,11,12 x 100, 102 = 144 = 12 squared. Wow! Make a double take on this. Further along the planet dimensions will be looked at and discussed to determine how they relate or correspond to each other in a sensible, or meaningful way.

The above endeavor speaks for itself and as discussed the value Phi + 0.5 proves to be an important number to work with in this aspect of the undertaking. Some of the percent comparisons with modern science input for planet dimensions are fairly close but it appears it had a problem with Pluto. Today's astronomers need to do a little more than wipe the fog off their telescope lenses, they need to listen closely to what the Great Pyramid has to say about the solar system and the planet dimensions, and astronomy in general now that it is talking to us. The axial tilt on the planets is a natural occurrence in the solar system influenced by gravity. The varying seasons on earth are a result of the angular tilt of its axis and has little to do with the distance from the sun. The maximum tilt from perpendicular to the ecliptic plane is close to 23 degrees and 26 minutes. The solstices are days when the sun's rays reach earth when it is at its farthest northern and southern declinations, or when earth is at its furthest distance from the sun on its orbit path on December 21/22, the shortest day of the year in the northern hemisphere, June 21, the longest day in the northern hemisphere. In the southern hemisphere it is the other way around. The equinoxes in the northern hemisphere occur on September 22 and March 21 when the days and night are of equal duration and that is when earth is 91,795,450 miles from the sun the AU mean radius distance. We were going so fast I didn't get a chance to set up my survey instruments, oh well, life goes on and we will travel on with the ancients to complete the survey of the solar system. In a manner of speaking the heavens will be traversed…in the end.

Orbit, Time Periods and Velocities of Planets in the Solar System

As for the geometry of the solar system, the planets, their moons, the sun and their orbital paths are entirely and uniquely ellipsoidal in the Phi Proportion and that is the way of the natural domain in all the Universe where there are no straight lines. An ellipse is simply defined as a closed curve of oval shape but there is no easy way to compute its circumference exactly. The mean radii from the sun to each planet are known, these are multiplied by 2 for the short dimension of the ellipse, and distances north and south, shall we say, are based on 1.16 times the mean radii times 2 which defines the long dimension of the ellipsoid orbit of each planet pathway, therefore the ellipsoid circumference distance each planet travels around the sun can be computed using a scary looking foot long formula as follows:

CE = Pi x 2 x Square Root of (½ Long Axis) Squared + (½ Short Axis) Squared

Oh, oh! To make it a little easier the long and short dimensions data of the ellipse can be fed into a programmable calculator or computer program that handles such problems and it will provide at best an approximations in order to acquire the circumference. The only force in the Universe that can compute an ellipse circumference with total accuracy is nature itself because it knows exactly what it is doing at all times. I should mention that drawing an ellipse manually is somewhat difficult. There is a rather complicated way to do it that will be discussed further along, but if you have a drawing program that includes the ellipse/draw feature you have it made.
Before we get into the next topic a review of some interesting values as they relate to the shape of an egg using the true dimensions of the Great Pyramid are available.

Base 762.08 feet
Height 484.68288 feet

Starting with an ellipsoid that has a long axis of 2 x 762.08 = 1524.16 feet and a short axis of 2 x 484.68288 = 969.36576 feet. The object is to determine its circumference.

When this data is entered into the ellipse calculator program the circumference value is 4012.60, divided by 2 = 2006.30. When this value is divided by 1+Phi the result is 1240.00 the elliptical circumference part of the egg. The next step is to compute the semicircle part of the egg, 484.68288, divided by 2 = 242.34144 x 2xAncient Pi = 1524.16. As it happens 2006.30 + 1524.16 = 3530.46 squared =1,250,000, divided by 100,000,000 = 0.125, reciprocal = *8.0 the number of Infinity which is about what one of the main lessons the Great Pyramid teaches. In other words it reminds us that the Universe is infinite in nature.

Keep track of this information as it will relate further along with reference to the Geometry of a Phi Based Egg when it becomes another intense topic.

To carry on with other important issues:

Sun to Earth AU = 91,796,000 miles.

Short Width Axis = 91,796,000 x 2 = 183,592,000 miles.

Long Width Axis = 91,796,000 x 1.16 = 106,483,360 x 2 = 212,966,720 miles.

The answer on the ellipse calculator was 620 million miles or so. Based on intuition and another leap of faith in the Royal Cubit Theorem the distance value for earth's orbit around the sun could only be Phi x 1,000,000,000 = 618,000,000 miles. The following list of the other planet orbit distances and velocities will be based on this value with reasonable accuracy.

The astronomers of the day claim the circumference of earth's orbit is around 584,000,000 miles based on an incorrect AU x 2Pi and its orbit velocity is some 67,000 m.p.h. yet they generate drawings of the orbit alignment as an ellipse. The percentage of their value to the correct one is around 0.94%. As far as I'm concerned they are out to lunch on the issue and the jig is up with this crowd because they are stuck in the dark ages when it comes to understanding astronomy.

Earlier in this chapter reference was made to there being 365.24 days in a year x 24 hours in a day = 8765.76 m.p.h. Therefore, the orbit velocity of earth around the sun is 618,000,000 miles, divided by 8765.76 = 70,500 m.p.h. rounded off. Using the above information it becomes known all is motion, time and velocity in the proportions of Phi as they relate to the workings of the solar system, God, Life and the Universe. Twentieth and twenty first century physicists and astronomers should know "The Theory of Everything" was well practiced during an ancient time on earth. The problem is, today's scientific community is supremely ignorant about applications of the Divine Ratio. According to the general theory of relativity a black hole is a region in space from which nothing, including light cannot escape because of the tremendous forces of electromagnetism that exist in one. In this age nothing is known about extracting and making use of this energy but it becomes apparent that the Great Pyramid could do so when it was fully functional. With regards to information loss due to black holes, the main one is, there are no records or information on how this Grand Science was employed during that ancient time on this planet. As discussed earlier, the Universe is perpetually and instantaneously in a state of expansion and contraction which is what holds it together and the conscious spirit is the closest thing to infinity there is. If today's astronomers and physicists wish to communicate with intelligent life elsewhere in the Universe, perhaps they should send the correct type of intelligible messages or signals in its language which are provided in the above and following. It has become obvious the Great Pyramid speaks the language of the Universe through the Golden Ratio. They might direct these signals to the constellations of Orion and Sirius to begin with. It might take a long while to get a response using today's equipment but it is worth a try and it could happen sooner than expected. The thing of it is, the entire Universe is very much alive and we need to learn how to live with it and within it, and at the same time be at one with it. The only way we can succeed with this is by understanding it on its terms and this will happen when we become fully informed on the details of the Golden Section, the Divine Ratio. It is a great loss to civilization that this has not been available in the education system and that is why it would have been a waste of time for me to have set foot in one of those places of so called higher learning. Without fully realizing it when I was younger this subject area was the game I was after, and the prize was to gain the larger understanding of life, and I am free to chase it down without any interference from the so called authority on "the Theory of Nothing." No thanks...

The above also shows the unique relationship between the Earth and Sun, thanks to input deduced from the true Great Pyramid dimensions. Mercury is part of the family and has something to say about this:

Knowing the Phi multiple value that approaches the infinite limit of *Phi Squared x AU 91,796,000 miles x *0.3819709702 = 35,063,407.18 miles, the distance from the sun to Mercury. The perimeter of its orbit according to the ellipse calculator is virtually 236,000,000 miles, divided by 1,000,000,000 = 0.236 = the square root of 5 – 2.0. The length of one day on Mercury is 24 earth hours x 0.3819709702 = 9.167 hours. One year on Mercury is 365.24 x 0.381970902 = 139.5 years and the orbit velocity of Mercury around the sun is 70,500 m.p.h., earth's orbit velocity x 0.3819709702 = 26,929 m.p.h. It stands to reason that Mercury the closest planet to the sun has the slowest orbit velocity.

With these revelations in hand for the first time in the history of this time zone it is now also possible to compute the motion and time factors, orbit perimeters and the orbit velocities of every planet in the solar system using a space lab supplied by the Ancient Sciences. The teachings of the Great Pyramid on these issues are as follows:

Moon orbit distance around earth = 1,618,000 miles, divided by earth's orbit 618,000,000 miles = 0.002618 x 1000 = 2+Phi. The moon circumference 6,754.13 miles x (Phi, divided by A ~ Phi), divided by 24,903.5 that of earth = 0.272 = the square root of 1+Phi – 1.0. Furthermore, 0.272 x 100 *27.2 days x 24 hours in a day = 652.8 hours. Therefore, the moon's orbit around earth is 1,618,000, divided by 652.8 = 2,478.6 m.p.h. an enduring Pi / Phi relationship has always existed between earth and it moon. The astronomers of the day claim the orbit time is 27.322 days, within 0.9955% and their velocity of 2,236 m.p.h. is .90% of what it actually is. All too familiar looking percentage comparisons by now.

Following is an eye opening story on the planet orbit dimensions and their velocities around the sun. The Phi multiples between the planets have been derived and the earth and moon orbit dimensions have been computed accurately, therefore the following additional information is presented:

Mercury orbit perimeter *236,058,059.6 miles. Orbit velocity = *26,929.6 m.p.h.

Venus orbit perimeter 236,058,059.6 x 1.927 = 454,883,880.8 miles. Orbit velocity 26,931.054 x 1.927 = 51,896.14 m.p.h.

Earth orbit perimeter 236,058,059.6 x 2.618 = 618,000,000 miles. Orbit velocity 618,000,000, divided by 8765.76 = 70,501.6 m.p.h.

Mars orbit perimeter 618,000,000 x 1.5413 = 952,523,400 miles. Orbit velocity 70,501.6 x 1.5413 = 108,664.1 m.p.h.

Jupiter orbit perimeter 952,523,400 x 3.391 = 3,230,006,849 miles. Orbit velocity 108,664.1 x 3.391 = 368,480 m.p.h.

Saturn orbit perimeter 3,230,006,849 x 1.85 = 5,975,512,671 miles. Orbit velocity 368,480 x 1.85 = 681,688 m.p.h.

Uranus orbit perimeter 5,975,512,671 x 2.0037 = 11,973,134,740 miles. Orbit velocity 681,688 x 2.0037 = 1,365,898.2 m.p.h.

Neptune orbit perimeter 11,973,134,740 x 1.5413 = 18,454,192,570 miles. Orbit velocity 1,365,898.2 x 1.5413 = 2,105,258.9 m.p.h.

Pluto orbit perimeter 18,454,192,570 x 1.309 = 24,156,538,070 miles. Orbit velocity 2,105,258.9 x 1.309 = 2,755,783.9 m.p.h.

Pluto orbit perimeter 18,454,192,570 x 1.309 = 24,156,538,070 miles. Orbit velocity 2,105,258.9 x 1.309 = 2,755,783.9 m.p.h.

A comparison made between the Pluto orbit perimeter distance generated by the ellipse circumference calculator program and the one above came to 0.99% and it was the same for the orbit perimeter of Planet 12, decent results.

Planet XB10 orbit perimeter 24,156,538,070 x 1.535 = 37,080,285,940 miles. Velocity 2,105,258.9 x 1.535 = 3,231,572.4 m.p.h.

Planet 11 orbit perimeter 37,080,285,940 x 2 = 74,160,571,880 miles. Velocity 3,231,572.4 x 2 = 6,463,144.8 m.p.h.

Planet 12 orbit perimeter 74,160,571,880 x 2xPi = 465,952,272,400 miles. Velocity 3,231,572.4 x 2xPi = 20,304,030.4 m.p.h.

It is seen each of the planet orbit perimeters and their velocities are reliant on the Phi intervals between them as shown in the Sun to Planets list above. This sort of information is a must in all science class rooms especially for the astronomy professors who should be providing straight goods on the subject matter to the students of this era one might assume.

Orbit Distance Planet 12 Investigation:

Planet 12 orbit perimeter = 465,952,272,400 miles x 5,280 feet in a mile = 2.46 E15 feet, divided by *15 = 164,000,000,000,000, divided by 100,000,000,000,000 = 1.64 – 1.0 = 0.64 x 100 = 64 and the square root of 64 = 8.0 the number of Infinity.

Planets 10, 11 and 12 have been reclaimed from the cosmic lost and found and they need names. This will be addressed in chapter 8. It has come to light that every number value in the above and following presentation makes total sense because the laws and rules of Golden Section Mathematics have been diligently applied.

There has been reference made to a 360 day year in ancient times, or 12 months x 30 days each = 360 but this isn't possible unless the earth orbits the sun at a rate 98.6 % of its present speed and that isn't the case today, never was and never will be. It appears that one year has 365.24 days in it and whoever master minded the design and construction of the Great Pyramid were well aware of this. After all this value appeared when the short width axis of Planet 12 was multiplied by 5280 = 3.6524 E 14. Later on, when the dimensions of the Milky Way are determined, it will be seen that 12 x 30.43666 days in a month = 365.24 days in a year. The claim that a 360 day year existed in ancient times is no more than a whimsical assumption promoted by less informed parties in more recent times.

The above fascinating information shows how the Phi proportions between the Sun and the planets in the solar system relate with regards to their sizes, orbit dimensions and velocities. This is the 'Language of the Universe' in action. Once I became acquainted with the basics I had no problems and found the computations to be correct. The real check is shown above in the Sun to Planets list where the total of the Phi Multiples add up to 1+Phi between Mercury and Pluto. It was seen how the regular Phi factor was employed from the Sun through Mercury and Venus to Earth and how the A ~ Phi factor was used from Earth to Mars then onto Pluto. I believe these results are trustworthy and show aspects of the solar system not known by modern science. They will surely be astounded with these findings as we move ahead out of the dark ages. A good start in the study has been made using a basic scientific calculator but once computers are programmed to manage computations such as these the process will become more refined and far more efficient.

Another list:
The *12 Planet Circumferences in miles:
The planet diameters are listed above.

Mercury: 9,851.7 / Venus: 11,284.3 / Earth: 24,903.5 Mars: 13,960.8 / Jupiter: 272,550.3 / Saturn: 22.362.0 Uranus: 94,788.5 / Neptune: 95,124.6 / Pluto: 279.6
Planet 10: 153.9 / Planet 11: 18.9 / Planet 12: 5.4

Total: 545,283.5 miles.

To start with: 545,283.5, divided by *12 = 45,440, divided by 2 = 22,720, divided by 10,000 = 2.272 – 1.0 = 1.272 squared = 1.618. Here we have the square root of 1+Phi, the value used to determine the height of the Great Pyramid multiplied by ½ its base dimension. Also, 545,283.5 x 5,280 = 2,879,096,880 feet, divided by 12 = 240,000,000, divided by 10,000,000 = 24 hours in an earth day. It is also interesting to find that 545,283.5, divided by the circumference of the sun (2,723,270.44) = 0.2, reciprocal = *5. One more detail for now, 2,879,096,880, divided by 3600 seconds in an earth hour, divided by 100,000 = 8.0, the number of Infinity. There it is again.

The above information provides more perspective of the solar system that has evolved since the Phi Harmonics intervals were employed earlier in the Sun to Planets list. In terms of the Golden Section Ratio this is the way life operates and it is a wondrous concept to deal with. There is nothing haphazard or abstract about it. With what has been learned the length of a year and a day for each planet in the solar system relative to earth time can be accurately computed with ease. To serve as an example the rotation velocity of earth spinning on it axis is achieved by multiplying its radius (3963.636) by 2Pi, divided by 24 hours in a day = 1,037.65 m.p.h. The same can be determined for the other planets now that their radii and the Phi intervals between them are known.

Pluto is the ninth planet from the sun, therefore, Planet XB10 is on the recount at 10, being Pi + 3.0, divided by 10 = 0.614 x 2.5 = 1.535 times the distance from the Sun to Pluto. The intention was to see if the planet circumferences would be in agreement with the Phi Ratios derived and all looks very reasonable. This is the way it works out as if all was planned out by some form of advanced intelligence that operates at a level beyond our comprehension. To many it means God and divine harmony and order are present in the Universe and in this sense it appears that such a force truly does exist. It is called the life force. The persistence of its energy which is the Golden Section in action has a mind and will of its own and it is not separate from us but is part of the living which is "all and everything." A visual of the planet orbits relative to the sun shown in total are like musical notes, or tones to the sense of sight and intelligence.

Review of Planet Diameters from Mercury to Planet 12:

Mercury: 3,136 mi.
Venus: 3,592 mi.
Earth: 7,927 mi.

Sub Total: 14,655 mi.

Mars: 4,444 mi.
Jupiter: 86,758 mi.
Saturn: 71,180 mi.
Uranus: 30,173 mi.
Neptune: 30,280 mi.
Pluto: 89 mi.

Sub Total: 222,924 mi.

Planet 10: 49 mi.
Planet 11: 6 mi.
Planet 12: 1.732 mi.

Sub Total: 56.732 / Overall Total: *237,635.732 mi.

Note: *15, divided by 1,000 = 0.00015 + 1.0 = *1.00015. This simple value expression will make total sense shortly.

The diameter of Planet 12 squared = 3.0 and 237,635.732, divided by Earth's diameter =3.0. The Grand Triune number cubed = 27and 237,635.732, divided by 27 = 8801.323, divided by 100 = 88.01323, divided by *1.00015 = 88, the number of Constellations. Mercury to Earth = 3 planets, Mars to Pluto = 6 planets and Planet 10 to Planet 12 = 3 planets, therefore, 3 + 6 + 3 = 12 and 9 x 3 = 27 and 27, divided by 12 = 2.250, square root of 2.250 = 1.5 x 10 = *15 and 12, divided by 15 = 0.8 x 10 = 8.0, the number of Infinity. Additional proof that there are 12 planets in the solar system. The over all total, 237,676 divided by the radius of the sun, 866,025 = 0.274, within 0.002 of 0.272, the square root of Phi - 1, but this is quite reasonable, the difference has something to do with the A ~ Phi value. The distance between Planet 10 to Planet 12 = 56.732 miles x 5,280 = 299,545 feet, divided by 100,000 x *1.0015 = 3.0. The diameter of Planet 12, 1.732 miles x 5,280 = 9,144.96 feet, divided by *3 = 3048.32 feet, divided by 1.732 = 1,760 Royal Cubits, the base perimeter of the Great Pyramid. Here are more examples of the Royal Cubit Theorem in action. The overall distance between Mercury and Planet 12 = 237,635.732 miles, divided by the Sun's diameter, 866,000 miles = 0.2744, divided by *1.0088 = 0.272 = the square root of 1+Phi – 1.0. Let this part of the discussion end with the clear statement there are 12 planets revolving around our Sun and each of the other sun-stars in the Milky Way.

It becomes clear on this issue by knowing the last three digits in the height of the Great Pyramid with a height of 484.68(288) feet. The number of sun-stars and planets in the Milky Way are as follows: *12,000,000,000 sun-stars x *12,000,000,000 planets = 144 E 20 x 2 = 2.88 E 20, divided by 1.0. E 21 = 0.288. And the square root of 12, divided by 2 = 1.732 = One Royal Cubit.

Not to overlook the fact that 484.68288 x 1,000,000,000 = 484,682,880,000 feet, divided by 5,280 = 91,796,000 miles = the correct AU which is the distance between the Sun and Earth.

Volumes of the Planets in Our Solar System…

Not long after the diameter dimensions of the sun and planets in our solar system were established as shown above there was a need to compute the volumes of each to complete the survey. Therefore it was elected to compute their volumes in cubic miles rounded off in an agreeable manner using the Ellipsoid Volume Calculator program that was available in my desk top. The volume of a sphere formula wouldn't do because it has been determined all heavenly bodies are ellipsoidal. The object of the exercise is to see how the sum of the planet volumes compare with that of the sun.

The following information provides the information needed:

Sun: 338,000,000,000,000,000 cu. mi.

Mercury: 161,000,000,000 cu. mi.

Venus: 24,100,000,000 cu. mi.

Earth: 260,000,000,000 cu. mi.

Mars: 45,700,000,000 cu. mi.

Volume of Ceres, other asteroids and moons. See below.

Jupiter: 340,000,000,000,000 cu. mi.

Saturn: 14,300,000,000,000 cu. mi.

Neptune: 14,500,000,000,000 cu. mi.

Pluto: 367,000 cu. mi.

Planet 10: 61,000 cu. mi.

Planet 11: 113.0 cu. mi.

Planet 12: 2.7 cu. mi.

Volume of all planets = 3.693 E 14 cubic miles, divided by the volume of the Sun = 0.0011 x 100 = 0.11% excluding the volume of Ceres, the other asteroids between Earth and Mars and the moons that orbit the planets.

The volume of the solar system, divided by that of the Sun = 25, reciprocal = 0.04 x 0.11 = 0.44 x 1,000 = 440 Royal Cubits = the base dimension of the Great Pyramid.

This discussion serves as a proof that all bodies in the solar system originated from the Sun.

Some interesting number values come to light when the volumes of the Sun and planets are converted to feet. One cubic mile = 1.472 E 11. Just multiply the first three numbers of the volume by this and round off to one place of decimal.

Sun: *0.5 x 10 = 5, the quintessential number of Life

Mercury: 0.24 x 100 = 24 hours in a day

Venus: *0.36 x 1,000 = 360 degrees in a circle

Earth: 0.383, reciprocal = *2.6

Mars: .673 x 4 = *2.7

Jupiter: *0.5 x 10 = 5, the quintessential number of Life

Saturn: 0.277, reciprocal = *3.6

Uranus: 0.21, divided by 8 = *0.026

Neptune: 0.309 x 2 = *Phi

Pluto: 0.54, divided by 2 = *0.27

Planet 10: 0.9 x 10 = 9.0, the triple trinity number before reset

Planet 11: 0.166, reciprocal 6.0, squared = *36

Planet 12: 0.4 x 2 x 10 = *8, the number of Infinity

Total: 5.46 x 1.464 = 8.0, the number of Infinity. In this case 1.464 is the multiple based on 2 x 1.732 – 1.0.

In summary, this more than reasonable input has served its purpose. What is learned from it is, with no exceptions the Geometry of Nature, Golden Section Mathematics or the workings of the Divine Proportion that was vigorously applied during those ancient times has the last word on the workings and proportions of all solar systems, galaxies and the Living Universe while the science of today remains well off the mark and in the dark about such matters. The only conclusion that can be made is the above listed information is correct.

ROYAL CUBIT THEOREM MGTS

BASED ON ANCIENT EARTH MEASURES

EQUATORIAL RADIUS 3963.636 mi.
LESS POLAR RADIUS 3950.0 mi.
DIFFERENCE = 13.636 mi.

POLAR RADIUS − 13.636 = 3936.3636 mi.

$$\frac{3936.36 \times 5280}{12,000,000} = \sqrt{3} \ (1.732 \ \text{ft.})$$

STATEMENT:

TRUE ROYAL CUBIT = $\sqrt{3}$

$$\frac{\sqrt{3}}{20} = 0.0866 \ / \ \text{PYRAMID INCH}$$

TEN MILLIONTH PART OF THE SUN'S DIAMETER

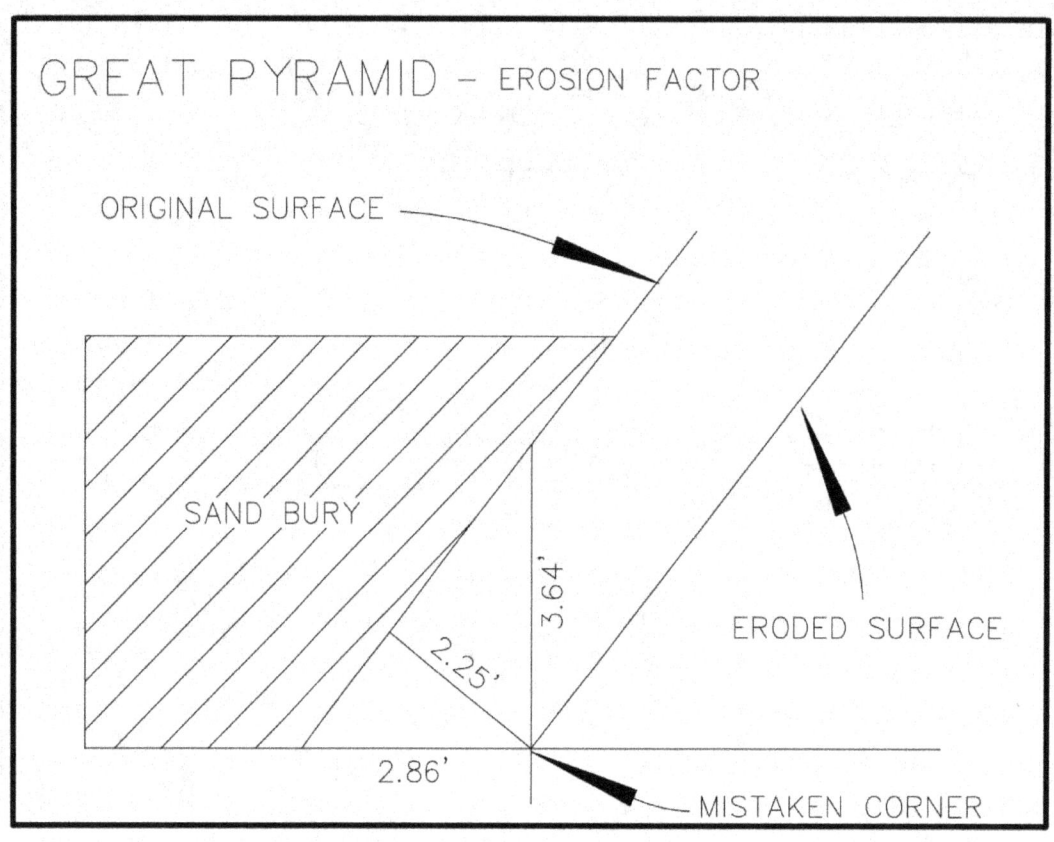

ANCIENT PI AND THE ROYAL CUBIT THEOREM

$b = \sqrt{1+Phi} \times \dfrac{a}{2} = 0.636$

$c = \sqrt{\dfrac{a^2}{2} + b^2} = 0.809$

$\dfrac{c}{\frac{a}{2}} = 1+Phi$

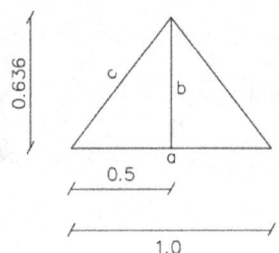

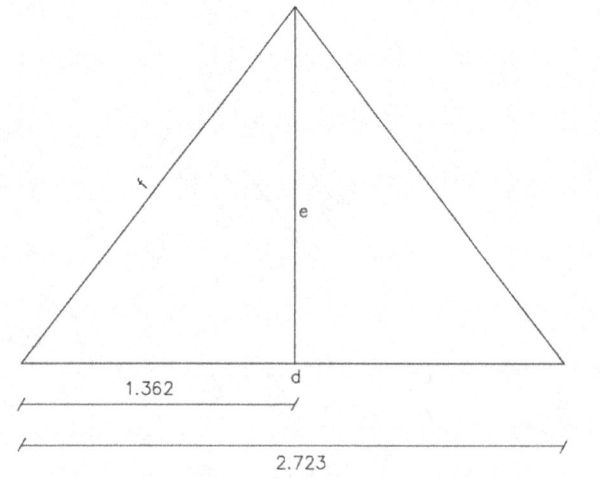

0.5

1.0

1.362

2.723

$\text{ANCIENT PI} = \sqrt{\dfrac{d}{3}}{2}\ \ = 3.144654$

$\text{Ancient Pi} = \dfrac{P}{2} = \dfrac{2.0}{b} = 3.144654$

or: $\dfrac{d}{\frac{\sqrt{3}}{20}} \div 10 = 3.144654\ \ /\ \ \dfrac{P}{2 \times API} = \sqrt{3}$

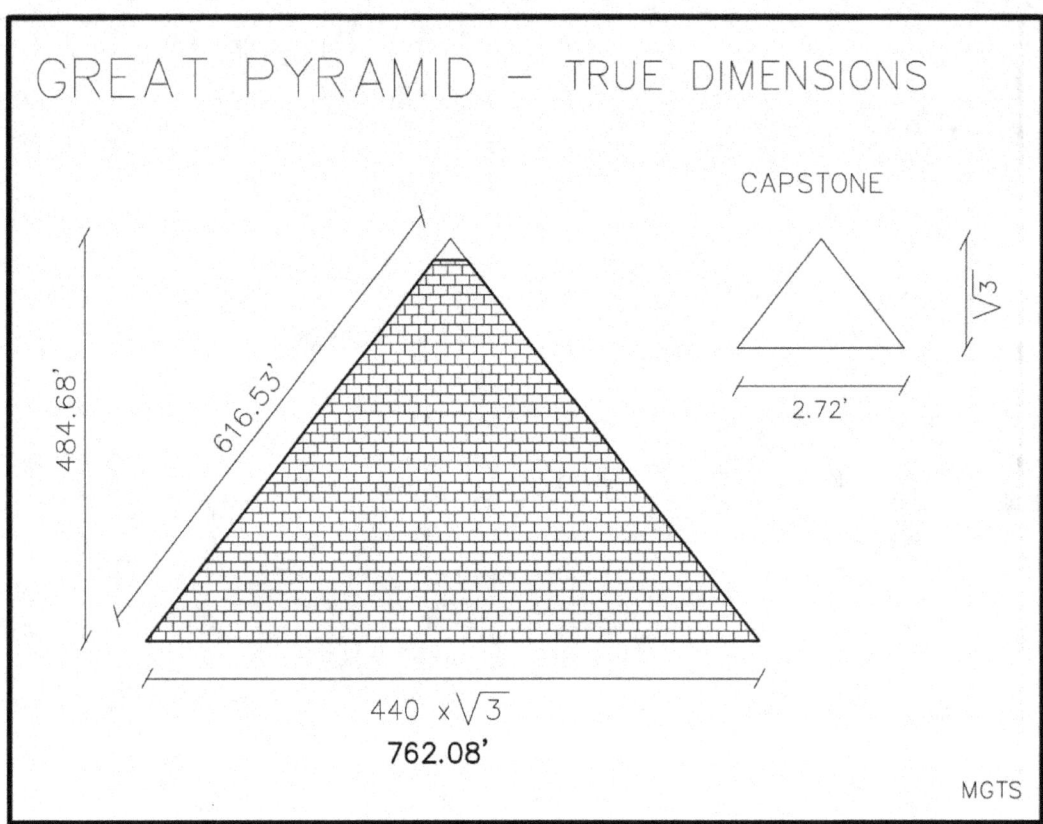

GREAT PYRAMID – TRUE DIMENSIONS

CAPSTONE

484.68'

616.53'

$\sqrt{3}$

2.72'

440 x $\sqrt{3}$

762.08'

MGTS

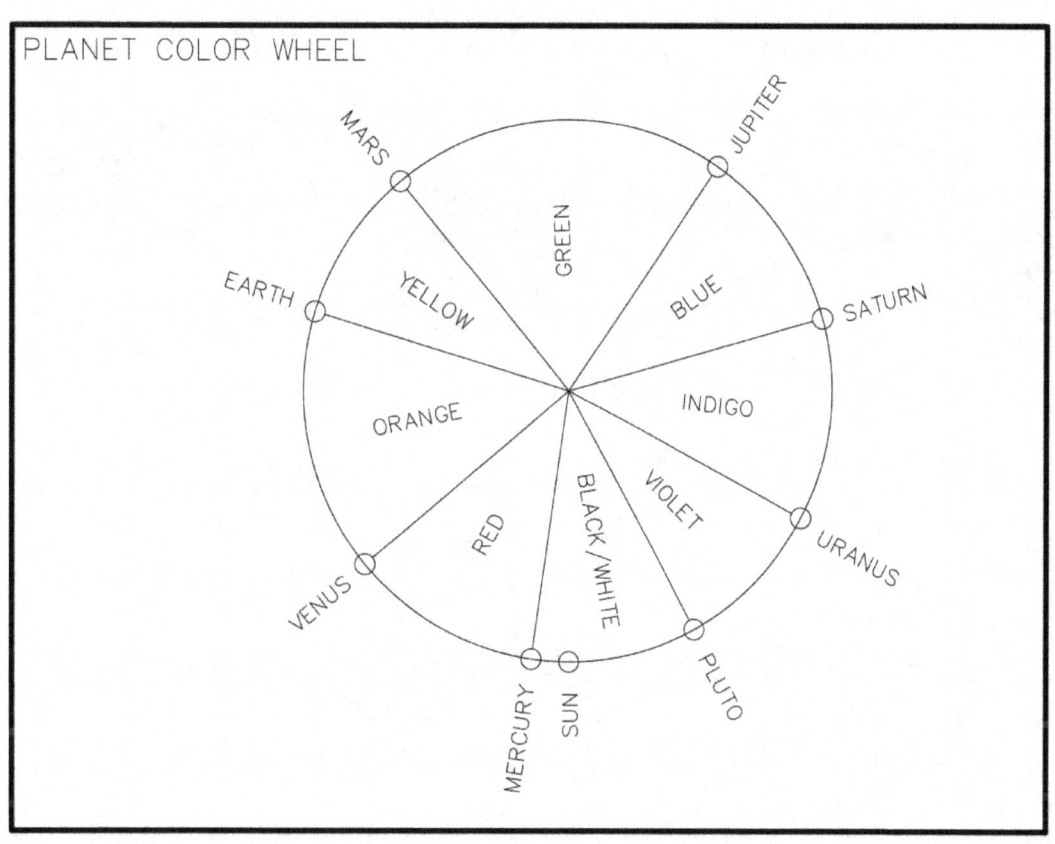

PLANET COLOR WHEEL

THE PLANETS AND RAINBOWS

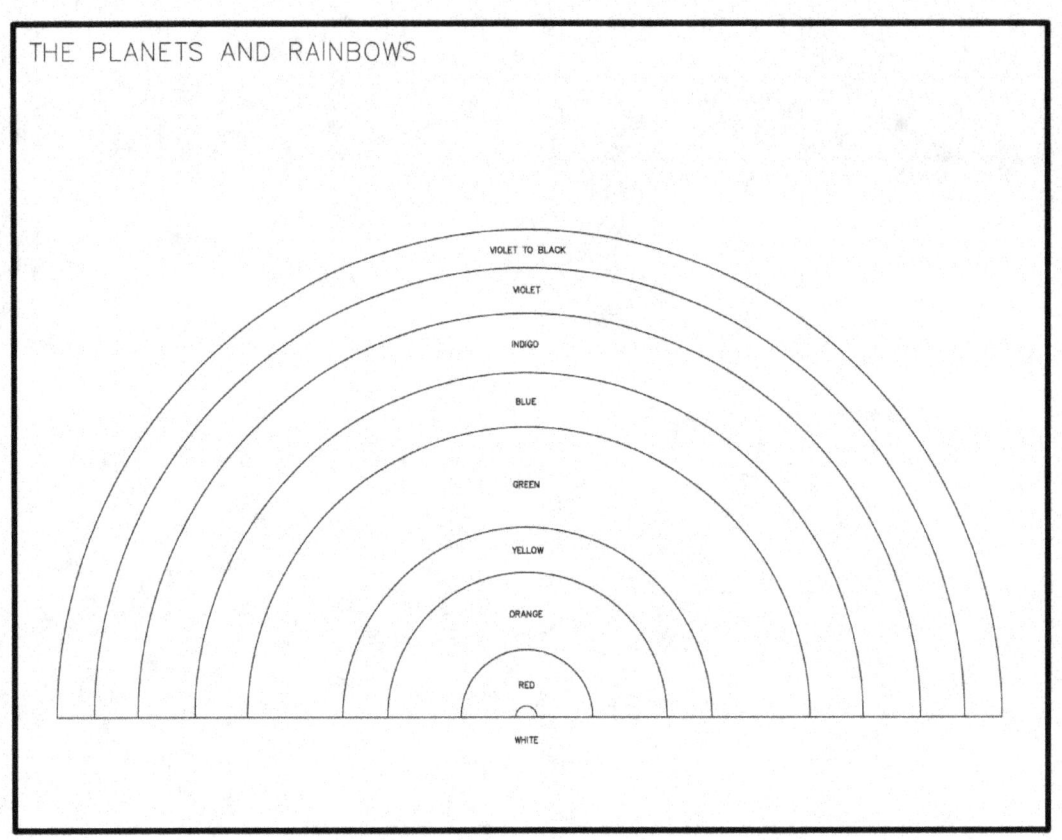

VIOLET TO BLACK

VIOLET

INDIGO

BLUE

GREEN

YELLOW

ORANGE

RED

WHITE

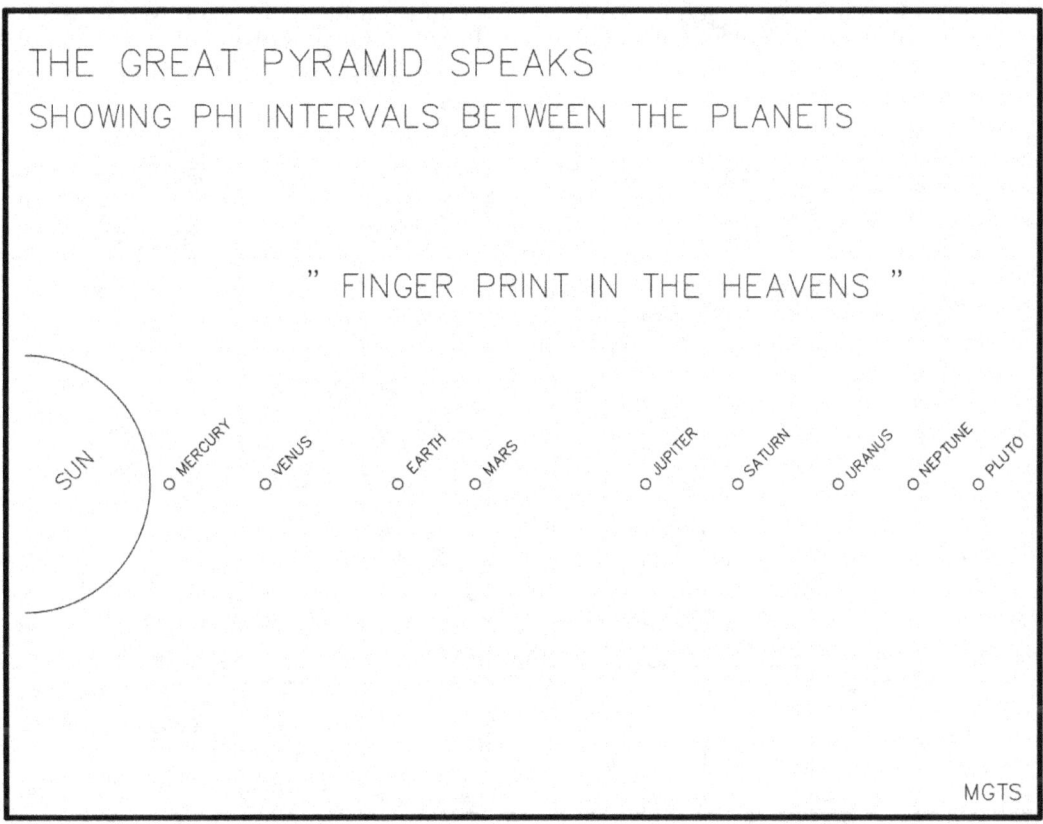

THE GREAT PYRAMID SPEAKS
SHOWING PHI INTERVALS BETWEEN THE PLANETS

" FINGER PRINT IN THE HEAVENS "

SUN

o MERCURY

o VENUS

o EARTH

o MARS

o JUPITER

o SATURN

o URANUS

o NEPTUNE

o PLUTO

MGTS

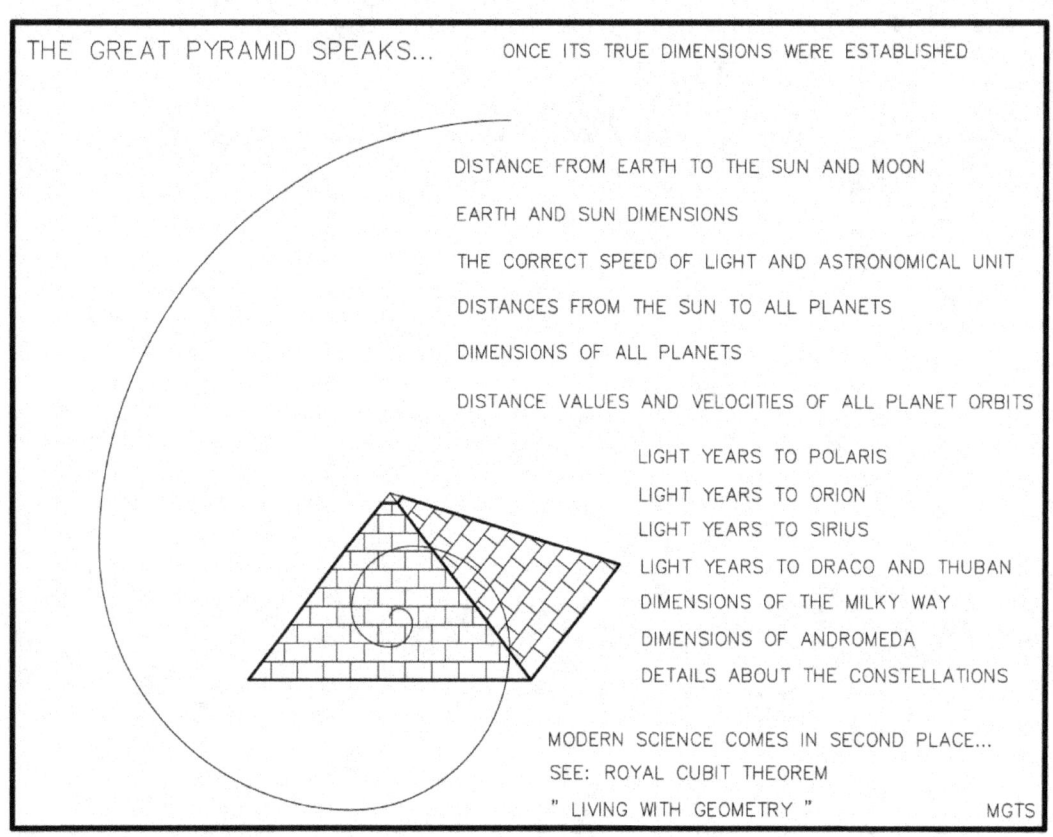

THE GREAT PYRAMID SPEAKS...   ONCE ITS TRUE DIMENSIONS WERE ESTABLISHED

DISTANCE FROM EARTH TO THE SUN AND MOON

EARTH AND SUN DIMENSIONS

THE CORRECT SPEED OF LIGHT AND ASTRONOMICAL UNIT

DISTANCES FROM THE SUN TO ALL PLANETS

DIMENSIONS OF ALL PLANETS

DISTANCE VALUES AND VELOCITIES OF ALL PLANET ORBITS

LIGHT YEARS TO POLARIS

LIGHT YEARS TO ORION

LIGHT YEARS TO SIRIUS

LIGHT YEARS TO DRACO AND THUBAN

DIMENSIONS OF THE MILKY WAY

DIMENSIONS OF ANDROMEDA

DETAILS ABOUT THE CONSTELLATIONS

MODERN SCIENCE COMES IN SECOND PLACE...
SEE: ROYAL CUBIT THEOREM

" LIVING WITH GEOMETRY "   MGTS

PLANET ELLIPSE ORBIT

GREAT PYRAMID PROPORTIONS

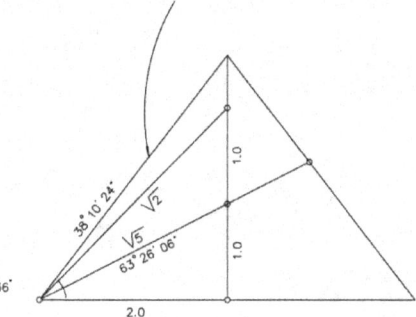

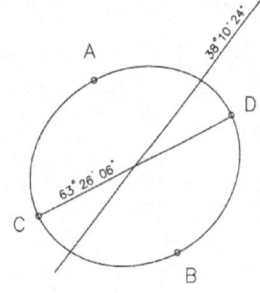

SHORT AXIS AB = 2 x 1 = 2.0

LONG AXIS CD = 1 x 1.16 x 2 = 2.32

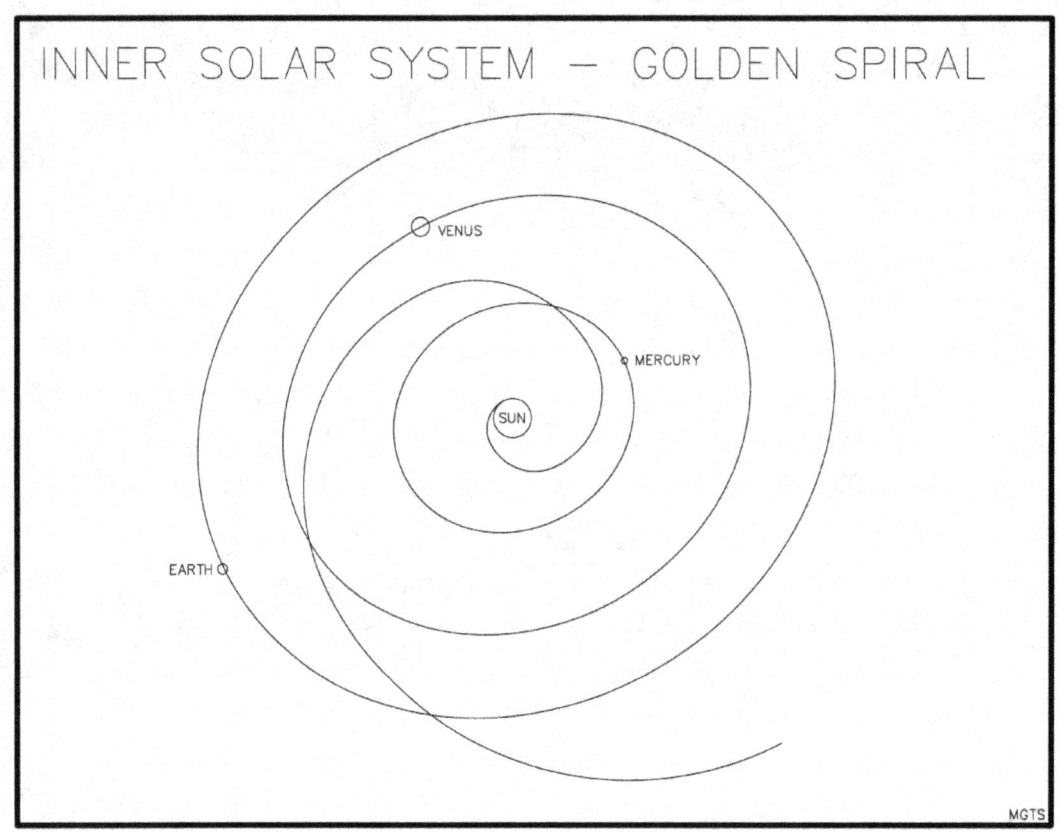

INNER SOLAR SYSTEM — GOLDEN SPIRAL

VENUS

MERCURY

SUN

EARTH

MGTS

GEOMETRY OF ALL SOLAR SYSTEMS

SEE THE NAUTILUS SHELL / LIVING GOLDEN SPIRAL

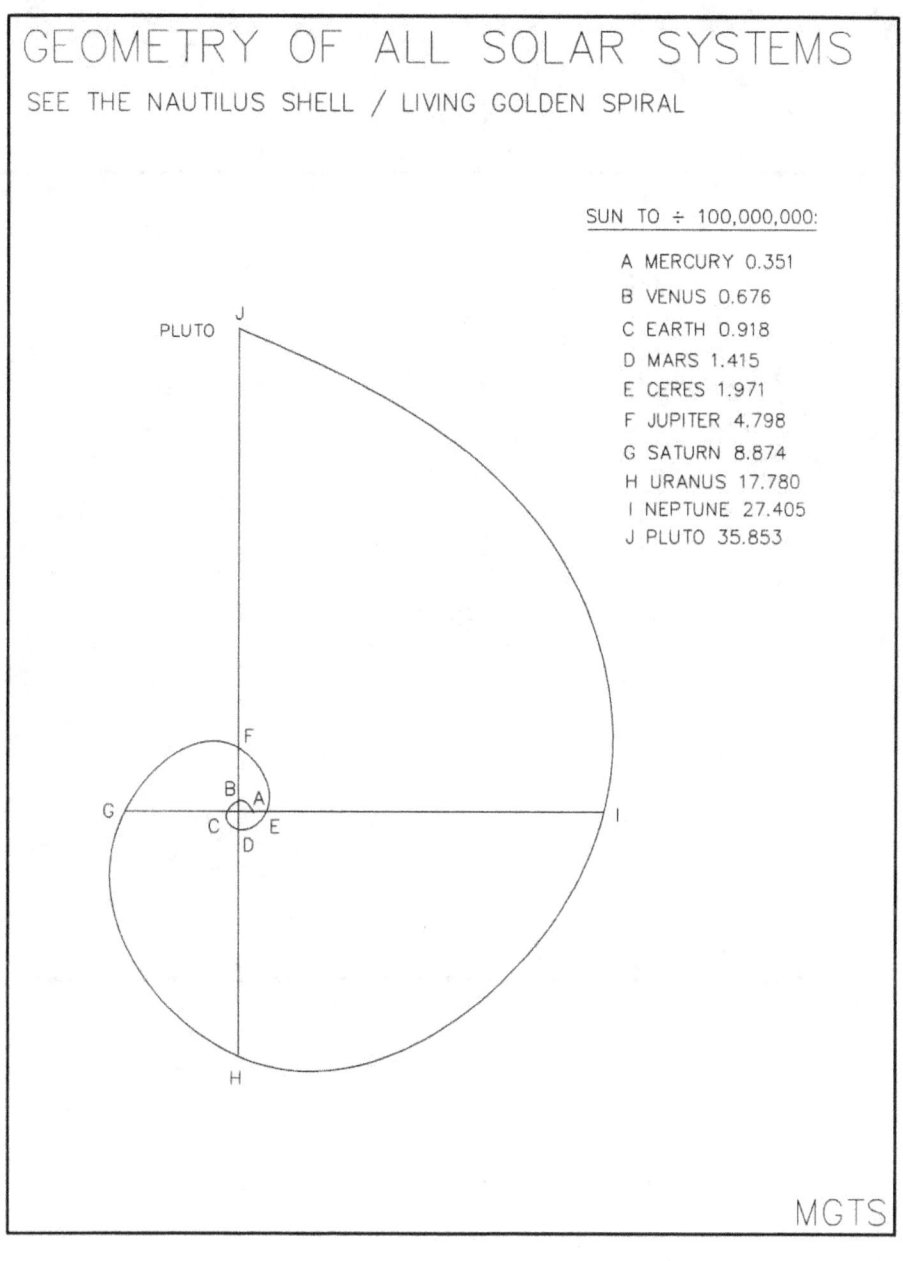

SUN TO ÷ 100,000,000:

A MERCURY 0.351
B VENUS 0.676
C EARTH 0.918
D MARS 1.415
E CERES 1.971
F JUPITER 4.798
G SATURN 8.874
H URANUS 17.780
I NEPTUNE 27.405
J PLUTO 35.853

MGTS

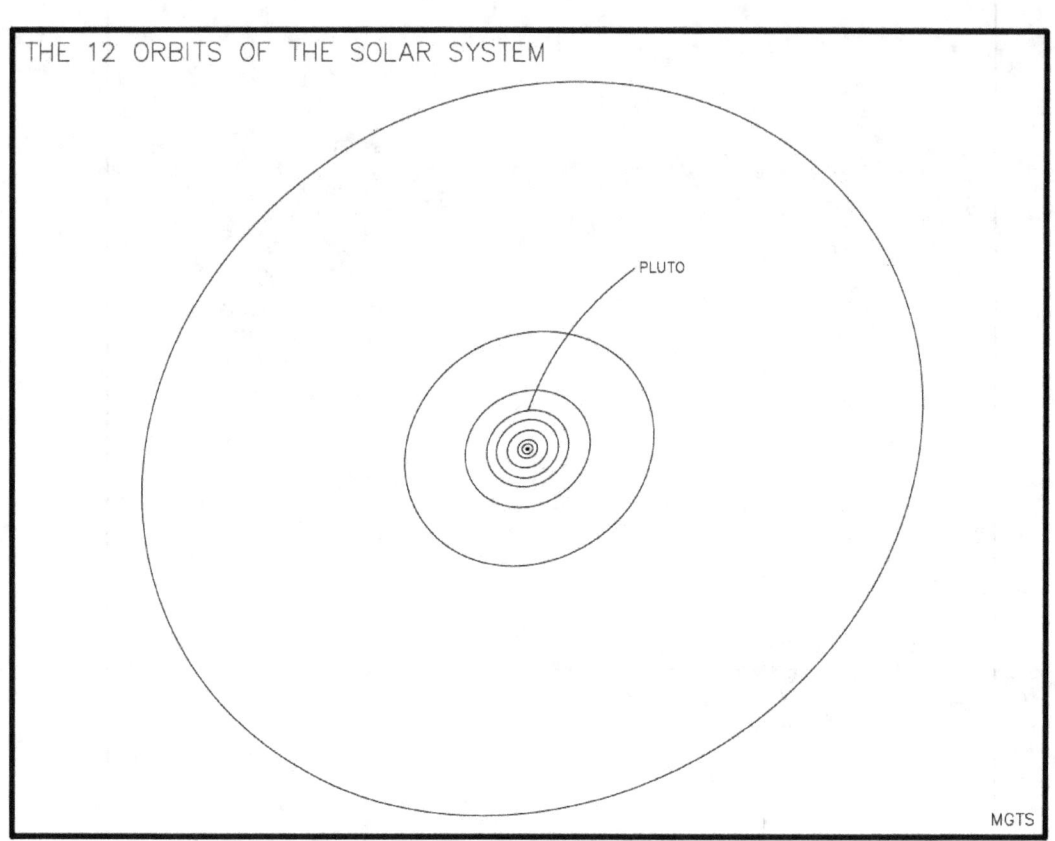

THE 12 ORBITS OF THE SOLAR SYSTEM

PLUTO

MGTS

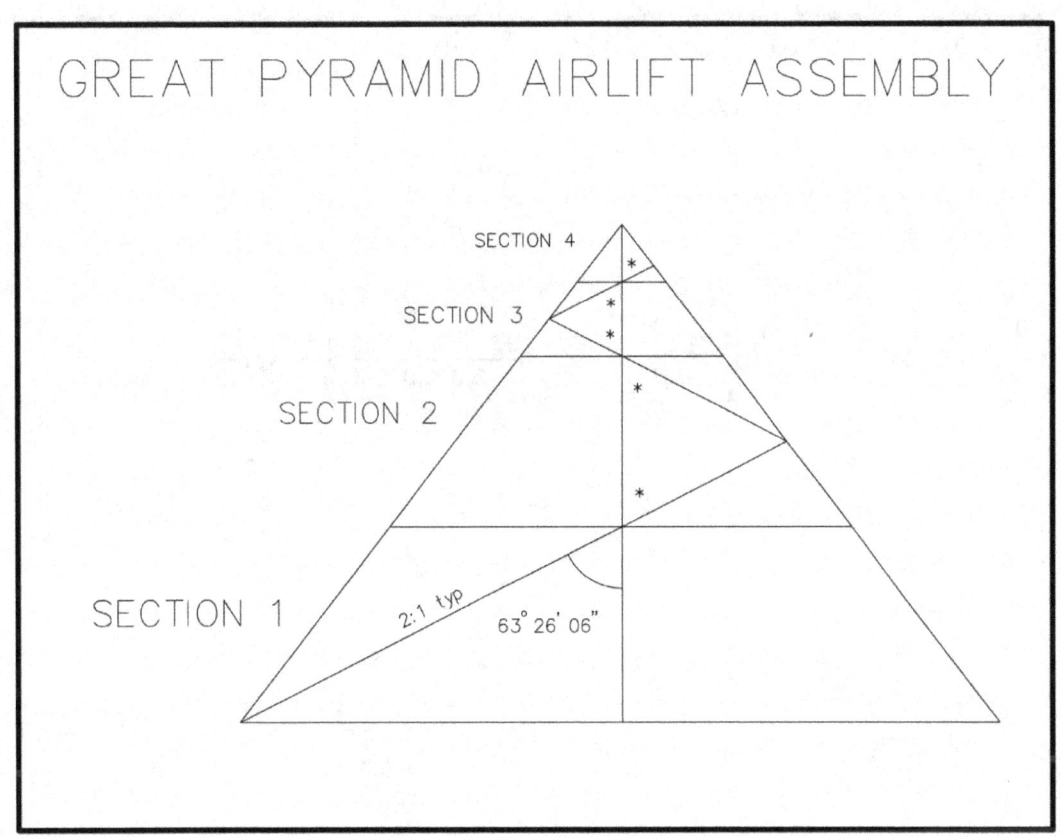

GREAT PYRAMID AIRLIFT ASSEMBLY

SECTION 4

SECTION 3

SECTION 2

SECTION 1

2:1 typ

63° 26' 06"

Royal Cubit Theorem — Based on Ancient Earth Dimensions

One Mile = 5,280 ft.

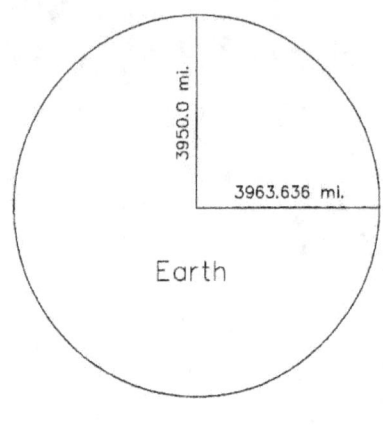

Earth

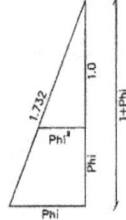

Equatorial Radius = 3,963.636 mi.

Polar Radius = 3,950.000 mi.

3,963.636 − 3,950.000 = 13.636 mi.

3,950.000 − 13.636 = 3,936.364 mi.

$$\frac{3,936.364 \times 5,280}{12,000,000} = \sqrt{3} = 1.732 \text{ ft.}$$

Note:

1.732/2 = 0.866 x 1,000,000

= 866,000 mi. = Diameter of the Sun

1.0/1.732 x 100,000 x 2 = 115,473.4411 lt. yr.

= Diameter of the Milky Way

Statement:

One Royal Cubit = $\sqrt{3}$ = 1.732 ft.

to three places of decimal

1.732 ft. = Capstone Height Great Pyramid

1.732/20 = 0.0866 = Pyramid Inch

= Ten Millionth Part of the Sun's Diameter

MGTS

CHAPTER 4

Dimensions of the Milky Way...Theorem Part II

Given: Height of Capstone ~ Great Pyramid = Square Root of 3, 1.732...
and 1.0 is the number of all...

Proposition:

1.0 divided by the Square Root of 3 x 2 x 100,000 = Diameter of the Milky Way
In terms of Light Years...it is that simple, or, is it? Please read on...

Not enough can be said about the Square Root of 3 and here again it shows the way to acquiring the dimensions of the Milky Way as it did for the true base length of the Great Pyramid.

Focusing on its two dimensional aspects for now, some simple computations are carried out:

1.0, divided by 1.732 = 0.5773672055 x 100,000 = 57,736.72055 Light Years = Radius
Diameter = 57,736.72055 x 2 = 115,473.4411 Light Years
Polar Radius = 57,736.72055 x 0,997 = 57,563.51039 Light Years
Note: 3 x 0.5773672055 = 1.732, the Royal Cubit
      and 1.0, divided by 0.5773672055 = 1.732
The interesting thing of it is 1.0, divided by 0.0866 x 10,000 = *115,473.4411, therefore the Royal Cubit Theorem is behind the story.

Now to get down to it...

Circumference = Roughly 362,227... On the calculator program...

The thinking on this value is, when it is divided by 100,000 the quotient is 3.62 to 3.63 and that value is close to the Phi value of 3.618 or 3.616, 3 + A ~ Phi... which is based on the slope side of the Great Pyramid, divided by 1000. Therefore it will be seen where it goes if the circumference is based on the radius, 57736.72, multiplied by 2 x Ancient Pi and what happens if 3 + Phi x 100,000 = 361,803.39 is assigned as a value for the circumference of the galaxy.

Radius 57,736.72055 light years x 2 Ancient Pi = 363,124.0286
1+Phi x 100,000 = value 361,803.39

Difference = 1320.6386, divided by 3.6158 = 365.24, rather amazing.
Another fascinating piece of information is the circumference of the Milky Way, divided by the number of miles in a light year (5,876,386,823,000) as shown earlier x 10,000,000 = 0.618 = *Phi x 2 + 1.0 = the square root of 5. Just to be double sure the circumference of the Milky Way

is 363,124.0286 light years.

When the value 361,803.39, divided by 100,000 then divided by 1+Phi the quotient is the square root of 5 and this has much meaning in the investigation. When the circumference of the Milky Way is divided by 100,000 x 3+Phi the result is 1.00365, the reciprocal = 0.996636, Interesting.

Multiplying the diameter by 0.997 rounded off to acquire a polar dimension of the galaxy seems reasonable enough because this was the value derived by the earth dimensions when the polar radius was divided by the equatorial radius. In the case of the Milky Way dimensions, when the polar radius is subtracted from its equatorial radius the square root value of 3 comes up like it did in chapter 3. As follows then: 57,736.7, its radius - (57.736.7 x 0.997 = 57,563.5)...here we go... 57,736.7 – 57,563.5 = 173.2 , divided by 100 = 1.732 and there it is. The tentative proposal is that the diameter or, width of the core, the black hole at center of the Milky Way is 1.732 x 1000 = 1732 light years making it1.50% of its diameter,approximately 2.2 million times greater than our sun's diameter. The horizontal long diameter or length of this black hole is 2 x 1732 = 3,464 light years making it 0.3% of its diameter, approximately 4.4 million times greater than our sun's diameter. Further along the value of this information is seen when it comes to computing the distances between galaxy centers. All stargazers including today's physicists and astronomers are welcomed to join in on the debate.

Note: The ratio of the length to width of the King's Chamber is 2:1, the same as the short width axis is to half its length in the Milky Way and there is probably a connection here. Details of the King's Chamber and the Sarcophagus will be discussed further along.

The Milky Way is a galactic unit within the Universe composed of sun stars and their planets, comets, asteroids, nebula etc. within its spiraling arms with the same characteristics as the rest of the celestial bodies, only being a galaxy, it operates on a much larger and more complex scale. It is also believed that the sizes of other galaxies larger and smaller than the Milky Way are based on square root of 3 ratios because it is a universal constant and in the case of the Milky Way this value uniquely relates to and is the basis of the mathematical system we exist in. The reason why the view of a galaxy in space appears longer and flatter and not as circular is because we see it from an obtuse angle side view and due to the outward centripetal forces acting from within it a halo of gases and cosmic dust accumulate at their ends. Simply stated, the plan view of any galaxy shows that it has a circular shape but it is actually ellipsoidal. The side view of the Milky Way is in a ratio of 2:1 and not by any coincidence the same slope as the ascending and descending passageways in the Great Pyramid. This delightful piece of information will be reviewed shortly when it becomes known as the Galactic Incline. Please see a conceptual not to scale drawing titled, Milky Way – 2:1 Galactic Incline that shows our galaxy from a side view perspective. With sound reasoning as mentioned above it is estimated the nuclear bulge dimensions at its center are black hole width or, diameter = 1,732 light years and its length = 3,464 light years. Perhaps the origins of life in our galaxy are found here and if this is so all the sun stars and planets in it must have the seeds of life in them because this is the source of all materials and energy in the galaxy. Taking it a step further it might be asked if life's origins are from the center of the Universe. The logic here is, if the Universe is infinite, it has no center and that develops into a larger question. What exactly is infinity and what does it mean? The size of our sun and the entire solar system is barely a pin point in comparison to the vastness of the Milky Way, like a point in space but it is not as if we are insignificant because life and consciousness are the grandest state of being, in which the concept of infinity is perceived.

Modern science, the others, make a very rough approximation of the Milky Way claiming its diameter is somewhere between 100,000 to 150,000 light years and because they have no true understanding of the topic or Golden Section mathematics its elliptical shape is not discussed. It is using an incorrect AU and the further the distance the larger the difference there is, as we know, and not all that is in the Universe works out to nice round numbers such as this. It is more the case that the dimensions work out nicely when the correct approach is taken, then they make sense with relation to the Golden Ratio. To propose a diameter closer to 100,000 light years this way I tried 9 x 100,000, divided by 1.732 = 519,630.48 x 2 = 103,9260.9 x 0.997 = 1036143.1 for dimensions and it didn't work out of course. Some other combinations turned out to be futile as well, therefore it was back to the 1.0 divided by the square root of 3 concept, being that it is the only thing that makes sense. Never the less today's scientists have been working hard on the case and the above estimate of its size based on the Square Root of 3 value compared to theirs is somewhere in between what they estimate it to be and we can go with it for now until some conclusions can be made. The claim is our solar system is in the outer reaches of the spiraling arms of the Milky Way in the Orion Arm next to the Sagittarius arm and the third planet from the sun, Earth, and as far as we know it is the only one in the galaxy or the entire Universe that has life on it. Well, then at least we have a sizeable backyard all to our own and our neighbors in the other galaxies must have large backyards as well if only one arm of a galaxy per is allowed to have life in it. The distances are so vast it would take thousands years at the speed of light for us to receive their messages and for them to pick up on ours. The problem with this is, the signals become inaudible noise after traveling a couple of light years. We don't have the answers to any of this hypothesis but there very well could be life elsewhere in the Milky Way or elsewhere in the galaxy's beyond but it hasn't been proven. What we do know is there is an abundance of life energy perpetually at work throughout the Universe and we should be pleased enough with that. What has been determined is the Milky Way consists of five spiraling arms named as follows: Perseus ~ Gygnus ~ Sputum ~ Sagittarius ~ and Orion, the arm in which our solar system exists and it is not a disproportionate blob in space, like all else in the Universe it is an organized living organism composed of interrelated working Golden Ratios. Chances of life elsewhere ought to be good in our galaxy it is thought because there are some 200 to 400 billion stars in the Milky Way and each one is a sun with planets revolving around them and the odds are at least one in a million, even a billion that one planet or a few of them might have conditions similar to earth where life support as we know it exists. Perhaps there is life in each spiral arm in the Milky Way. Some of us might still be living with the concept that the Ancients came from planets in the Sirius and Orion constellations or other locations. It is an intriguing question no one has the answer to. Another question that comes to mind is what is our location relative to the center of the Milky Way. The answer, is given by the azimuth of the Great Pyramid slope as it is from the sun to the planets, 63 degrees, 26 minutes, 06 seconds. The assessment for the circumference of the Milky Way, 363,124 light years is how long one full rotation of the Milky Way takes and all the sun stars and their planets and all else in it revolve around its center. The center must be a mass of energy like a sun only vastly larger and more powerful, from where all energy influences in the galaxy generate. When it is seen in perspective our hair ball in a space suit headed for Alpha Centauri A on his brief four light year trip has less of a problem and now that more about the Milky Way is known it almost seems a smaller more manageable place.

360 degrees, divided by 5 = 72 degrees, divided by 2 = 36 degrees and like the square root of 3, not enough can be said about the pentagram, the number 5, the square root of 5 and how it relates to the square root of 3 when it comes to the geometry of nature. This is seen in the drawing for Golden Spiral - the Pentagram and the drawing titled Golden Spiral - Cosmic

Expansion and Contraction in Chapter 2. Please see the drawing titled The Milky Way which shows its five arms spiraling outwardly counter clockwise from its center and the relative location of our Sun in the Orion Arm. Its location is determined by the azimuth of the Great Pyramid Slope right of celestial north from the center, intersecting with the zero north azimuth from the midway point from center to the right or east and it appears to be more or less in agreement with today's picture of where it is. The triangle dimensions in light years are: Base = 57736.72, Height = 36720.55, Side Slope = 46709.52. The gradient of this line is that of the Grand Gallery 0.5 %, 2:1, or 26 degrees, 33 minutes, 54 seconds. Since the true slope angle is determined a vertical fix on the sun relative to the center of the galaxy can be made, that is if what the Great Pyramid tells us is to be depended upon. It hasn't let us down yet, so let us see where it leads to. What is discovered here is a Phi Based light year triangle that has our sun at its apex is exactly in proportion with a side view of the Great Pyramid and our solar system revolves around it. There are three other quadrants in the Milky Way where this takes place due to the principle of symmetry and this is enough to evoke interest in any party who is looking for a mystery to solve. That includes our physicists, astronomers and the science fiction sector etc.

The following self explanatory computations convert light years into feet to provide ratio comparisons with the dimensions of the Great Pyramid:

365.24 x 24 x 3600 x 186,216.56 = 5,876,386,823,000 miles in a light year x 5,280 = 31,027,322,420,000,000 feet in a light year and 5,876,386,823,000, divided by 5,280 = 1,112,952,050, divided by one billion = 1.113 rounded off. This we know. It is also important to know that the Light Year Pyramid is in the same proportions as the Great Pyramid. One half the base of the Light Year Pyramid is the reciprocal of sq. rt. 3 x 100,000 = 57,736.72 light years, divided by 2 = 28,868.36 light years. Therefore its height is, 28,868.36 x 1.272 = 36,720.55 light years and the slope is the square root of the sum of the squares of one half the base and its height = 46,709.54 light years. That is all for now. Remember, we are dealing with astronomical numbers.

The Grand Gallery, Ascending and Descending Passages...

A side view drawing of the Great Pyramid titled, Grand Gallery-GP is introduced. The dimensions of the Great Pyramid were divided by 100 and when the drawing was generated it worked out that 1+Phi x 100 was the magic number to use from center on the vertical to the top of the Kings Chamber, Point A, and on the slope of the Grand Gallery. Indeed the top of this chamber is within a hair of one third of the height of the structure. Flinders Petrie had determined the floor of the King's Chamber was 1692 inches, or 141 feet above the base of the Great Pyramid but the problem with this is there was no sure way of determining the base level after 5000 years plus of erosion, sand bury and molestations by looters etc. According to this proposal the vertical distance from the base to the floor of the King's Chamber is 142.603 feet differing from the Petrie estimate by 1.603 feet. This value is determined as follows: 161.803 - 19.2 = 142.603. It is suggested to ride with this proposal for now and see where it goes and if by chance the following input is incorrect the sky isn't going to fall because, if nothing else it could lead us in the right direction. The logic can be seen on the drawing when 1+Phi x 100 is divided into the height and when 2.9954945, divided by 484.68 = Phi precisely. The focus is on what I believe to be the true dimensions and geometric structure of the Grand Gallery, the Ascending and Descending Passages, where the Grotto is located and what their physical meanings are. The King and Queen's Chambers, Grand Gallery and passageways are shown. The airshaft locations have been covered. The object is to analyse these dimensions and see how they corresponde to the access details and determine what their geometric makeup and proportional relationships are in terms of Golden Section mathematics. I didn't enjoy dickering around with those bogus pyramid inches in order to convert to feet but what I got out of it after dividing their pyramid inch numbers by 1.006106, then by 12 for feet the height of the Grand Gallery is 28.024, it has a width of 6.86 feet and its length is 138.19 feet, based on Phi squared + 1.0 x 100, along the A-1 to B line. The estimate by others of 156.6 for its length, a difference of 18 or so feet, is likely because the measurement was taken to the north side of the one thousand year old well eroded well shaft that was dug. Theories abound about this strange feature which is around 210 vertical feet in length. Some think it was an escape route for the workers during construction when the rock plugs were placed but more than likely it was dug by looters looking for treasure long after the construction was completed. Around 820 A.D. a well equipped gang of Arab looters broke into the structure below the north side entranceway using battering rams. Fire was used to heat up the rock as cold vinegar was poured over it to help break it up and this caused much damage at the passageway junction after they reached it. Eventually they found their way down into the lower passageway and into the grotto where nothing of value in the form of gold or gems was found. It was this type of destructive activity that made it difficult for the investigative surveys to be carried out accurately in the later years, though none of these parties had any understanding of Golden Ratio Mathematics. To carry on with this part of the story the well shaft must have been dug from the bottom up while the looters lusted for material riches they had assumed were in the upper levels inside the structure. More than likely a large pile of shattered rock was dumped on the ground outside the entrance as it would have been laboriously transported up grade along the descending passageway to that area. Little did they realize or appreciate the fact that the gold of understanding life and the workings of the Cosmos was hidden in the ratios of the structure. The truth of it is, beyond doubt, to the seeker who needs answers, the Great Pyramid is truly an enormous nugget of gold, when the value of this type of knowledge is considered.

To carry on with the investigation and examine the dimension proportions to see what meaning they have, the first item of interest is the height, 28.024 divided by 1.732 = 16.18 = 1+Phi x 10. Then the length, 138.2 divided by the height = 4.932, divided by 1.618 = 3.04832 x 1000 = 3048.32, the perimeter of the Great Pyramid. The perimeter divided by Ancient Pi = 220 x 2 = 440, the number of royal cubits in the base. The length divided by the width = 20.144 divided by 2.236 = 9.0. The height divided by the width = 4.085, divided by 1.732 = 2.36, divided by 10 = 0.236 = Phi cubed. We are told by the previous researchers that the width of the ascending and descending passageways is 3.5 feet. What makes more sense is a width of 3.568 feet, divided by 0.618 = 5.7736, reciprocal = 0.1732, meaning the true width is a 10 thousandth part of the radius of the Milky Way. The results of a simple computation shows that the volume of the Grand Gallery divided by the combined volumes of the upper and lower passage ways including the distance to the Grotto = 3.6524 x 100 = 365.24 days in a year. The length of the Grand Gallery divided by the total lengths of the upper and lower passages including that of the length to the Grotto = 0.24 x 100 = 24 hours in a day. The volume of the ascending passage divided by the volume of the Grand Gallery = 0.0866, the true pyramid inch. or 10 millionth part of the sun's diameter and the Royal Cubit Theorem makes its presence known. Finally, BC + DE + EF = 1143.0, divided by 10 = 114.3 = FO. Herein are yet other proofs that the proposed dimensions being used for the Great Pyramid are correct. We will leave off on this topic for now and move on to other areas of interest. A question mark is left aside the point that is 323.6 feet, equal to DE on the drawing, above the base on the vertical at center where there might be an area of interest such as another chamber or power center made accessible by a spiral stair well above the antechamber that has eluded the investigators these many years. It is an intriguing thought at least.

It was this angle of 26 degrees that bothered me, another source claims it is 26.3 degrees and it became known it wasn't very clear to anyone what the grade was or what anything meant in the Great Pyramid with relation to the Grand Gallery, the passageways, the chambers and grotto etc. I had no other choice but to carry out some research by proposing to begin with, that if the grade of the Grand Gallery was 50 % or 0.5 to 1.0, divided by Phi the quotient is Phi, divided by 2 + 1.0 and that was exactly what I was looking for, a Phi ratio relationship that makes sense. In a flash the slope angle then became 26 degrees, 33 minutes and 54 seconds, in other words it was a 2:1 slope I was dealing with, and the complimentary angle is 63 degrees, 26 minutes and 06 seconds and that is a "Bingo" because this is the same angle seen on the Elements of the Golden Section drawing and it means the same thing, it is the Phi angle used to determine the slope and height of the Great Pyramid. In the drawing it is Angle OAC. When the proposed slope of the Grand Gallery and ascending passage from the top of the King's Chamber is intersected with the base line the hypotenuse length = 361.803 feet and the opposite side from the center of the base = 323.606 feet and the remaining distance from there to the base corner = 0.574. When this value is divided by 1.732 = 0.3314 x 100 = 33.14 - 10 x 3 = 3.14, Pi, is the result. There appears to be a proof to the above proposal emerging then. * Note: 2:1 = 2 atoms hydrogen + 1 atom oxygen = Water.

Then I was to discover that yet another group had determined the slope was 2:1 but they didn't make the connection as presented here because they didn't have the correct dimensions of the Great Pyramid to begin with. When the Royal Cubit Theorem is applied, 0.5, divided by 0.0866 = 5.773672 a multiplication of 10,000 provides the radius of the Milky Way in light years. The only conclusion that can be made is the slopes of the Grand Gallery, Ascending and Descending Passages in the Great Pyramid are officially on a 2:1 incline which is at 26 degrees, 33 minutes and 54 seconds, or on a grade of 0.50 % and the supplementary angle is 63 degrees, 26 minutes, 06 seconds and this has great significance with regards to the alignment of the solar system because this is the azimuth of the line from the lower left side that passes though the point 2.0 units above the top of the Grand Gallery at center. In this case some extra gold digging, it could be said, has paid off handsomely. The distance from the floor of the Grotto at F to the bottom of the structure at O is 114.3 feet. From O to the floor of the Queen's Chamber the distance is 80.8 feet, half way between the base and the top of the King's Chamber and from that elevation to the bottom of the Grand Gallery, the floor level of the King's Chamber the vertical distance is 61.8 feet which is in the ratio of Phi squared to the distance from the base to the top of the King's Chamber. The ratio of the height of the King's Chamber to the distance of 161.8 feet is 0.118, which is Phi + 0.5 = 1.118 - 1.0 = 0.118. Then when CD is divided by EF = 1.118 = Phi + 0.5. The question is, how does this mathematical arrangement lead us to know how the cosmic energy is processed in this chamber. There may be no coincidence in seeing that the base line intersects at the junction of the passage ways, when the distance to the intersection is divided by ½ the base dimension and the quotient is divided by the square root of 3 the result is 0.433. Example: 286.251, divided by 381.04 = 0.75, divided by 1.732 = 0.433. Another none coincidence is, the perpendicular distance from D to the base line = 31.4 feet and when this is divided by 10 = 3.14, Pi is the result. Then there is the case where the distance from the top of the King's Chamber to the bottom of the descending passageway at the opening, distance AD = 379.5, divided by Phi = 614.0, divided by 100 = 6.14 = 3 + Pi. Furthermore, the distance from the top of the King's Chamber to the top of the Grand Gallery = 0.088 and this has significance relative to the base dimension as shown in the following computations: 762.08, divided by 0.088 = 8660 / 0.088, divided by 762.08 = 0.00011547344, quite amazing, and on track with the above proposal. Then it is determined the slope distance from the base at center to the top of the King's Chamber at center = 414.0 feet which is 100 x 's Pi + 1.0. The distance from the base at center to the floor of the Queen's Chamber is 389.5 feet, divided by 1.732 = 2.25. The distance from the base at center to the floor of the grotto, divided by 3, then divided by Phi = Pi - 1.0. Also, the vertical distance from the floor of the grotto to the base, 114.3 feet, when divided into the base length = 6.667. When 114.3 is divided by the height, 484.68 the result is 0.236, sq. rt. 5 - 2.0. The vertical distance from the floor of the grotto to the apex = 598.98, divided by 7 = 85.569, divided by Phi = 138.461 and this value is very close to Phi squared x 100 = 38.192 + 100 = 138.192, within 1.002 %. When 85.569 is divided by 114.3 the result is 0.75, divided by 1.732 = 0.433. Therefore, there is some critical mathematical significance attached to the grotto and its location while others have made the claim it merely has something to do with drainage. This is not at all the case as every theoretical and measurable dimension within and without the Great Pyramid tells us something about the cosmos in Golden Ratio code form. The distance to the bottom of the Grotto at centerline is 114.3 feet divided by 100 = a multiple of Pi - 2 down from the base at center and it "represents" Alpha Centauri B, the south pole star. To the right, or east of this point at 67 feet is represented the red dwarf star known as Alpha Centauri C, or Proxima. The only way this arrangement can be appreciated is by viewing the constellation from space, looking north toward earth, or by the side on view drawing of the Great Pyramid itself. The location of Alpha Centauri A is to the north west in this view and its relationship to Proxima and B in the celestial arrangement will be discussed further along in this chapter. In this section it will also be seen how the grotto details and descending passage help determine the distance to and position of the star constellation, Sigma Draconis, or Draco, the Dragon.

For a change of pace:

Search for a legitimate function of Phi for computational purposes. Please examine the following statements:

Given:

Ancient Pi x 2.0 = 6.289 / Square root of 3 = 1.732... / Square root of 5 = 2.236...

1 + Phi or 1.618, divided by 1000 = 0.0001618 + 1.0 = 1.0001618

1.0, divided by 1.0001618 = 0.9998382 ~ (proposed function of Phi)

From floor of grotto to apex = 598.045 x 0.9998382 = 597.948,

597.948 x 6.289 = 3760.495, divided by 365.24 = 10.296

10.296, divided by 0.618 = 16.66

* 1 2/3 Phi x 10 = 10.296

Hence: 3760.495, divided by 10.296 = 365.24 days in a year

****

From floor of grotto to apex = 598.045 x 0.9998382 = 597.948,

divided by 2 = 298.974 x 6.289 = 1880.247, divided by 365.24 = 5.148,

divided by Phi = 8.33

* 8 1/3 x Phi = 5.148

Hence: 1880.247, divided by 5.148 = 365.24 days in a year.

In the case of the Royal Cubit Theorem:

The half pyramid inch where 1.732, divided by 40 = 0.0433 x 1000 = 43.3
x 0.9998 = 43.29 x 6.289 = 272.25, divided by 365.24 = 0.7454,

divided by 2.2360679 = 0.3333...

* 1/3 x 2.2360679 = 0.7454

Hence: 272.25, divided by 0.7454 = 365.24 days in a year

The pyramid inch where 1.732, divided by 20 = 0.0866 x 1000 = 86.60

x 0.9998 = 86.58 x 6.289 = 544.50, divided by 365.24 = 1.4908,

divided by 2.2360679 = 0.6667

* 2/3 x Square root of 5 = 1.4908,

Hence: 544.50, divided by 1.4908 = 365.24 days in a year…rather interesting.

There is every reason to believe the grotto and descending passage relate to star constellation energies that were apparently made use of in the structure. This subject can only be touched on because the full details are unavailable. What might be determined is that when the star and sun energies combined with gravitational influences of the various constellations, earth and all other energies involved during certain cosmic events was directed, or found its way through the airshafts and passage ways etc., the intended results within the King and Queen's Chambers were acquired at a time very long ago, by those who understood exactly how it worked and if this was the case it might be asked indeed, if a form of Star Travel and Heaven only knows what else was in the works when it came to use of the Sarcophagus. The word pyramid means fire, energy or light in the middle and that relates to where the King and Queen's Chambers are located within the structure. It can only be surmised that during these cosmic events the crystalline capstone made of high grade quartz, silicone dioxide would become fully functional for its intended purposes and radiate, or emanate refined cosmic energy that spiraled down from there to become focused in these chambers. It occurs that whoever was in the Sarcophagus at the time would have been enormously influenced by these forces, and we might ask what the object of the exercise was and what was the end result. Amongst other thoughts, the recipient of such energies might develop super human capabilities as well as become supremely knowledgeable in the workings of the Universe. An outstanding question might be, what does this energy processing have to do with the creation of life in human form? Perhaps only an adept would be permitted to use the Sarcophagus for its intended purpose whatever that was. The other thought that occurs is when the components of the interior details within the Great Pyramid are compared to human anatomy they appear to relate as follows:

Grotto = Sexual Organs, or Lower Chakra, the area between the grotto and chambers = Intestines,

Queen's Chamber = Solar Plexus, below Stomach, perhaps the female component,

King's Chamber = Heart, the area between the chambers = Lungs and other vital organs, the air shafts, perhaps the male component,

Antechamber = Left Pulmonary Artery of the Heart,

Grand Gallery and passages = Blood Vessels, Oxygen Supply, Sense of Smell

at the 161.803 foot level above the top of the King's Chamber = Throat Center,

Breathing System = Expansion and Contraction of the Living Universe, the same way we breath.

Capstone = Crown, Mind, Brain, Nerve Center, including the third eye, or all seeing eye. The phrase " All seeing Eye " as it relates to the capstone is appropriate because without question, it certainly tells us a great deal as if it has vision and mental capacity. I believe it " Sees the Universe " and now it can tell us all about it.

Please see a drawing titled: Great Pyramid and the Chakra's in which the vertical spacing of the Phi based value 85.568 which is derived by dividing the distance between the floor of the grotto and apex by 7 is shown. It can be seen how the floor elevation of the King's Chamber corresponds with the heart and lungs area while the floor of the Queen's Chamber corresponds with the solar plexus. For a striking picture color in the chakra centers as referred to in Chapter 3. I believe this concept to be more than a vague possibility and since this is the case I am left wondering how the central core blocks might be arranged in terms of a spinal chord and what might be found, or that which exists at the seven chakra centers.

Therefore the Great Pyramid can be seen as a living, breathing, thinking, organism with vision capacity not unlike a human, or other living things that are created in the image of, or are a reflection of the living Universe itself, and if this is so, it is in a state of consciousness as well. Its source of nourishment is cosmic energy and its true meaning and terms of existence are perceived to be that of the living God in form, made of Stone, designed and engineered by some group with incredible capacities who performed under an influence we know nothing of. Its arms and legs are seen in the Golden Spiral workings of the life force throughout earth and the Universe. Take note that a golden spiral is seen in the anatomy of the inner ear, therefore this marvel has a sense of sound as well. The answer to that ageless question appears to be; that God power is the life force active throughout the Universe which interacts with the integral values from 1 to 9 including all resets, expressed in terms of infinity by the Number 8, and the Solar Number 7, empowered by the glorious workings of the Divine Ratio especially through the square roots of 3 and 5. For thousands of years words by others have spoken of this notion with regards to the Great Pyramid using somewhat different terminology, but meaning the same thing, and it does seem to help find a definition for the indefinable. It also appears as though the above mentioned geometric determinations, which are not merely mathematical coincidences, are proofs beyond any doubt that the above proposition with reference the interior design and how it relates to the outer geometry of the structure is correct, and a few more pieces of the puzzle have fallen into place. Therefore, I rest my case on this issue for the time being.

The Queen's Chamber

Some raw measurement information of the Queen's Chamber is available. The south side of this enclosure is on the vertical center and its floor level is very close to the half way distance between the true base and top of the King's Chamber. The distance from the center of the Grand Gallery to the south end of the Queen's Chamber on the level through an access proves out to be 123.6 feet. The length of the Grand Gallery, 138.19 divided by 123.6 = 1.118 = sq. rt. 5 divided by 2. The access to this chamber is at the junction of the Grand Gallery and ascending passage 138.19 feet from its top.

According to the information provided by others the dimensions in feet of the Queen's Chamber are as follows: H = 15.3, W = 17.19, L = 18.9, and to the apex 20.3, but though they are close none of these are precisely in the Phi Ratio agreement with each other and this doesn't make sense. These measurements are suspicious because air and water loss shrinkage factors need to be taken into consideration, especially in an enclosure that has been dehydrating for thousands of years. To make a long story short the Royal Cubit Theorem was applied and the following dimensions were derived: H = 15.415, W = 17.3205, L = 19.052. The shrinkage factor is around 0.99 % and this is reasonable. After all it has been proven the entire structure shrunk for various reasons that were addressed earlier. One tenth of a foot 0.10, divided by 5000 years = 0.00002 = rate of shrinkage per year. It can be seen that by dividing the width by the height the quotient is 17.3205, divided by 15.415 = *1.1236 = the square root of 5 - 1.0, divided by 10 + 1.0, and the width divided by the square root of 3 = 10. When the length is divided by the height 19.052, divided by 15.415 the quotient is the square root of 5 - 1.0. When the width is divided by the length the quotient is, 17.3205, divided by 19.052 = 0.909, which when divided by 1+Phi, divided by 2, 0.809 = 1.1236. Therefore, it is determined that the square root of 5, the square root of 3 and their variations interacting in terms of the Phi Ratio relationships is the mathematical theme in the workings of the Great Pyramid. It is observed that the width of the Queen's Chamber equals the width of the King's Chamber and the length of the Queen's Chamber is virtually equal to the height of the King's Chamber and the length of the Queen's Chamber is around 55 % the length of the King's Chamber. The perimeter of the Queen's Chamber to that of the King's Chamber is 0.70 x 10 = 7, the solar number. This may not give us much to go on at this stage but it must be part of the puzzle. As seen, many of these mysteries will be solved by finding the proportions between one true dimension to those of the others and it will take time.

As a check using the Royal Cubit Theorem, the square root of 3 and the pyramid inch are divided into each of these revised dimensions of the Queen's Chamber and both interesting and realistic results come into focus:

H = 15.415, by 1.732 = 8.90, by 0.0866 = 178.00

W = 17.3205, by 1.732 = 10.00, by 0.0866 = 200.00
L = 19.052, by 1.732 = 11.00, by 0.0866 = 220.00 ...
this serves as a proof of the Royal Cubit Theorem and the shrinkage factor theory

Therefore, the above listed information on these issues appears to prove out and is in accordance with the mathematics of the Great Pyramid, the language of the Universe, yet the true usage of this chamber and the one above it remain a mystery for the time being, although progress is being made. Once all the facts are in, the code locked within those ancient triangular stone walls, like shimmering silhouettes, originally white, now turned to gold, that have been mocking time for thousands of years will be more completely deciphered and it will be all systems go for a new age of science based on what was known in very ancient times. I did say this wasn't a math course and it isn't really. It is just a matter of simple arithmetic along with an understanding of the Golden Ratio. The best term to use for it is, " Number Crunching ", it can't be avoided. There is more coming, we aren't done with it yet. The old adage applies here, no pain, no gain.

From a surveyor's point of view the true incline of the Grand Gallery would be easy enough to acquire using a well adjusted engineer's level for elevation differentials, or simply by reading the vertical angle on a survey instrument from the bottom end to the top end or vice versa to a point on a sight pole that has the same height as the instrument, and though it would be more difficult but not impossible to do it that way in the narrow Ascending Passage an electronic digital grade beam could have been used as an alternative method to determine the grade in more recent times. I believe this section is in its original position and settlement is not a factor. It is interesting to see that 161.8 x 0.5 % = 80.9, divided by Phi = 130.9 = Phi, divided by 2 x 100 + 100. It is unfortunate the previous researchers missed this detail because the incorrect estimate of 26 degrees or so and the incorrect length of the Grand Gallery and passage ways has left many a Great Pyramid enthusiast in the dark for many years and that has been counter productive. Basic training shows that correct survey information is very important and the archaeologists who visit these sites should put this into practice and hire personnel who can handle this type of work.

When Napoleon Bonaparte first scrambled up through the narrow ascending passage all hunched over then boldly strode up the Grand Gallery and into the King's Chamber back in the 1800's it is doubtful he had any concern over the percent grade of it or how long it was, but his face turned ash white in that enclosure because he must have seen or felt something unusual, or so it was reported. On his death bed some years later he was about to tell a close friend about what happened in there, then he stopped in mid sentence and sighed, saying, what's the use, no one would ever believe me. We are left wondering if an influence there had opened his mind's eye to the expanse and details of the Heavens and the workings of the Universe, or perhaps, he was somehow informed on what his future would be.

I have never visited the Great Pyramid in person and may never do so, however, that is what the writings and drawings in this chapter are primarily about because a break through with its true dimensions has been made and it can finally talk to us, and I believe this hasn't happened since long before biblical times. This information wasn't available in ancient Greek culture and historical records of that time are hazy. There is no answer to this question that I know of, but we have the goods now and the impact will be felt around the world once the information is released, and this might turn out to be timely in our process if it is handled properly. I make frequent visits to my 10 foot pyramid when I need answers and this gets me by. It is a structure made simply of wood but its proportions are exactly those of the Great Pyramid and it is aligned with magnetic north. Starting as an experiment it became my think tank, so to speak. Coupled with the desire to find the answers and mental focus I believe that it has enabled me to solve a number of mysteries as old as time. One big question I have is why was the Great Pyramid made inaccessible. What bothers me is the destructive manner in which it was accessed and the false motive such as material greed that was behind it.

What I have learned is, it isn't all that necessary to visit this magnificent structure to take measurements in person as many others have done so already. Once the recorded information is sifted through and deciphered it takes a thorough knowledge and appreciation of the Golden Section to progress through the facts and the misinformation in order to understand and make sense of it. It could be said too, that I am a surveyor who possesses a certain degree of intuitive, deductive reasoning power with many years practice in taking measurements and this has proven to be an assist in the inquiry. The analyses of the Grand Gallery, the descending and descending passages is a prime example of this. As suggested before, if the accurate angles and distances were known with reference to the airshafts that are aligned with certain stars the distances to them could be readily computed based on their lengths inside the Great Pyramid or by certain functions we need to know that are locked up in the code, the distances between them could be determined by simple triangulation. Many have labored long and hard to generate drawings of the interior chambers, air shafts and passageways etc. in the Great Pyramid but it has been all been for naught because the correct dimensions were unavailable but now they are.

Light Years that have Meaning...

The ratios of the Great Pyramid match with those of the Light Year Pyramid in the Milky Way. The simple breakdown of this being the case is outlined below:

a.) 365.24 x 24 x 3600 x 186,216.56 mi./sec. = 5,876,386,823,000 miles in a light year x 5,280 = 31,027,322,420,000,000 feet in a light year x 57,736.72 = 1,791,415,827,000,000,000,000 feet in the base of one of those awesome light year pyramids in the Milky Way.

b.) Divide 762.08 feet, the correct base length of the Great Pyramid into the base length of the light year pyramid = 2,350,692,614,000,000,000, divided by 10,000,000,000,000,000,000 = 0.23507... x *(8, the number of infinity,divided by 2 = 4, divided by 1000 = 0.004 + 1.0 = 1.004) = 0.236 = Phi cubed...to make it simpler, 0.23507 x 1.004 = 0.236. The same value makes itself known when the vertical distance from the base of the Great Pyramid at center to the floor of the Grotto, 114.3 was divided by 484.68, its height = 0.236. It is also meaningful that 0.236 x 2+Phi = Phi and (Phi + 1+Phi) = 2.236, squared = 5.0, the quintessential number of life.

c.) Conclusion: The above is a proof that the reciprocal of the square root of 3 x 100,000 = 57,736.72 light years, the radius of the Milky Way and the implications are huge. Just around the corner another joyous revelation will be at hand when this radius is multiplied by 2 x Ancient Pi.

It needs to be accepted it is the Golden Numbers and Grand Ancient Sciences behind a Phi Based Pyramid with meaningful dimensions that has educated us about the workings of God, Life and the Universe. There are three other Light Year Pyramids in the geometric composite of our galaxy and these together working in harmony with each other involve the number 5 in terms of a natural pentagram in their makeup. A name for them could be, the Pyramids of Light Years, or simply, the Pyramids of Light and each of them relate in a Phi Proportion way to the Great

Pyramid on earth. Incredibly the proportions of the Great Pyramid are fully expressed in terms of light years throughout the geometry of our galaxy forming a perfect golden rectangle with a width to length ratio of the square root of Phi and a double Golden Section pentagram having a ratio, arc over radius = Phi, within it. Since this is the case there is far more to the Great Pyramid than ever imagined before and somehow earth and our sun are factored into the picture in a big way. Is this because it is the only planet in the Universe that has, or ever had life on it to begin with and that is why it is so special, or was or is something going on in the Milky Way we don't know about? To say it again, the one thing that is known for sure is, some colossal science we don't understand was at work here during some ancient era in the Universe and on this planet that we know nothing of. It is as if it was somehow possible the cosmos was charted out and traveled throughout by humans long before our time, and the big questions are, how, why and when was this done.

Now that it is known there are two pentagrams within the Milky Way geometric makeup the interior angles subtended by the equal arc lengths around the circumference are 36 degrees each, or 360 degrees, divided by 10 = 36 degrees. Two five pointed stars, one pointing up and the other pointing down are not shown. There are enough lines to contend with on the drawing as it is, however, a five pointed star fish in our oceans seems to relate very well to the geometry of our galaxy and the pentagram in an apple core seen when it is cut in half at a right angle to its vertical plane appears to do the same. When the circumference is divided into 10 equal parts, 361,803.39, divided by 10 = 36,180.339. It is interesting to see that 36 degrees x .9973317 = 35.903941 degrees x 0.0174533, (one degree in radians) x 57736.72, (½ the long axis) = 36,180.265, a very decent approximation, within 0.9999979%. And this speaks well of the 0.997 factor times the equatorial diameter used to compute an estimate for the polar diameter of any celestial body. Interestingly, this value was acquired by dividing the Phi Based circumference into the circular circumference, ½ the long axis x 2 Pi and another mental note is made.

The three other locations of great interest are where the lines of the pentagram intersect with the corners of the rectangle, and aside from that there are two interior ellipses with circumferences of * 314,000 light years that also coincide with these points and two more points at the intersections of the pentagrams east and west. It isn't clear what this is all about but it might be important. Perhaps they are transportation pathways or like great circle routes between these locations that we might travel some day, or that were traveled by others in some very remote time in the past. This is very interesting because it might be telling us there are three other suns and solar systems identical to ours with a chance of life on the third planets from their locations in these systems. It would be jumping to conclusions of course, but even if there is this possibility they are a long way apart and we will never see or know about them using our present day technology. We will need to travel much faster than the speed of light to get anything out of it in our era. Therefore, I believe we need to know exactly how the Sarcophagus and energy relationships in the Great Pyramid work because understanding this may be the key to future space travel. Could there be Great Pyramids on these planets identical to the one on earth? Perhaps the ancients are from these planets, or the 4 races of human kind on earth originate from them. Make way for a new brand of science fiction that has a certain deal of credibility to it. Then, please have a look at the drawing titled, Geometry of the Milky Way which shows its circumference, the golden rectangle, the pentagrams and the six Phi Based light year pyramid triangles within it. The perimeter of these triangles works out to, 57736.72 x 2 + 46709.52 x 4 =302311.52 x 3 = 906934.56, divided by the square root of 3 = 523600, divided by 1,000,000 = the square root of 5 + 3 and a mental note is made. There are some similarities between this drawing and the one for the Enneagram but the angles are different of course and they represent different meanings.

Interestingly, the six triangular faces seen in the geometry of the Milky way is reminiscent of a quartz crystal and what is seen in this drawing might relate to an ancient energy technology we should know about. A six sided light year double tetrahedron could be constructed from this detail and of course a light year pyramid could also be put together but none of this makes very much sense really. I was looking for something more workable that would relate to what is seen in this drawing. Then I cut along the six exterior lines on the perimeter of the triangles and ended up with a geometry not unlike a diamond or at least a crystal type shape of some sort. I drew this independently of the others to study it and it occurred to me that I needed some bristle board paper because I saw an opportunity to construct a form from what I saw. Please see the drawing titled, Milky Way Pyramid Crystal. The form embodies the triangles that are in proportion with the Great Pyramid and the Golden Rectangle. At each end there are pyramids which are enlarged by a ratio of one over ancient Pi in order that a double terminated pyramid form at each end is the result. These can be identified as capstones. The points of these pyramids are seen where the pentagram lines intersect left and right on the Geometry of the Milky Way drawing. Within this form are many Phi Proportions that are meaningful to the study and the operations of the Milky Way. Fold on the outside vertical lines and on each of the horizontal lines where the diamond shaped pieces meet, find some glue, put it together and add it to the collection. First, give it a coat of speckle stone and spray it with our favorite color and you will have the machinery of the galaxy in your hand that looks like a good sized gold nugget. Well, it certainly is loaded with Golden Ratios at least and it becomes a representative platonic solid that represents the geometry of the Milky Way. Then give it the energy emission test. With one hand direct one of the pointed ends to the palm of your other hand rotating slowly and a coolness will be felt on it, then direct a point toward your chakra centers moving it up and down and each one will be stirred as if you were using a quartz crystal sphere. It wouldn't do any harm to balance it atop your head while in meditation mode. When I took my model with me for a session in my pyramid I used it in place of the crystal sphere and, wow, it was as though my being was being encased in a cocoon of energy with jolts of electricity surging through me as I slowly deep breathed and moved it up and down over my chakra centers. Then when I rotated it clockwise and counterclockwise at each of my energy centers the sensations were much more pronounced than when I used the sphere. Also, I believe that having the capstone in place atop the structure makes a pronounced and positive difference with regards to higher energy levels. After the session I felt totally solid and mentally alert, ready for any challenge I might be confronted by. The symptoms of a sore throat that had been bothering me instantly vanished shortly after my visit. I can only conclude that the use of such a form in this manner inside a Phi based pyramid can be only be beneficial. The form is made of only paper, and I thought, what would be the case if it was made of high grade quartz. The next time I visited there for a session I held the crystal sphere in one hand and was busy with the other hand using the form as described. Of course my wheels are turning now, my next project will be to build one that I can fit myself into. It will be around 12.14 feet long if the pyramid base is 6 feet. What has been constructed is a scale model of our galaxy which is not unlike a space craft on a journey across the infinity of space and time powered by an energy we have much to learn about. It is reasonable the depth, or vertical width of the Milky Way energy zone is the height of the light year pyramid times two and this is seen in the form that has been assembled.

The above assessment of the Milky Way dimensions and relationship with the Great Pyramid and our solar system are correct as far as can be determined because, as shown, when its dimensions are defined using 1.0, divided by the square root of 3 in terms of light years they are given by those of the capstone and a distinct correlation with the Great Pyramid has been found in its geometry. Should this concept stir up other individuals to find their own answers and spawn another theory that works then let them be heard. What has been discovered here suits the old cliché, the answers were right in front of us all along but we couldn't see for looking. By dividing 1.0 by the square root of 3, in a flash the microcosm has expanded into the macrocosm in terms of light years and a chart of the Milky Way and Universe were opened up for exploration, therefore, without doubt, the benefits from studying the lessons of the Golden Section are truly immense. A thought on the capstone is, it might not have been necessary for it to have been a gold trimmed high grade silicone dioxide pyramidion affair after all, whereas it might have been a simple structure of consisting of an alloy of bronze and tin, or gold rod anchored in a special way at the apex. The answer to this question may never be known, but the importance of it is what its dimensions tell us and I believe there is agreement on this by now.

It is time to roll up our sleeves and get to work because it is now possible to assign start coordinate values to the center point of the Milky Way such as 0.000 / 0.000 / 0.000 that will provide a relative fix on the points that define the astronomical geometry of the Milky Way, on our sun and solar system within it and beyond. The azimuth of 38 degrees, 10 minutes, 24 seconds and distance between our sun and the center of the Milky Way is known and this provides the required baseline of 46,709.52 light years on a 26 degree, 33 minute, 54 second angle, or a grade of 0.50 %, or 2:1 slope to begin with. Therefore, 46,709.52 light years x 0.50 % = 23354.76 light years is the elevation of our sun above the center of the Milky Way and it is that simple.

Note: 23354.76, divided by 36720.55, the height = 0.636... that familiar number...
and, 2 x 115473.44, the base = 230946.88,
divided by 36720.55, the height = 6.289308847 = 2 x Ancient Pi.

The blank spaces and unknowns can be filled in with some help and fine tuning from what more the Great Pyramid tells us. We have the right idea with our star dome planetariums at this time, all we need to do is derive the true dimensions and it will all make sense, then our celestial inheritance and destiny will be known. By doing this we will become much more knowledgeable about the Universe and I believe the ancient code locked up in stone for these past thousands of years within the Great Pyramid will show us how to travel to certain destinations at the speed of thought and there will no longer be any need of space ships and telescopes because we will have advanced beyond today's limitations and into those higher and more noble levels of enlightenment.

Near Journey's End…

Our journey through the solar system and Milky Way is coming to a close for now but we aren't coming back empty handed after having a look around and making good use of the Royal Cubit Theorem. Before we head back to earth let us take a peak at some special locations beyond its limits. I am quite sure many of you will find your own way around out there with this Theorem and the true dimensions of the Great Pyramid at your disposal. Modern Science claims it is 430 light years from earth to the North Star, Polaris but we know they have been using the wrong AU value all along, therefore what they have to say about anything beyond the status of the weather is doubtful. That may sound unkind but we have an opportunity to pull all the way out of the dark ages because we have the advantage now by knowing the true dimensions of the Great Pyramid and what it can tell us.

Polaris:

The surveyors, map makers, astronomers and the general public would be nowhere without a true north reference. Nothing would make sense if there was no bearing or azimuth relative to Polaris from positions on earth. What is needed to acquire the azimuth or bearing of a line on earth is the horizontal angle from a base line A, back sighting B and fore sighting to Polaris when it is at elongation and an accurate chronometer to record the time of the measurement so that the longitude east or west of Greenwich can be determined. The vertical angle to Polaris will furnish the latitude of the set up point. The distance in light years from Earth to Polaris is based on the revised height of the Great Pyramid multiplied by the inverse of 1.118 = 0.894 x 484.68 = 433 light years, within 0.995 % of the estimate by modern science. Remember, the square root of 5 = 2.236, divided by 2 = 1.118, subtract 0.5 = 0.618 and this was the assist for determining the planet dimensions further back. In simple terms 1.0, divided by 1.118 = 0.894. This calls for a proof, therefore, the diameter of Polaris will be discussed a little further along, then it will be seen how its distance from earth in light years multiplied by 2 x Ancient Pi relates to the number of days in a year. The dimensions of the Great Pyramid in feet are in terms of light years and the Royal Cubit Theorem provides the distance to Polaris when 1.732, divided by 4 = 0.433 x 1000 = 433 light years. Presented at the end of this chapter is a drawing titled, Great Pyramid and the Airshafts. A Phi based pyramid was drawn with a height of 433 light years divided by 100 = 4.33. The object was to get a handle on the true slopes of the air shafts and what is shown in this drawing with regards to where they connect at the chambers appears to be on the mark. Please see the notes which are self explanatory. The code provides the light year distances and azimuths from Earth to Sirius, Orion, Sigma Draconis, Thuban and Alpha Centauri, or Rigil Kentaurus, the ancient Egyptian name for it. Today it is known as the Constellation of Centauris. Please see the drawing titled, Great Pyramid and the Constellations for a view of the celestial arrangement in the heavens.

The researchers have been claiming the air shaft in the Queens Chamber to Sirius was on an incline of 39 degrees for years when it is actually 40 degrees and their claim that the vertical angles for the airshafts from the King's Chamber to Orion and Thuban is 32 degrees is shot down as well because it proves out to be 30 degrees. One or two degrees in terms of astronomical measurements in space makes a huge difference. Some benefits of this addition are the distances in light years from earth to these constellations are now known and the true inclines of the much talked about air shafts have been established. Please see the following list of light year distances from earth to certain stars: Alpha Centauri A = 4.0, Alp cma Sirius = 8.66, Alnilam in Orion = 1381, Sigma Draconis = 18.926, Thuban = 309.4 as we probe further into deep space.

The distance to Alpha Centauri A, the closest star to us we are told, the one our hairball in a space suit is headed for, is seen in the dimension of 426 feet from the base at center on the west side to the vertical of the Great Pyramid on the 2:1 slope, or on azimuth 63 degrees, 26 minutes, 06 seconds, starting at the lower west side, same as descending, ascending passage ways and Grand Gallery, only on the opposite side of the structure. The distance 426, divided by 100 = 4.26 and 4.0 is the number of light years from earth to Alpha Centauri on the 2:1 incline to the south west. As discussed in the previous chapter the ancient code locked up in the Great Pyramid informs us 0.26 x 2 x 100 = 52 weeks in an earth year and 0.26 x 0.5 x 100 = 13 the sun + 12 planets in the solar system and the AU 91,796,000 miles = 1.56 E-5 light years x 10,000 = 1.56, divided by 2 x (*1.732 squared) = 0.26 from earth to the sun. Note: the above expression 2 x (*1.732 squared) = *6 the double trinity number. Alpha Centauri is the closest star to our sun on the 2:1 galactic incline. At the bottom of the drawing Alpha Centauri B, the south pole star and Alpha Centauri C, Proxima are seen showing their relationship in the celestial arrangement of this constellation. Note: If the Grotto point is Alpha Centauri B, it is 3.05 light years from our sun and 2.94 from earth, closer than Alpha Centauri. Also note: The azimuth and distance from Alpha Centauri A to B is 106 degrees, 42 minutes, 01 second, 3.978 light years. Proxima is due east of B at 0.67 light years and this speaks of a 2/3 proportion. When the locations of the above mentioned constellations are introduced into the drawing it gets more interesting and the earlier question about the value for the slope distance 616.528 feet has been answered. This is the distance in light years from Polaris to Alpha Centauri A. The geometry of the drawing more or less matches with the distorted star maps by others but they don't understand how the 2:1 slope between Alpha Centauri A and our sun works. As it turns out, it is another baseline that can be worked with for mapping purposes. In these drawings are seen what the relationships are between our favorite star, the constellations, our sun and the center of the Milky Way. Remember to view this triune of stars from space, not from earth, just as it is shown on the drawing of the Great Pyramid presented in this discussion. A way to understand this more clearly is to draw a sketch of the star arrangement on a sheet of paper with little circles joined by lines as seen from earth. Alpha Centauri A will appear to be south east of Alpha Centauri B. Then flip the sheet over in a sunny window and the image will appear the other way around like the view from space, same as the drawing presented here. The center of our galaxy is at 90 degrees clockwise from points B to A and is 2.01 light years south east of this star on azimuth 153 degrees, 26 minutes, 06 seconds. Now we know exactly where it is, not just somewhere near the middle of the Milky Way as put forward by others.

As previously discussed, a picture says a thousand words. Please have a look at the drawing titled, Cosmic Chord where the west side pyramid angle from the location of the Great Pyramid on the globe cuts a chord on the arc from the location of the Great Pyramid producing a dimension equaling the side slope distance, 616.5 feet x 10 = 6,165 miles on a line parallel to the one from earth's center to Alpha Centauri A. As a foot became a light year in the above, 10 feet becomes one mile. There doesn't seem to be any coincidence in this and the unique angular values shown on this drawing say it all. Incredibly Angles ABC = CAB = the value of the azimuth of the side slope and half the subtended angle = the Pyramid Angle. Though it may seem curious, the catch is, it works out this way only if the azimuth starts at a vertical distance of 1.0, divided by (sq. rt. 5 - 2, divided by 10) = 1.0, divided by 0.0236 = 42.35 miles x 5280 = 223,600 feet down from the Great Pyramid at center and this has to do with the code cracking process because the square root of 5 is involved and the values are arrived at mathematically, making it a valid proposition. It becomes more understandable when it is seen, 223,600, divided by 100,000 = 2.236, divided by 2 = 1.118 - 0.5 = 0.618, and when DE, divided by AE = 0.618 shown in the following self explanatory drawing titled, Great Pyramid Earth Measures Theorem. Details not shown on this drawing are: Angles BCD = CBD = 38 degrees, 10 minutes, 24 seconds

Angles DEC = DEB = 51 degrees, 49 minutes, 36 seconds

BD = DC = 4846.8 miles

Angles ABD = ACD = 90 degrees

Angle BAC + BDC = 180 degrees...In a rather extraordinary fashion applications of the Golden Ratio are apparent in the geometry of this drawing and the correct answers are provided. What can be seen is 7,620.8, divided by 1.732 = 4400, divided by 10 = 440, the number of royal cubits in the base of the Great Pyramid x 1.732 = 762.08, the number of feet in the base. The latitude and longitude of the Great Pyramid is 30 degrees north and 31 degrees east of Greenwich. The interesting thing about this is, the chord terminates at latitude 30 degrees north and 72 degrees, 39 minutes west of Greenwich, approximately in the middle of the Bermuda Triangle, north east of Cuba and north west of the deepest part of the Atlantic Ocean known as the Milwaukee Deep. There is nothing available in this edition that offers an explanation for the mysterious types of events that have occurred in this region over the years such as airplane and ship disappearances etc., however, the intriguing results of a recent oceanographic survey has yielded images of many structures inaccessibly deep down on the ocean floor in this area that are suspected to be man made, and the story about ancient road ways left over from the Atlantis era along the coast of Brimini is well known. The popular opinion is, there is a link between Ancient Egypt, the Great Pyramid and Atlantis but at this time proof and hard evidence is lacking to substantiate this claim. Refer back to the drawing and discussion about, the Giza Phi connection where it has been determined there is an ancient global ley line passing through this area from the north west down through Central America into South America to Nazca. When 10 feet becomes a mile it appears the riddle of the true dimensions of the Great Pyramid is solved, along with the previous input which supports and agrees with the presentation here-in. In a sense then, it serves as a final proof of the true dimensions. The drawings explain the geometric arrangement better than the words it would take to describe what is at play here, because it is beyond the word amazing, better words to use might be, just plain amazing. The term mind boggling was already used for the Giza Phi connection drawing and we are left wondering if there might be an active energy vortex of some sort deep below the floor of the ocean in the Bermuda Triangle that has something to do with the Great Pyramid and its connection with the lost continent of Atlantis.

There has been ongoing speculation for several years by both modern science and science fiction parties that there is or was life existing on planets that revolve around the above mentioned constellations. In particular, in the case of Alpha Centauri A because spectral analysis of its light reveals that it has similar qualities as compared to our sun light. One might think though, since it is 4 light years closer to the center of the galaxy it might be much hotter there. What we do know for sure is, the decoding of the Great Pyramid provides a glimpse of a truly fantastic science that has a purpose we need to know much more about.

Diameter and Circumferences of Alpha Centauri A, B and C = Proxima Centauri

Alpha Centauri A

We are about to embark on an adventure to explore what the lost code tells us about star dimensions as they relate to the diameter of our sun-star and the Golden number values. It is believed to be a matter of proportion, therefore, after some research the opening statement is put forward on the diameter of Alpha Centauri A: The distance, 4.0 x reciprocal of Ancient Pi, or the square root of 1+Phi x 866,025. In other words, 4 x 0.318 = (1.272) x 866,025 = 1,101,584 mi. Modern Science says 106,100,000 and that works out to 0.936%. What is seen is 1,101,584, divided by 2 x 6.289 provides a double cubit value of 3,464,000 mi. and that seems reasonable with regards to the Royal Cubit Theorem. It is put out tentatively that the diameter of Polaris is 3.0 x 10 = 30 x 866,025 = 25,980,750 mi. and this correlates with the modern science that claims it is 60 times the radius of the sun and it works out that 433 x Ancient Pi x 2 = 2723.27, divided by 3 1/3 x 2.236 = 365.24 days in a year.

Alpha Centauri B ~ 1,101,584, divided by 1 + Phi (1.618) = (D) 680,831 mi., divided by 2 = 340,416 x 6.289 = 2,140,000, divided by 1,000,000 = 2.14 = Pi - 1.0.

Alpha Centauri C, or Proxima Centauri ~ 1,101,584 x the reciprocal of the sq. rt. Phi - 1.0 = 0.272 = (D) 299,631 mi., divided by 2 = 149,812 x 6.289 = 942,168, divided by 3 = 314000, divided by 100,000 = 3.14 / Pi.

Relationship of Diameters Alpha Centauri Constellation

A, divided by B, divided by 1000 = 1 + Phi / B, divided by C = 2.272 = sq. rt. 1 + Phi + 1.0

The code in the Great Pyramid provides information about the constellations of the Zodiac and what we need to know about their locations in the Cosmos. The above examples serve as a start for what lies ahead in the mapping department of the Milky Way and beyond. With regards to what the researchers have had to say about these details, look up the distances in light years to these constellations they have on file in the web sites and come to your own conclusions. It's the same old story, the science of guess work has been at play all along, however, progress is being made in spite of these influences.

Journey through the Stars ~ Astronomy made Simple ~ the Navigator's Guide

No seat belt or countdown to blastoff, or any other discomforts and inconveniences of space travel need to be endured and it won't take very long because we will be traveling at the speed of thought of course. Please view the attached four drawings titled: Great Pyramid to, or Earth to: Thuban ~ Sigma Draconis ~ Sirius ~ Orion

It becomes apparent in these drawings, that in each case, the constellation positions, or locations in the cosmos relative to Earth in our solar system are defined, or explained in Phi based mathematical terms when the base and height dimensions of the Great Pyramid are interpreted as the key to the code which provides the answers needed, and the procedure serves as a proof that the distances shown, including the airshaft inclines directed to these constellations from the Great Pyramid are correct. This information becomes known using basic terms of geometry when the facts are in, yet these details have eluded astronomers and the science departments in our education system for centuries if not thousands of years. In fact, none of this was known during our recorded history and that is because training in the applications of the fundamentals of the Golden Section was never introduced because the science of it was somehow lost and this has seriously altered our ability to understand the workings of the Living Universe, or anything else that is important. Not to despair, this sort of information provides a new beginning and I dare say some more bingo's can be added to the growing collection. Note: The distance to Thuban in light years is based on Phi, divided by 2 x 1000 including a touch more such as Phi x 1.001, to allow for the astronomical nature of its position in deep space. Divide each distance by Phi, 0.618 for further entertainment.

A Flash Tour of the Constellations the Great Pyramid is Aligned with:

Center Milky Way to: Orion / 300 degrees, 11 minutes / 1380.7 light years
Sirius / 336 degrees, 40 minutes / 10.0 light years
Thuban / 59 degrees, 41 minutes / 313.4 light years
Sigma Draconis / 66 degrees / 22.5 light years
Orion to: Sirius / 119 degrees, 56 minutes / 1372.5 light years
Thuban / 110 degrees, 6 minutes / 1558.8 light years

The Zodiac Constellations:

Imagine a drawing titled, The Zodiac Wheel showing Aries at the top through all the other constellations going counter clockwise to Taurus at 30 degree intervals. More details on the Zodiac arrangement will be discussed in a future edition. In the following there is some very interesting reference as to how the Zodiac was understood and dealt with in that ancient and glorious period long before our time.

It has been determined the diameter of Alp cma in Sirius is 866,000 miles, the diameter of our sun x *1.732 = 1,500,000 miles, divided by 100,000 = 15 that intriguing number that was looked at back in chapter 3. By proportion then Sirius is no less than 1.732 times brighter than our sun. Since the diameter of Anilam the middle star in Orions Belt is 30 times greater than our sun, 866,000 x 30 = 25,980,000 mi. and by no coincidence the computation, 25,980,000, divided by 1,500,000 = 17.32 = 10 Royal Cubits the width of the King's Chamber and 17.32 ft. x 2 = 34.64 ft. = 20 Royal Cubits = the Length of the King's Chamber and the Royal Cubit Theorem makes it presence known again. A further investigation based on its true dimensions reveals that the height of the Great Pyramid, 484.683 ft., divided by 2 x (2 x Ancient Pi) = 1524.16 ft., or 2 x 762.08 ft. = the same, divided by 88, the number of constellations = 17.32 ft. and its perimeter, 3048.32, divided by 88 = 34.64 ft. For certain this Grand Monument is divinely tuned in with all that exists in the Heavens by way of those Golden Section Values which speak to us in the Language of the Universe. It occurs, if there are any other human inhabitants in the Milky Way, no matter where they might be located therein, the constellations of the Zodiac would appear the same as they do on earth. Again it is asked, was a form of space travel made possible in ancient times, and what does the Sarcophagus in the King's Chamber and the Queen's Chamber have to do with it? We need to know about this. One consideration is, the velocity of reflected energy such as sunlight coupled with the interaction of piezoelectricity and electromagnetism from within the structure and through the concave faces might be accelerated, or perhaps this occurs at the apex of the structure. We really need to look into this. Earlier on in the study I experimented with that 12 inch pyramid oriented toward magnetic north with witching rods and a gold ring on a thread atop the apex and proved there is an active energy field in that area of the structure. The conclusion was, it must have something to do with electromagnetism that is generated due to the Phi ratios of the model, then I considered how powerful those forces would be when it came to the real thing that would be using all the above listed energies. The simple experiment of sharpening a dull razor blade at the one third height of a Phi based pyramid stimulates the imagination and is quite thought provoking.

At the end of the chapter please see the drawings titled, Constellations of Sirius, Orion, Draco / the Dragon = Sigma Draconis, Aries and Polaris, Ursa Minor, Little Dipper, Little Bear. The key to drawing these patterns is the lines in the Great Pyramid as shown, using the 5 and 10 factors in the baseline and vertical line at center, and employing the Phi Ratio for the dimensions between the stars. This is a major break through in the study. What is needed are reliable graphics of the constellations based on photographs with good resolution, referenced to the north azimuth and square to the horizon. The vertical and horizontal lines, parallel lines and on occasion the pyramid side slopes and 2:1 slopes can be seen along with the relative Phi intervals between the stars. With a working knowledge of the Golden Ratio it all falls into place for drawing purposes. If a drawing program is not available the use of a protractor, scale and electronic calculator will help get the job done as well. Beware that the graphics provided by other parties of the star patterns may be distorted somewhat. It is a good idea to take in a visual of these constellation patterns at night and make sketches of them. Use binoculars or a telescope to view the constellations that are more distant from earth.

Cassiopeia's Chair

Included are two drawings, the first one is titled, Cassiopeia's chair to help explain the process of determining distances and azimuths between the stars from the code and the one after that titled, Cassiopeia's Chair / Light Years. In the Cassiopeia's Chair drawing a schematic of the constellation is shown above along with the dimensions for AB, BC, CD and DE. These are the computed distances of the lines labeled a, b, c and d in the to scale drawing of the Great Pyramid below where the dimensions are divided by 100 to convert to light years. At first the baseline and vertical line at center were divided by 5 and the connecting lines were drawn as shown. In the case of distance CB the vertical line 1.905 between the base and the 2:1 intersection point was divided by 5 = 0.381, then by 10 = 0.381+0.1905 = 0.5715. Then the lines that defined the alignment of the constellation were saved and the others were deleted. This detail showed the constellation arrangement based on the azimuths of those lines and their distances in light years and, voila, a to scale drawing of Cassiopeia's Chair was generated. This is a result of having a graphic of this constellation from another source, viewing it at night and making a sketch of it then making comparisons with the lines made available in the pyramid. It would help to read the horizontal and vertical angles to each star from a baseline on earth and I intend to carry this out at the time of the summer solstice on June 21, 2008 at 10:00 P.M., but what is seen in this drawing is not only a strong resemblance, but what appears to be very much a true likeness of this popular group of stars achieved in a way never achieved like this before in our history. Therefore, a decent start has been made in this area of the inquiry and additional support from the code within the Great Pyramid is in evidence. The numbers and ratios in the constellations are a story in themselves irregardless of the legends and folk lore attached to them.

To serve as proofs of the undertaking the first observation to be had in the Cassiopeia's Chair / Light Years drawing is the A ~ Phi factor in the first leg from A to B, and the distance A E x *1.013, divided by 5 = 3.0. *1.013 = 1.0, divided by (1.732, divided by 2 = 0.866 x Pi - 2.0 (1.14)). Worth special mention is AC, divided by 5 = 1.75 because its reciprocal = 0.5715 x 2 = 1.143, that mysterious and unique value that just happens to be half way between Pi - 2.0 and Ancient Pi - 2.0. Incidentally, the reciprocal of 0.5715 + 1.0 = 0.636, which when divided into 2.0 = 3.144654088..., Ancient Pi. This value was mentioned on page 86 in chapter 8 and what that does is clear up the mystery on where this number comes from. I ran out of room in the drawing to show the other details. The distance 0.5715, divided by 5 = 0.1144 x 10 = 1.144 = Ancient Pi - 2.0. The total horizontal distance from east to west, 14.48, divided by 5 = 2.896, its reciprocal = 345, the dimensions of that 3-4-5 unit right angled triangle mentioned earlier. The vertical distance from the base to the intersection point where the 2:1 slopes meet at center = 1.905, divided by 5 = 3.81 x 100 = the ½ base of the Great Pyramid divided by 2. The base divided by 5 = 0.762, the full base length of the Great Pyramid, divided by 10. Additional information based on the dimensions ~ AB, divided by BC = 1+Phi ~ BC, divided by CD = 0.666..., reciprocal = 1.50 ~ CD, divided by DE = 0.80, reciprocal = 1.25. And some more yet ~ cD = 3.048 is the perimeter value of the Great Pyramid divided by 100. This is the distance from the base at center to 1.143 above the 2:1 intersection point, divided by (8) = 0.381 / dE - bB = 1.785 = sq. rt. A ~ Phi + 1.0 ~ dE, divided by bB x *0.9976 = sq. rt. 2.0. BD, divided by 5 = 1+Phi. AB - (Y-Y) = 1.33, reciprocal = 0.75. No doubt the dimension relationships of all the constellations possess a good number of Golden Proportions in their geometric networks. After all they express their existence in terms of the language of the Universe and once this is mastered by us, we will be able to communicate with them, and learn about their true meanings.

Light Years to Cassiopeia's Chair

The approach to answering this question will be by employing the Royal Cubit Theorem as it was in determining the distance in light years to Polaris and in several other cases. The five stars in this constellation are at different distances and azimuths from earth but the distance to the closest star named Shedar, the brightest one which is point D in the drawings can be readily computed. Each constellation has a main star. For example, the main star in the Leo constellation is Regulus. The object of the exercise at this stage is to see what the code tells us about the distances to the other stars in Cassiopeia's Chair relative to Shedar and earth. The following computation provides the distance in light years from earth to Shedar.

Square root of 3, divided by 4 = 0.433 x 1000 = 433, divided by 2 = 216.5 light years from earth to Shedar.

The diameter of Shedar is tentatively, 30 x 10, divided by 2 = 15 x 866,025 = 12,990,375 mi., and though it is half the diameter and distance from earth to Polaris that is why it shines brighter than the north star in the night sky. This is a good thing to know, the size of the sun viewed from planet 12 must look like a large star. It would be simple enough to work out a proportion on this now that the facts are known.

Azimuths from earth to Shedar and the other Stars in this Constellation:

Note: All computations following are determined for the moment of the summer solstice in the northern hemisphere when they are at their highest altitude.

Amazingly the angle given in the Cassiopeia's Chair drawing to A, the western most star is from the top of the Great Pyramid to its base on the west side then to the point where line Y-Y meets the vertical line at center at y. Its azimuth value = 13 degrees, 59 minutes, 30 seconds. The note on the drawing indicates XZ = 1.143 and XZ, 1.143 x sq. rt. Phi, divided by 10 = 0.08985. "A legitimate function of Phi". The horizontal distance from the western star to Shedar = 10.67, divided by 216.5 = 0.049284065, divided by one degree in radians (0.0174533) = 2 degrees, 49 minutes, 26 seconds. The sum of these angles providing its azimuth from earth = 16 degrees, 48 minutes, 56 seconds. Azimuth to B, 3.81, divided by 216.5 = 15 degrees, 0.0 minutes, 0.0 seconds. To C = 16 degrees, 0.0 minutes, 30 seconds. To E = 17 degrees, 49 minutes, 26 seconds. It can be seen how simple it is to determine the azimuths from earth to the stars in any constellation once the data on the main star is known. This one has only 5 stars and others have several more and that is why this is a good one to start with, in an attempt to keep it simple to begin with. Notice how the horizontal distances can be used effectively for angle computations. That is because when it comes to light year distances between the stars in a constellation the chord and arc are virtually equal. Of course today's surveyors are using their global positioning system units that are tuned into satellites that orbit around earth and the science of astronomy is becoming a lost art. However, if I am not mistaken, another piece of history is in the workings as we progress through this section and it has everything to do with the ancient code that has been locked up in the Great Pyramid these past thousands of years. A theme that sounds like a broken record by now, but consider for a moment how truly a wonder it is how information from a so very ancient source is telling us what we need to know today... Surely this knowledge wasn't used for the sole purpose of constructing monuments like the Great Pyramid that simply tells the story of the workings of the Universe and how each feature of it relates to the other in terms of the Golden Section and Phi Ratios. Since the code provides a road map of the Universe, which vastly outclasses the capabilities of today's GPS units, we can only speculate that it might have something to do with a form of space travel throughout the heavens that was taking place during a very ancient time, but the means of accomplishing this are totally unknown to us at this time. Perhaps, somehow, some day, we can only hope, the code in the Great Pyramid might let us in on that secret too. It is certainly a very big question. It might be thought that since we are able to travel through the Universe at the speed of thought, or in the mental sense at this time, the faculties of the mind might some day kick in and the solution to this question will manifest itself. Since the turn of the 20 th. Century air speed has gone from zero to faster than the earth revolves on its axis. It will be a matter of time and scientific evolution it might be supposed.

Altitude ~ Latitude of the Star Members in Cassiopeia's Chair:

Polaris is at 90 degrees latitude above earth in the northern hemisphere, therefore the vertical angle to it at the equator is 45 degrees and since Shedar is half way between the vertical angle to it at the equator is 22 degrees, 30 minutes at the time of the summer solstice. The latitude of the area where I conduct my survey / engineering activity is close to 52 degrees, therefore the vertical angle to Shedar at that time would be around 38 degrees. Refer to the drawing titled, Cassiopeia's Chair / Light Years to examine the horizontal and vertical distances indicated between the stars on the lines attached and below the drawing. Using Shedar as a reference point at 22 degrees, 30 minutes it can be seen how to determine the latitudes of the other stars in the constellation relative to the equator of earth. For example, the latitude of Star E...6.059, divided by 216.5 = 0.027986143, divided by 0.0174533 = 1.0 degrees, 36 minutes, 13 seconds + 22 degrees, 30 minutes, 0.0 seconds = 24 degrees, 06 minutes, 13 seconds, some 2.0 degrees, 39 minutes, 51 seconds above the Tropic of Cancer which is said to lie at 23 degrees, 26 minutes, 22 seconds, the highest point on the globe where the sun strikes the earth at the time of the summer solstice. Try the one for C which is using the Great Pyramid height, 0.572 + 4.275 = 4.847, divided by 216.5 = 0.022387991, divided by 0.0174533 = 1 degree, 16 minutes, 58 seconds + 22 degrees, 30 minutes, 0.0 seconds = 23 degrees, 46 minutes, 58 seconds. Then for B, 4.275, divided by 216.5 = 0.019745058, divided by 0.0174533 = 1 degree, 07 minutes, 53 seconds + 22 degrees, 30 minutes, 0.0 seconds = 23 degrees, 37 minutes, 53 seconds, which is 0.0 degrees, 11 minutes, 11 seconds above the value given for the Tropic of Cancer. This is worthy of a comment. The latitude given for the Tropic of Cancer, divided by the computed latitude for B = 0.9919 %, very similar to the comparison made with Polaris. It is as if the computed latitude for B might be the correct one for this point on the globe. One might think it would be more convenient to have the latitude of Star B equal to the Tropic of Cancer, and that makes more sense. The latitude for the Tropic of Capricorn given is the same as the one for the Tropic of Cancer but see what develops when 2 x 23 degrees, 37 minutes, 53 seconds = 47 degrees, 15 minutes, 46 seconds. First, the angle is converted to decimals then multiplied by 0.0174533 rounded off to two places = 0.83 x 3950 mi., the polar radius = 3278.5 x 5280 = *17310480 ft., the north to south arc on the globe, divided by *1.732 = 0.999 = 1.0, Wow!. Then the arc on the globe, divided by half the polar circumference, 17310480, divided by(3950 x 6.284, divided by 2 x 5280) =

65529552 = 0.264 and its reciprocal = 3.786 = sq. rt. Phi + 3.0, 2 x 3.786 = *7.572, Chord = Phi, divided by 2, divided by 10 + 1 = 1.0309, and the Royal Cubit Theorem gets a piece of the action again! I rest my case on this issue for the time being. We might as well finish the project by computing the latitude for A / 4.275 + 4.847 = 9.122, divided by 216.5 = 0.042133949, divided by 0.0174533 = 2 degrees, 24 minutes, 51 seconds + 22 degrees, 30 minutes, 0.0 seconds = 24 degrees, 54 minutes, 51 seconds. The angle difference between Shedar and star A is 2 degrees, 24 minutes, 51 seconds. When the angle is converted to decimals, then x 's 0.0174533 x 216.5 = 9.122 = a to A, a check is provided. It would be so nice if the mathematical symbols could be typed in using these word perfect programs, it would make life more bearable for me and the reader I'm sure. At least I can do this in the drawings. I trust you are still with me on this, because favorable results are coming in...

Aside from using Polaris to acquire the azimuth of a line on earth there are 26 other stars that can be used in various constellations and Cassiopeia's Chair is one of them, especially star B. Tables that provide star positions throughout the year are available from the astronomical almanac. Star B in Cassiopeia's Chair is a popular one that is often used by surveyors, or was until GPS usage became the rage in more recent times. Sirius and Orion are also popular ones for surveyors who are working in the northern hemisphere.

Light Years to the Star Members in Cassiopeia's Chair

Light years to the star east of Shedar, E...

From the lower right side of the Great Pyramid where line c intersects the base the distance is 0.762, divided into the base length = 10....10 x 0.762 = 7.62 from Shedar / 216.5 + 7.62 = 224.12 light years from earth. Line d on the left side is also at 0.762 east of the west side of the base and the distance from C to D includes the distance BC + OP = 7.158. It is interesting to find that CD, divided by OP = 5.0, and OP, divided by 1.143 = 1.25. A glance at the drawing titled, the Golden Section tells what this is all about.

Light years to the first star west of Shedar, C...

Where line b intersects the vertical line at center the distance is 0.5715 from the base. The depth of the Grotto at center = 1.143 divided into the height x 0.999 = 4.236, the sq. rt. 5 - + 2.0. This value, 4.236 becomes the multiplying factor for the distances to the stars on the west side of the main star Shedar. Therefore, 0.572 x 4.236 = 2.42 from Shedar / 216.5 + 2.42 = 218.92 light years from earth.

Light years to the second star west of Shedar, B...

The total distance from where line c intersects the vertical line at center to where the 2:1 slopes intersect the vertical line = 0.572 + 1.333 = 1.905 / 4.236 x 1.905 = 8.07 from Shedar / 216.5 + 8.07 = 224.57 light years from earth.

Light years to the most western star in Cassiopeia's Chair, A...

The total distance from the base at center to where line d intersects = 0.572 + 1.333 + 1.143 =

*3.048 / 4.236 x 3.048 = 12.91 from Shedar / 216.5 + 12.91 = 229.41 light years from earth. This makes sense because star A appears to be the faintest one of the group because it is the furthest away. The diameters of the stars in Cassiopeia's Chair will be reviewed at a later date.

The three dimensional arrangements of this particular constellation have been discussed and the means of how to derive the same on all the others once we have our ducks in row is at our disposal. There are many more to do including the 12 Zodiac Constellations. The golden goodies list of details for Sagittarius, the Archer - Shoot for the Stars fills an entire page. The full story on this will be revealed in future, I promise. There is something intriguing about this constellation because of the way the distance values to three of its stars works out. The distance to Arkab Prior and Arkab Posterior divided into the distance to star cloud Nunkin works out to a 1.6 ratio. Of course it occurs that each star in the night sky seen in the constellations are suns that have *(twelve) planets revolving around them and it is difficult not to anticipate, or speculate on whether or not there might be life on them yet we know so little at this stage. Those thoughts didn't do that Church member in the middle ages much good when he was burned alive at the stake for thinking them aloud. Attitudes on this sort of thinking today are more casual it might be said, but nothing more has been learned about it since that time. *We have a very long way to go, but understanding the Golden Ratio will come to the rescue.*

What is envisioned is an awesome survey control that can be developed for mapping the heavens to begin with based on the findings in this study, and possibly navigation throughout them in future to the many points of interest in the Living Universe. Further along it will be suggested that the center of the Milky Way can be used as a start point with assigned coordinate values ~ Elevation / 0.000, North / 0.000, East / 0.000 because all that has been discovered about the heavens to date in this article can be mathematically connected to it. At this stage it appears to be a workable concept once a staggering heap of details are processed and stored by specialized programs in the computer banks. I am sure modern science would like to participate in this venture but they need some basic training in the Golden Ratio department first and I trust this presentation will help steer them in the right direction. We are left with the consideration that all this knowledge was well known, long ago during an ancient time on earth and the records of it somehow got lost. It appears though, since the process of decoding the Great Pyramid properly is presently in progress there is hope that the rest of the information needed can be retrieved from the cosmic lost and found. After all, that is where Planets 10, 11 and 12 were rediscovered. In closing with this section it needs to be considered what the case would be if the code had been deciphered during the Golden Age of Greece and that knowledge was passed on to us from the earlier beginnings of western culture. For certain we would be a good 3000 years ahead in the study and not have suffered through those false religious doctrines and a primitive understanding of science which have been in the way of our progress all along. If it wasn't for this scenario religion and science would have been on the same page in our era along with the true concept of God being established.

Following the drawings for Cassiopeia's Chair is another, titled, Alpha Centauri Constellation which allows the opportunity to zero in on its three star arrangement as compared to the glimpse of it seen of it in the drawing titled, Great Pyramid and the Constellations. This self explanatory drawing shows the parallel lines involved and it is very interesting to see that DE = AB = AD. One other detail to explain is, AO = 381.04 + BC, 67.0 = 448.04, divided by 100 = 4.48, divided by 5 = 0.890, its reciprocal = 1.116 - 0.5 = 0.616 = A ~ Phi. Known is the distance between Alpha Centauri A and B, 3.81 light years, a one hundredth part of the Great Pyramid Base. The side view in the drawing shows that the distance from Alpha Centauri B to Proxima = * 0.67 light years. * 1.0, divided by (724.774, the base corner to apex, divided by 484.58, the height = 0.67. The value 724.774 was discussed earlier and the latter appears to be the answer to part of that mystery, however, the feeling is, there may be a star of great importance in one of the constellations somewhere out there in the heavens that matches this dimension. The reciprocal value of 724.774 x 100,000 is within 0.998% of the distance from earth to Arkab Prior and Arkab Posterior in Sagittarius which might provide a lead on this question for now. What is suggested to get a handle on where Proxima is in plan view is as follows: The square root of 0.67, squared x 2 = 0.95 light years at azimuth 135 degrees. A proof of this is 3.81, divided by 0.95 = 4.0, the reciprocal = 0.25, that magic number seen in the Golden Section drawing where 0.5 squared = 0.25 seen in the right angled triangle where the square root of 0.25 squared + 1.0 squared = 1.118 - 0.5 = 0.618. Don't forget, the way it is viewed from space is the reverse of how it is seen from earth. The alignment will basically match with the star charts by others. A to B, then a dog leg to proxima. This part of the inquiry is somewhat beyond the preliminary stages and it will develop more fully as time goes by and it may be found as it has to date that the 3, 5, 10 and possibly 12 factors in this exercise play a significant roll in our coming to an understanding with the constellation patterns. Further along an appealing drawing of the Aquarius constellation will be presented using the 5 and 10 factors along the base and perpendicular line at center.

All aspects of the Living Universe show that there is something not unlike an intelligence behind its geometry that wins out with respect to order vs. chaos and we can learn from this. It is interesting to see that Thuban is the third star south west from the north east end of the dragons tail and of great interest, the azimuth from Alpha Centauri B, the Grotto, south pole star to Thuban is 63 degrees, 26 minutes, 06 seconds as it is from Proxima to the first star in Draco, and the dimension from the first star south west of Thuban to its location is 2.236 = sq. rt. 5 on the west pyramid side slope angle 51 degrees, 49 minutes, 36 seconds, azimuth of 38 degrees, 10 minutes, 24 seconds. Also of interest, the distance from the first star in Draco to Thuban, divided by 5 = 1.785, sq. rt. A ~ Phi + 1.0, and the distance from C to Thuban = Pi - 2.0..., the implications are enormous that the lines in the Great Pyramid provide the geometry of the constellation patterns. The previous drawings of Sirius, Orion, Polaris and Aires were drawn using 5 and 10 factors in the baseline and vertical for a good reason and these along with the others serve as exercises in the inquiry. This becomes evident with the analysis of the Alpha Centauri Constellation and it is details such as this that are exactly what is needed to further understand the structure and workings of the cosmos and the reasoning behind it. It is found that A ~ Phi, 0.61653, divided by 88, the number of constellations = 0.007 x 1000 = 7.0, the solar number. This number of magic 88 will be dealt with more specifically in the follow-up. Simply stated, the results of this edition in the above and following are in the process of bringing to light reliable information which is based on the elusive code based on the true dimensions and ratios that have been locked up in the Great Pyramid these past thousands of years, long before the Old Testament was written by those unenlightened scribes. Eventually this information will become the accepted norm within the hierarchy of the world scientific community once it ventures into a reevaluation of the facts about God, Life and the Universe then it will be free to proceed with its re-education program. The problem is, since Newton's time, input from others for around 15 generations have been making rather a mess of it because they never had the true dimensions to work with from the beginning. This tragic tangle of misinformation is being left behind as we move ahead with confidence restored in what this magnificent gift of truth and light made of stone is telling us and we are no longer hindered by these negative influences.

The deciphered code of the Great Pyramid is to science and all of mankind from every walk of life irregardless of their ethnic background, race or religious following, more so than the scriptures are to the Christians and the Torah is to the Jewish faith., and it is testimonial to the endorsement of the life force by that greater power, or what is known to be God, as previously discussed. Now that the code is in the process of being deciphered it will be simple enough for all peoples in the world to interpret what it is telling us, provided, the Universal language is understood and that is but child's play. Isaac Newton spent the remainder of his adult life until his death attempting to find a hidden code in the words of the Holy Bible without success as did others before and after his time, when all the while, the key to it was in Exodus, Isaiah 19:19, 20 ~ the Great Pyramid is a "Sign" = within its proportions is the lost Code, and it is a "Witness" = to the workings of life and all that exists and moves in the Universe, but because the scribes did not understand it they failed to educate anyone on the topic. Another quote from the Bible ~ Jeremiah 32:20 ~ " God has set signs and wonders in the Land of Egypt, even unto this day, and in Israel, and among other men; and have made Yourself a Name, as at this day." This doesn't give any clarity on what the Great Pyramid was all about either but, no doubt, Newton took this in with his bible studies and that probably led him to the Great Pyramid site. If only he and his following had at the time an understanding of that special language defined by the Golden Ratio and the code had been broken when he was involved with the dimensions of it, his influence would have been far more profound and beneficial to this era, alas, as discussed, the mark was missed due not only to erosion of the structure but also that of human capacity when a primitive approach to numbers was employed keeping the western world populace and those under its influence in the dark for ages to come. If Newton had reviewed the works of Fibbonaci and others previous to his time he might have scored a bingo or two. He gave thought to an apple falling from a tree, and considered the force of gravity, but never pondered over the geometry of a snail shell. It goes back to 762.08, divided by 756.36 = 1.0075625, and 1.732, divided by 1.719 which equals the same, when a working knowledge of the Golden Ratio would have solved the riddle. If it wasn't for that we would have been, literally, light years ahead in the study with our understanding of the Living Universe. It is also bewildering how these details were missed by geometers during the Golden Age of Ancient Greece.

Andromeda:

Andromeda is the closest galaxy to us which is said to have a diameter of 70,000 to 80,000 light years and is 2.2 million light years distant from earth, but Modern Science is using the wrong AU, therefore, it would seem their estimates are not very reliable. Now it is known the distance of our sun from the center of the Milky Way is a constant, 4.67, or 4.7 light years on an azimuth of 38 degrees, 10 minutes, 24 seconds, the side slope of the Great Pyramid. If we are going to survey the heavens the information needs to be organized properly and we need to have our ducks in a row when the time comes to travel through them. We can't afford to travel to some destination in space and be off the mark by 91.8 million miles. Judging from what today's level of understanding is on the topic of astronomy along with its inaccurate estimate of the Milky Way diameter the Royal Cubit Theorem more than suggests that the diameter of Andromeda is 115,473.4411 x 0.75 = 86,605 light years and its distance from the center of the Milky Way is 1.732 x 2 = 3.464 x 1,000,000 x 0.75 = 2,598,000 light years, within 0.847% of the estimate by the astronomers of this age.

It might occur to those who register with these observations and revelations since the diameter of Andromeda is 86,605 light years, the distance from Earth to Sirius is 8.66 light years, the diameter of our Sun is 866,000 miles and the reciprocal of 1.732, divided by 2 = 0.866 x 100,000 = 115,473.4411 light years is the diameter of the Milky Way an 866 constant makes itself known in the study of astronomy.. More information on this revelation and the light year distances to galaxies and their diameters east and west of the Milky Way will become available in the follow-up edition. It has been determined that the Milky Way is the fifth member of a chain of 12 galaxies from east to west beginning at an as yet undefined point, not necessarily the center of the Universe having nothing to do with the big bang theory. In progressive terms the diameters and distances from our galaxy to those in the east are based on a multiple of 0.75 x the Milky Way diameter going east. For example, 115,473.44 x 0.75 = 86,605 x 0.75 = 64,954 x 0.75 = 48,715...And times 1.333 going west, 115,473.44. x 1.333 = 153,926 x 1.333 = 205,183 x 1.333 = 273,510...then using the same multiples 0.75 x 3,464,000 light years going east and the distances to the ones west are based on a multiple of 1.333 x 3,464,000 light years. The arrangement for the infinite number of galaxies in the Universe are spaced and sized in the Phi Ratio similar to what has been discussed in the sun to planets list above. Coincidentally, 5.77 - 4.67 = 1.1 x 2 x one million = 2.2 million. In this case the distance of the earth to the sun is as the radius of the Milky Way is to its center. Keep in mind 1.0 - 0.75 = 0.25 and the reciprocal value = 4.0, the number of faces on the Great Pyramid and this point of interest will be touched on shortly. The Milky Way is one of a galaxy cluster, or galactic system connected to ours that are arranged similar to the planets in our solar system in a counter clockwise Golden Spiral commencing at an undefined beginning, perhaps the point of infinity. This seems reasonable enough. Notice the correspondence in numbers between the proposed diameter of Andromeda, the distance from earth to Sirius and the diameter of the sun with regards to the 866 factor, 1.732, divided by 2 = 0.866, interesting. As we know by now, the square root of 3 and the Royal Cubit Theorem are at play in all this. The next question that reveals itself is, how many galaxies are in the cluster the Milky Way is in, 10, 12, 20, 100 or an infinite number? The position of the proposal for now is, the azimuth from the center of the Milky Way to our sun is 38 degrees, 10 minutes, 24 seconds, or on an astronomical bearing of N 38 degrees, 10 minutes, 24 seconds E and this is what is used to compute the distance to Andromeda which is in a position to the east of this line when moving clockwise on the golden spiral alignment of the cluster. For example draw a counter clockwise spiral and make an x on it for the Milky Way and make another on the curve to the right, clockwise, for the position of Andromeda. Then visualize, or rough plot other galaxies to the right and left along the spiral at Phi intervals without getting bogged down with computations. Therein is a schematic of galaxy cluster members, the younger ones would be to the right and the older ones to the left it might be assumed. It might also be safe to assume the sizes of the galaxies diminish to the east at a rate of 0.25 per and increase to the west at the same interval. An intuitive guess is, there might be 3 x 4 = 12 galaxies in each cluster when the possibility of there being that number of planets in each solar system throughout the cosmos is considered. Besides that the sq. rt. 12, divided by 2 = 1.732, one of our favorite numbers. Today's astronomers report on a galaxy that has a diameter of 6,000,000 light years and is 29 billion light years from earth but we don't know if that is part of our cluster or not. To serve as a comparison 50, divided by 1.732 = 28.86836 x 100,000 x 2 = (5773672 light years), divided by 6,000,000 = 96.2 % and this fits in with the percentage differences found along the way in the discussion.

It will come as no surprise that 365.24 days in a year is the result of using the radius of the square root of 3, divided by 4, shown on page 81. A drawing of Andromeda is not provided, however what would be seen in it are Phi based light year pyramids, more locations of interest at their apexes, circle routes and pentagrams etc. like those shown in the geometry of the Milky Way, all scaled down by 75%. It is obvious now and with no exceptions, this is the geometric arrangement of all galaxies throughout the Universe now that we know their dimensions are based on the square root of 3. Other galaxies must be at right angles left and right of the above mentioned azimuth and at vast distances from each other in the infinite expanse of space relative to the square root of 3. The problem is, these distances are unknown except for rough approximations. It might be assumed there are approximately three quarters the number of star suns and planets in Andromeda in comparison to our galaxy because it is 75 % its size and exciting thoughts of there being life somewhere there and in other galaxies as well filter into the mind. From this proposal a somewhat more clear picture can be determined mathematically on how the other galaxies in the Universe are positioned. Like the planets around our sun they are on multiples of the Golden Spiral at Phi interval distances to infinity, commencing from the center of the Universe which is at the point of infinity and the same can be said about the sun star position arrangement from the center of our galaxy and others beyond. We are challenged to find out the workings of these arrangements and first learn how to travel them in a mental sense then physically...somehow. Indeed, there is nothing haphazard about the natural processes of the Universe because they all make perfect sense and there is a particular order to it.

At the end of this chapter is a drawing titled, Great Pyramid Cosmic Matrix showing the angles, azimuths and proportions discussed. Some photo copies will be needed for the following. Place this graphic along with those titled Planet Orbits - Solar System, Planet Ellipse Orbit and Inner Solar System - Golden Spiral above and to the right of the drawings titled, Geometry of the Milky Way and The Milky Way on a table top free of clutter and orient them toward astronomical north, then while viewing them allow the visual interpretation and thought processes to take hold and it will turn out to be an enlightening exercise when connections are made on how the cosmic arrangements in our galaxy work. Another simple thing that can be done is make a photo copy of the matching scale drawing titled, Great Light Year Pyramid that shows the 2:1 incline of the planet orbit and place it overtop the one for, Geometry of the Milky Way in a sunlit window and the concept of the orbit incline discussed is readily seen. Imagine for a moment an octahedron of life energy that has no mass or substance with the same proportions as the Great Pyramid at the center of the unlimited Universe. The top side has 4 directions and the bottom has 4 more, each one providing directions and proportions of and to locations in space in terms of the Golden Section that exist or are in the process of coming into being, we would like to know about, and 8 of course is the number of infinity. The point of infinity cannot be determined but the concept appears to be a reasonable thought to toy with at this time. This topic will be touched upon again further along in chapter 6 when the shape of the Universe is discussed and why an egg is shaped the way it is. There is much work to be done yet.

The Ancient Zodiac:

It makes sense that the time clock of the Ancient Zodiac would have been based on 360 degrees in a circle, divided by 12 = 30 degrees with the center point being that of the Milky Way and each counterclockwise cycle would be a 1/12 part of its circumference with Aries at the top = 12 o'clock and Libra at the bottom = 6 o'clock. When 30 is divided by 360 the quotient is 0.08333... and this is the conversion factor from feet to inches, 1.0 divided by 0.08333 = 12 inches. Note, with regards to that elusive Pyramid Inch the answer is quite simple: 1.0, divided by 20 = 0.05, reciprocal = 20, 1.732 divided by 20 = 0.0866. That which is 0.0866 light years per second = the speed of thought = the velocity of space travel in that ancient time. Please see the second to last drawing titled, Pyramid Inch that provides a final resolution to this matter. This is what Isaac Newton missed because first, the erosion factor was overlooked and he had no knowledge of Golden Section Mathematics. Keep in mind that the only valid dimensions we are dealing with are the Royal Cubit = 1.732 ft. and Imperial Foot and how they relate to the Royal Cubit Theorem. What I have learned is whether its those bogus pyramid inches, imperial inches or metres they must be converted to imperial feet in order to make sense. The beliefs by others on how long it takes a Zodiac cycle to complete vary from one source to the other but it could never be determined if the correct dimensions of the galaxy were unknown, and it is obvious this has been the case. Since the radius and circumference of the Milky Way are known that Zodiac interval in earth years can be determined quite simply. The computation for this is straight forward, circumference = 363,124 light years, divided by 12 = 30260.33333 light years, divided by 5.876 (that which provides a quotient in terms of millenniums) = 5150 earth years per Zodiac cycle. And though it will blow a number of enthusiasts and scholars away one proof of this statement is, 12 x 5150 = 61,800, divided by 100,000 = Phi and that is hard to beat, and by no coincidence another proof is, 5150, divided by 5 + Pi x 1.732 = 365.24 days in a year and this is also in harmony with the presentation. Another way to look at it is, 363,124, divided by 12 = 30260.33333, divided by 5150 = 5.876. Also, another way to determine the number of light years in a cycle is, Ancient Pi = 2.0 divided by 0.636 = 3.144654088 divided by 180 = 0.0174703 x 30 degrees = 0.524109015 x 57736.72055 = 30260.33.

It is interesting to see that 12 x 61800 = 741,600 as does 12 x 12 = 144 x 5150 = 741,600. Also 5150 x 365.24 = 1,880,986 days in a Zodiac cycle, divided by 61,800 = 30.436666 earth days in a month. Another proof, 12 x 30.4366666 = 365.24 days in a year. As we know, the Gregorian calendar has a different arrangement for the number of days in a month but the number of days in a year work out the same. The number 741,600 is of interest in terms of numerology because 7 + 4 + 1 + 6 = 9, the full triple trinity number before the reset. It becomes more interesting when 7416 x 2 becomes 14.832 billion years and this is proposed to be the age of the Milky Way. The give away is 14.832 billion, divided by 0.618 = 2,400,000,000, or 1.4832, divided by 0.618 = 24 and perhaps this will become known as the cosmic day after the previous era came to closure, or the beginning of time in the Milky Way. The questions that come to mind are, just how old is earth and our solar system and just how did life come to be on earth? Surely it is older than 4 to 5 billion years. Modern science comes up with 14.4 billion for the age of the Milky Way, within 0.973 % of the before mentioned value but they never did know very much about the Golden Ratio. Furthermore, 14.832, divided by 24 = 0.618 x 7 = 4.326, or 4.33 rounded off to three places of decimal x 2 = 8.66 and these number values are known by now. As stated it happens that when 5150 is divided by the square root of 5 = 2303 = 2 + 3 + 0 + 3 = 8, the number of infinity appears and this has meaning.

Ratio comparison between the Solar System and Milky Way:

Given: The numbers are known by now...the correct speed of light = 186,216.56 mi./sec., 365.24 days in a year x 24 hours in one day x 60 seconds in one minute x 60 minutes in one hour = 31,556,736 seconds in one year etc.

Diameter of the Milky Way = 115,473.4411 light years.

As follows: There are 5,876,386,823,000 miles in a light year x 5,280 = 31,027,322,430,000,000 feet in a light year x 115,473.4411 light years in the diameter of the Milky Way = 3,582,831,689,000,000,000,000 feet = (a).

Diameter of the solar system = 69,211,253,750 miles x 5,280 = 365,240,000,000,000 feet = (b).

(a) divided by (b) = 9,809,527.129

The explanation offered is as follows: The number of years in a zodiac cycle 5150, divided by 100 = 51.50, reciprocal = 0.019417476 + 1.0 = 1.019417476 x 9,809,527.129 = 10,000,000, virtually.

Therefore, it has been determined after some more tedious number crunching that the diameter of our solar system is infinitesimally close to a one ten millionth part of the Milky Way diameter. At any rate it speaks of the 5 x 2 = 10 factor influence on our mathematical system as there doesn't seem to be any coincidence that we have 5 x 2 = 10 fingers and 5 x 2 =10 toes. There are 5 fingers on each hand and 5 toes per foot and the number 5 and its square root certainly has an influence in the study. As stated, our solar system is less than a spec of dust in comparison to the size of the Milky Way. On the one hand it shows that we may be lost somewhere in space. It is nothing to be overly concerned about and it provides an insight and appreciation into the vastness of outer space, yet if "below is as above" and infinity exists the human mind can handle it. Hence, the metaphysical definition of the number 8 is quite valid.

The Human Body ~ A Marvel of Mathematical Machinery and Mind…a brief note

One human body, a human being = 1.0, the number of all, a universe within itself. The structure of the human body including the details of its anatomy right down to the last detail are in the Phi Proportion. Therefore the square roots of 3 and 5 and these values squared are spoken for. We have 10 fingers + 10 toes = 20, less two thumbs and two big toes = 16, the square root of which is 4...sq. rt. of = 2...sq. rt. of = 1.414, and 16 - 4 = 12. The human body has one bipolar brain, all its organs, arms and legs etc. are dual and it has a brain and a mind, therefore multiplying or dividing by 2.0 is natural to the thought processes and acquiring the square root and squaring a number etc., is a sign of intelligence. A human has a heart, mind and soul and that speaks of the number 3, a trinity. Another form of a trinity occurs when 2 becomes 3, when a man and a woman engage harmoniously and an offspring is born as a result of that union. Multiplying by 3 and acquiring cube roots of numbers is a further a sign of intelligence. Therefore, it can be seen how the numbers one through to ten come to be. The numbers 1.0 to 5.0 have been explained, to carry on 2 x 3 = 6, 4 + 3 = 7, 2 x 4 = 8, 3 x 3 = 9, 1 + 2 + 3 + 4 = 10, 10 + 1 = 11. Other features that can be added to the collection are, it is on average 3.0 feet from the center of a man's chest to the end of his middle finger with his arm fully outstretched, and one pace is generally accepted as being 3 feet, 3 x 2Pi = 18.84, divided into 3048.32, the base perimeter of the Great Pyramid = 161.8, divided by 100 = 1.618, 1+Phi and 1.0, divided by 1+phi = Phi. Also, the height of the Great Pyramid, divided by 100 = 4.8468 x Phi, divided by 4 + 1.0 = 5.6 feet, the average height of a man at the time it was constructed, and we can't forget that the length of a mans arm from the elbow to the middle finger plus a palm width is a Royal Cubit, the square root of 3. Therefore, the human being is not to be counted out as being a cosmic creation. When the mind is engaged properly with the pertinent details it is found that within the dimensions and proportions of the human body and those members of the animal, vegetable, and mineral domains are found the keys and capabilities to solving and understanding the wonders and mysteries of the living Universe. A note on this is taken because the mystic, or metaphysical meanings of numbers is an important feature of the study. In the last chapter, number 8, a solid number theorem will be presented that will reveal the mystery of how the numbers 1 to 9 come into being.

How long do we have?

Another question is, how many cycles are left before the Milky Way becomes a black hole somewhere in space? To answer this question it seems prudent to use the trusted 1+Phi factor that has never let us down. Therefore 1.618 x 1,000,000,000 = 1,618,000,000, divided by 5150 = a total of 314,200 revolutions, divided by 100,000 = 3.142,Pi, which indicates the completion of a cycle. The tentative number of complete revolutions it has made to date, perhaps by the year 2012, will be 1,483,200,000, divided by 5150 = 12 x 12 = 144 x 2 = 288 x 1000 = 288,000 and this is of interest because the 12 factor appears as a result of the computation. Then when 288,000, is divided by 314,200 = 0.9166 x 100 = 91.66%, the same can be seen when 1,483,200,000, divided by 1,618,000,000 = 0.9166 x 100 = 91.66%. This may indicate that there are 314,200 - 288,000 = 26,200, or 26180, divided by 10,000 = 2.618 full cycles x 5150 = 134,827,000 years remaining in the cyclic life span of the Milky Way which might provide enough time to get our act together in order to determine the correct ways and means to explore the Universe and find a new home for human life when the time comes to do so. That is, if our species doesn't become extinct by its own doings beforehand. At this time our science is in the process of attempting exploration of space but it has a very long way to go. There is talk of colonizing in space but the futility of this proposal is seen when we realize it will be another 84 years before Voyager I reaches the limits of our solar system. A thought to share on this is, since the Great Pyramid has made details of the Universe available to the point where we can begin to chart it out, this could very well fit in with our plans on this issue. The astounding thing about it is, the science behind the Great Pyramid knew so much, it leads us to believe there were indeed, highly evolved beings who once traveled throughout the Universe at some time in the very remote past. It is interesting to find that Pi to two places of decimal, 3.14 = 3 + 1 + 4 = 8, the number of infinity and 1+Phi = 1.618 = 1 + 6 + 1 + 8 = 16 = 1 + 6 = 7, the solar number and for eternity there has always been an agreeable union between these two mathematical life functions. It is somewhat disturbing to realize that 91.66% of the life expectancy of our galaxy has expired, though 134,827,000 years, divided by 20 = 6,741,350 generations of humanity where from the answers to our needs might come from those who might pursue the topic. Another question might be, will the Milky Way have another cycle of time equal to the first one described, I think not. It might be theorized or we might entertain ourselves with the thought that those who constructed the Great Pyramid might have departed to a life supporting planet in another distant galaxy that has most of its cycle left. And again it is put forward, the intrigues about the energy processes that once took place inside the chambers and sarcophagus within the Great Pyramid could be very important to us.

Further reference to the above values in terms of the square root of 5 are just ahead when the true age of the Great Pyramid and how the science behind it fits in with the cosmic plan is considered. Before leaving this section note: 0.7416, divided by Pi = 0.236, the square root of 5 – 2. Also Phi, 0.618 cubed = 0.236. It is this type of information that strengthens the debate on the mathematical structure of the Milky Way. Furthermore, in search of other proofs please refer to the light years conversion to feet on page 119. The circumference of the Milky Way = 363,124.0286 light years, divided by the number of miles in a light year, 5,876,386,823,000 x 10,000,000 = *Phi x 2 + 1.0 = the square root of 5 and Phi x 1,000,000 = the orbit distance of Earth around the Sun. And we are reminded that 88, the number of Constellations, divided by 2 = 44 x 10 = 440 Royal Cubits, the base dimension of the Great Pyramid x 1.732 = 762.08 ft., divided by 2 x the square root of *1+Phi = 484.68288 ft. x 1,000,000,000 divided by 5,280 = 91,796,000 miles the Astronomic Unit based on its height.

Hell will freeze over before the Establishment enlightens the masses on this vital information because it is stuck in the dark ages yet the results of this study will enable Religion and Science to get back on the same page where they belong then the spirit of man will be freed from its material confines. The true dimensions of this grand structure have helped to explain the existence of God and how Life and the Living Universe operate. A drawing entitled, The Great Pyramid Speaks is made available and it shows what progress has been made to date. The greatest exploration of the Universe we have ever seen is at hand since the Great Pyramid began to speak, and so far what has been brought to light has covered the solar system, the Milky Way and got our big toe outside of it and I dare say, history has been made in the process. The day will soon arrive when number crunching becomes a thing of the past for this study because programmable calculators and computer programs will be equipped with the required functions that deal specifically with what this Ancient Wonder tells us, and then we will be, to start with, at least mentally traveling over vast astronomical distances at the speed of thought, knowing where we are going and what we are doing. We had better get started on it now, because there is no time to waste. An appeal is extended to all those who participate in consciousness of life to join forces and pool resources in order to develop a plan for the future welfare of life on earth as we know it.

Year 2012 ~ Pisces Cycle ending, beginning the Age of Aquarius, Transition Period

Aquarius, the Water Bearer…

Of course it would like to be thought of by optimists as the beginning of a New Age, a revival of the human spirit, the promise of a glorious new beginning on earth etc., however, it won't be that simple if social trends continue on they way they have in the past and there is no reason to believe the starving millions and terminally ill on this planet will suddenly cease to suffer because a new zodiac cycle begins. And it isn't likely grand installations such as the Sphinx and Great Pyramid will be constructed to commemorate the event. As promised earlier a drawing of the Aquarius Constellation titled, Age of Aquarius 2012 to 7176...Dancing with the Stars...is presented at the end of the chapter. To draw this one the base and vertical of the Great Pyramid shown below were divided by the 5 and 10 factors to obtain the lines and it appears to have worked out very well. The notes on the distances AB, AC and DB, divided by 5 are very interesting and this serves to verify that the geometry of the constellations are Phi based. Following are drawings for Taurus, Gemini and Leo where the same observation is noted. There is an indication that this is the way to go in the study to further understand the geometry of the constellations. It is noted that the lines created this way in the Great Pyramid also apply to the obelisks. A drawing titled, Phi Based Obelisk is available at the end of this chapter, and following this one is the Obelisk of Tuthmosis that has a sq. rt. of 3 scheme, and its slope is the base length of the Great Pyramid divided by 80, divided by 10 = 8, the number of infinity. The next one is of Washington Tower including its dimensions and a revealing fact about it. In-depth research on this fascinating topic is presently in progress. Be sure to add the obelisk to the list constructions for a type of solid that has great meaning. The next drawing after these ones titled, Constellation Pattern Lines is an attempt to find the range of lines between the stars in the constellations. The Alpha Centauri Constellation is shown and the horizontal baseline is divided by 20 equaling 0.391 intervals and the vertical line is divided by 11, which also equals 0.381 intervals. The vertical baseline includes the 1.143 value to the grotto below the horizontal baseline line at center and 1.143 above the 2:1 intersection point on the vertical line indicated by the triangle. The range of these lines have so far met with the requirement to draw the constellations that have been dealt with to date. There may be others in future projects that need lines beyond the range shown in this drawing, therefore, more study in this area will carry on.

Remember in math classes we were told that two upright obelisks side by side with a horizontal line above was supposed to be the symbol for Pi, and that was the only thing of importance about it. In fact this symbol was Greek in origin and since then the true meaning of the obelisks was lost. There is much more to it that we have been missing out on. Consider the ancient Egyptian hieroglyph for the number one that was discussed earlier in chapter two and how it explains the value for Phi. Take for example the two huge obelisks at the gateway to the Temple of Luxor, constructed during an ancient time in Egypt. One of these was mindlessly shipped off to France, in other words it was stolen. Irregardless of this sad state of affairs the distance between them was half the height of the matching pair of obelisks, which means, as discussed, they represented the value for Phi as well as Ancient Pi. The way Ancient Pi is expressed in this arrangement is as follows: First of all consider that the height of the obelisks equals one unit each, which adds up to 2, and the distance between them is 2 x 0.5 equals one unit, then the total becomes 3.0, that sacred number of trinity, therefore, 1.0, divided by (Phi - 3.0, divided by 10) = 3.144654088, Ancient Pi. This is seen in the Elements of the Golden Section Drawing and it is clear now that the entire matter was totally misunderstood by the earlier western investigators who had no clue of Golden Section Geometry and unfortunately that has hindered western culture's understanding of the grand science of an ancient yesteryear. In that time Ancient Pi was used for astronomical evaluations which dealt with the infinite and what we know as modern Pi was used for finite dimensions such as the circumference of a planet. This was discussed in chapter 8 when it was determined that the geometers of those times had a total handle on Phi, Ancient Pi and Modern Pi. Take note: Ancient Pi, 3.144654088... x 0.999 = Modern Pi, 3.1415 = 3.142, good to three places of decimal...see page 17, chapter 2.

A reasonable drawing of the Sphinx is made available, that mythical sculpture with the body of a lion and the head of a man. It is the Sphinx, the obelisks and Great Pyramid found half buried in the sands of time during the latter part of the middle centuries wherein the secrets of the lost code of the Ancient Sciences and the elements of the architecture of the Universe are in evidence, and we are at long last hot on the scent and coming to an understanding of them by now. In the case of the Sphinx its dimensions are an ancient time storehouse of scientific knowledge being that its body length divided by the height of its head = 8.0, the number of infinity, its length divided by its height = 3.636 and that number + 10 = 13.636 is seen in the Royal Cubit Theorem when the earth dimensions computation for the square root of 3 took place, when this number was subtracted from the polar radius, then converted to feet, then divided by 12,000,000 and that is just the beginning of this story.

The Great Pyramid is not only a spectacular marvel of engineering, and literally a time capsule, and an astronomical instrument, it is the greatest educator we know of because it has been providing us with step by step revelations about what we need to know about the Living Universe since it began to speak. At the earlier stages it was difficult to say if its teachings went beyond the astronomy of the Heavens and that there might be some great prophecy contained in its code we should know about. On the surface the information in the code didn't appear to refer to beginnings or endings except those that are in terms of infinity, and reference to a Big Bang Theory is not in evidence, and there doesn't appear to be any indication about life elsewhere in the Universe, however it does provide an indication of where it might be once the powers of deductive reasoning are employed. For now, the adventure of learning about the applications of the Golden, or Divine Ratio as it relates to life, energy and the workings of the Universe is of primary importance to the state of being because it needs to be understood by us. I just wince every time I see those Great Pyramid prophecies and false dates concocted by others because I know they never knew correct dimensions in the first place, however according to research there is a prediction by that the peak of a sunspot cycle will take place in 2012 and this can spell trouble on earth in the form of magnetic storms, serious electrical power outages, volcano eruptions, huge land slides, hurricanes, earthquakes, and tsunami's etc. as the electromagnetic field around earth becomes disrupted. According to reports over the past decade these activities are already in progress. The Mayan calendar indicates an end time or a new beginning in 2012 and a certain percentage of the world population are in a flap about it. To some it will be the predicted time of Armageddon as outlined in the scriptures, the second coming, the weeping, wailing and gnashing of teeth, or the day of judgment etc. Giving it some serious thought though, when facing the facts about the folly of humanity, if the abuse of the environment, wastefulness, lawlessness, greed and corruption, alcohol and drug abuse, disease, the mental instability of the masses, war, world terrorist activities etc. accelerate and this state of affairs is extrapolated to a time just four years hence, which isn't very far off, it very well might indicate a stressful period for the living on earth in that time period, especially for mankind. From what we gather times were not so good for western man around 2000 years ago at a time when his weaknesses beset him with many difficulties, about the time when the human spirit was pinned on the cross of matter. Since then war, material greed and the down side of religious fanaticism have taken their toll. In the dark ages Nostradamus got in on the act, predicting a final World War III in 2012, when the global human population will be dramatically reduced, and the experts on world energy reserves are saying oil production will be at its peak then, and that could mean trouble because the supply will probably become severely diminished and possibly run out. The world population is predicted to be some 7 billion by then and the solar number can be read into this as if it might mean that is the limit for stress on the earth's resources. When 2012 is divided by 4 as the four cardinal directions and the four base lengths of the Great Pyramid come into play, the quotient is 503 appears and this adds up to 8, the number of infinity. In ancient Mayan culture a dot stood for 1.0 and a bar represented 5.0. A base number of 20 instead of 10 digits was used, therefore 4 x 5 = 20 and 20 squared, or 20 x 20 = 400, divided into 2012 = 5.03 which is the same number x 100 and it also adds up to 8. Aside from that two pairs of hands with 5 fingers each and two feet having 5 toes each comes to 4 x 5 = 20 etc. Interestingly, when 2012 is divided by the square root of 5 the quotient is 8.998, or 9.0 rounded off, but when it is divided into 2012.4 the quotient is exactly 9.0 and this is the triple trinity full number before the reset at 10. The mathematical statement, 2.236 x 100 x 9 = 2012.4 explains it. This value is interpreted as being in March 2012, about the time of the Spring Equinox. Others believe the events of this year, what ever they might be, will take place during the winter solstice of 2012. What ever the case might be, what might be read into all this is, there could be a global crash of some sort at that time and man will have to pick up the pieces and carry on somehow after the smoke clears. It might be asked if some of the problems are avoidable. I believe they are if life is truly valued by us, and if we know what to do in order to support it instead of abusing its privileges. It is possible there will be consequences, or a negative karma because of man's collective separation from the natural domain for the past thousands of years. The Green Movement and themes such as An Inconvenient Truth are a sign of the times taken up by those who are concerned and want to do something about it. Even the revelations contained herein with regards to what the Great Pyramid tells us, are perhaps timely in our process. They are of help because it shows that the answers are available if there is a willingness to dig below the superficial and mine out the truths we need in the midst of all the failures and falsities that prevail in our social, religious and scientific environment. About the confusing allegories of an apocalypse referred to in the book of revelations, the very word apocalypse means an unveiling, a taking away of that which obscures and makes it clear, mysteries and their meanings, and this is taking place before us not unlike a miracle, in the process of the Great Pyramid being decoded.

One important thing that can be read into the study is, that to begin with, the proportions of all buildings, public and domiciles should be constructed in the Phi Proportions to be in agreement and in tune with the life enhancing energies of the Cosmos and this practice has been long lost for untold ages. Actually, it is the workings of the Golden Ratio that should pervade our thoughts at all times and be a strong influence on our designs, how we live and what is produced by us. Rather than sitting around on our hands, waiting for the worst to happen, by doing this in all ways possible it will serve to help save our social and physical environment and keep us embraced with the Universal God Force, that which is connected with life and consciousness. Necessity is the mother of invention and man can be resourceful, therefore what lays ahead could lead to intelligent alternate energy concepts or new food sources that have been there all along. Our position will be strengthened by these practices as we face future challenges. Whether or not it is all doom and gloom in 2012 remains to be seen, but we don't need a prophecy to tell us that we need to change our ways. In a sense, starting about now, our species is in a race against time to manifest outcomes that yield positive results. There could be dancing in the streets when peace and harmony prevail because we will have figured out what to do about the situation. That the ancients might revisit earth and help us out isn't very realistic, it's up to us. Coming in for a landing…back to Earth in an Ancient Time Zone, back at the site of the Great Pyramid where the story first began…

Age of the Great Pyramid:

A timeless question, and yet another mystery that has boggled the minds of many a scientist and scholar because they never had anything to base it on, but that has changed now. The general consensus of its age among the researchers according to carbon dating is, it was constructed some time around 3000 B.C. This provides a value of approximately 5000 years from today and it seems to suit most parties on the question, however, it doesn't mean the design for it didn't exist long before it was constructed and that other pyramids like this weren't constructed long ago, perhaps at the Giza site itself. When 5150 is subtracted from 2012 the result is year 3138 B.C., the year it was constructed, more than 2000 years before Greek and Roman Culture emerged. Since 5150 - 3138 = 2012 a correspondence with the Gregorian Calendar is provided. One way to look at it from a numbers point of view is that if the year 2012 indicates a major turning point in global history because the value 9 appears before the recount, when the square root of 5 is divided into it. Quickly do this with your calculator, 2012, divided by 2.236 = 8.998. Again it is mentioned, when 5150 is divided by 2.236 the quotient is 2303 which in terms of numerology equates to 2 + 3 + 0 + 3 = 8, the number of infinity. There is an indication that there was a high cycle in history on earth at that time, meaning that when the Great Pyramid was constructed it was a period of profound enlightenment carried out by unknown parties who fully understood the workings of the Universe. If we are entering the time of Aquarius marked by the year 2012 which in terms of numerology = 2 + 0 + 1 + 2 = 5, the quintessential number, according to this reckoning the Great Pyramid was constructed at the beginning of the Pisces era which was 3138 B.C. It is therefore determined, the age of the Great Pyramid will be 5150 years old by the Gregorian calendar year 2012. The number for year 3138 can be broken down into separate parts, 3 + 1 = 4, 3 + 1 + 3 = 7, the solar number, 3 + 1 + 3 + 8 = 15 = 1 + 5 = 6, the double trinity number, and 3138, divided by 2.236 = 1403 = 1 + 4 + 0 + 3 = 8, the number of infinity. Its age in 2012 = 5150 = 5 + 1 + 5 + 0 = 11, the number of love and purity. Taking it a step further, 11 = 1 + 1 = 2 = the principle of correspondence, " As above so below, as below, so above ", the mirror image of 1.0 = 2.0 and that is the way of the Universe, therefore notions of the Big Bang Theory are preposterous and so much mental rubbish because 1.0, the whole always existed. Its age in another 5150 years = 10,300, divided by 2.236 = 4606 = 4 + 6 + 0 +6 = 16 = 1 + 6 = 7, the solar number.

The age of the Sphinx has not been truly determined as yet. It could have been a project that was undertaken 5150 years prior to the construction of the Great Pyramid at the beginning of the Aires zodiac cycle. That would go back to 3138 + 5150 = 8288 B.C. and judging from the erosion of it there is an indication that a tropical climate prevailed for a long period since it was constructed. According to the archaeologists the people at that time lived humbly as hunters and gatherers with some knowledge of agriculture and livestock herding and they flourished in the Nile River valley because of the water supply and added benefit of the annual silt deposits that supported agriculture. In terms of numerology the value 8288 equates to 8 + 2 + 8 + 8 = 26 = 2+ 6 = 8, the number of infinity. The discussion on this topic is ending for the time being but it is believed that using the cycle interval of 5150 years relative to the age of the Great Pyramid has a value for determining historical periods on earth. Today's researchers are welcomed to think what they will about earlier people in this area but it is quite possible they are considerably off the mark with their assessments of human life at that time.

The Great Pyramid ~ a History Book that is conscious of the Future

An almost startling coincidence is seen when 3138 x 0.618 = 1939, the beginning of the WW 11 and when the key numbers we have become aware of are employed in certain ways along with the value 3138 other coincidences with regards to the year dates of historical events and certain personages are determined.

There is More:

3138 x 0.618 x 1000 = 618 - 8 = 610, divided by 1000 = 0.61 = 1914, the beginning of WW 1.

3138, divided by 1000 x 1.618 = 1618 + 2 x 9 = 18 + 1618 = 1636, divided by 1000 = 1.636 = 1918, the end of WW 1. Subtract 1.0 = 1917 = the Russian Revolution.

3138, divided by 1.618 + 3.0,divided by 10 = 0.030 + 1.0, divided by 1000 = 0.001 = 1.649 = 1903, about the time man took to the skies in heavier than air flying machines, the first airplanes. 3138, divided by 1.618 - 2 x 1.0 = -2.0 , divided by 10 = - 0.02, - 3.0, divided by 1000 = - 0.003 = 1.595 = 1967, the 100 th. year anniversary of the Dominion of Canada, and plans were well in place for putting men on the Moon by the USA.

3138, divided by 1.618 + 1.0, divided by 10 = 0.01 + 9.0, divided by 1000 = 0.009 = 1.637 = 1917, the year John F. Kennedy was born.

3138, divided by 1.618 - 2 x 9.0 = 18, divided by 1000 = - 0.018 = 1.60 = 1961, one of those rare invertible year numbers when President John F. Kennedy was assassinated.

3138, divided by 1.618 - 2 x 9.0 = 18, divided by 1000 = 1.60 = 1961, the year Princess Diana was born.

3138, divided by 1.618 + 2 x 2 = 4, or 4 x 1.0, the number of all, divided by 1000 = 1.622 = 1935, the year Elvis Presley was born. 3138 + 1935 = 5073 = 5 + 0 + 7 + 3 = 15 = 1 + 5 = 6, he was blessed with the union of Fire and Water, the double trinity number, and was therefore the embodiment of a powerful soul and there is no disputing this.

3138, divided by 8 x 5 = 40, divided by 10 = 0.040 + 1.0, divided by 1000 = 0.001= 1.659 = 1891, the first gasoline powered automobile was constructed.

3138, divided by 1.618 x 1000 = 1618 - 10 x 5 = 1568, divided by 1000 = 1.568 = 2001, the September 9, 911 events that occurred when the World Trade Center was destroyed by terrorists and the free world changed forever. Many lives were senselessly lost along with much public property. "One interpretation", When multiples and divisors are based on the 10 factor to increase or decrease the value of numbers and 1000 is looked upon as a towering figure, look at the sum of 1.568 x 1000 = 1568 = 1+5+6+8 = 20 + 1568 = 1570, divided by 3.14 = 500, meaning 1000, the (2) towers were cut down as the result of a cyclical change in world events.

3138, divided by 1.25 x 0.618 + 1.0 = 1.773 = 1770 + 2 x 3.0 = 1776, the United States of America came into being.

3138, divided by 1.25 x 0.618 = 1.773 = 1770 - 1.0 = 1769, birth date of Napoleon Bonaparte.

3138, divided by 2 x 8 = 16, divided by 10 = 1.6 + 2 x 3 = 6, divided by 100 = 0.06 + 1.6 = 1.66 / 3138, divided by 1.66 = 1890 - 1.0 = 1889, birth date of Adolf Hitler.

3138, divided by 1.618, divided by 1000 = 1618 - 5.0 = 1613, divided by 1000 = 1.613 = 1945, end of WW 11, when the 3 rd. Reich was defeated and atomic bombs were dropped on Hiroshima and Nagasaki. A turning point in the technology of war when man's attitude toward it changed forever as the Cold War developed.

3138, divided by 1.732 = 1812, the war of 1812
3138, divided by 1.732 = 1812 + 20 x 5 = 100 = 1912, the sinking of the Titanic

3138, divided by 1.636 x 1000 = 1636 + 5 x 10 = 1686, divided by 1000 = 1.686 = 1861 = beginning of the America Civil War.

3138, divided by 1.618, divided by 1000 = 1618 + 12 x 5 + 8 = 1686, divided by 1000 = 1.686 = 1861 + 2 x 3 = 1867, the Dominion of Canada came into being.

3138, divided by 2 x 9 = 18, divided by 10 = 1.8 = 1743 - 20 x 5 = 1643, Isaac Newton was born.
3138, divided by 1000 x 1.618 = 1618 + 52 x 5 = 260, + 2 x 3 = 6 = 1884, divided by 1000 = 3138, divided by (1.884) = 1666, the fire of London in an apocalyptic year following the Bubonic Plague of 1665, Isaac Newton made revolutionary inventions and discoveries in calculus, motion, optics and gravitation, Italian astronomer Cassini discovered polar ice caps on Mars.

3138, divided by 1.0 + 2 x 3, divided by 10 = 1.6, + 7 divided by 100 = 0.07 = 1.670 = 1879, the year Albert Einstein was born.
3138, divided by 1.618 + 26 x 5 = 130, divided by 1000 = 0.130 + 2 x 1.0 = 2.0, divided by 1000

= 0.002 = 1.750 = 1793, the year French science introduced the bogus metre as an earth measure. The history on this subject is known by now and it will be left at that.

3138, divided by 1.618 + 7.0, divided by 10 = 0.07 = 1.688 = 1859, birth date of Sir Arthur Conan Doyle, doctor of medicine and author who inspired readers in every language around the world with the adventures of Sherlock Holmes and other famous characters such as Brigadier Gerard and Professor Challenger. Observation and scientific deduction was the motto of the Sherlock Holmes series of adventures, and to present times this proves to be the correct approach to solving any problem, including the Riddle of the Great Pyramid. Conan Doyle was a Free Mason of the highest order and a profusely gifted individual who had bridged the gap between science and words. Note how the solar number 7 fits so well with his being, entering into the stage of this world.

3138, divided by 1.618 x 1000 = 1618 + 60 x 5 = 300 + 8.0 x 10 = 80 + 8 = 388 + 1618 = 2006, divided by 1000 = 2.006 = 1564, the year William Shakespeare was born. Volumes of his works have been published and much has been said of his unequalled contribution to the English language. His was the science of words. Note how 8.0, the number of infinity appears to play a role in his coming into this world. William Shakespeare was also a Free Mason who carefully kept it a secret because of the negative influence of the Church during his time.

3138, divided by 1.0 + sq. rt. 2 = 2.414 = 1300, a complete map of Europe was generated, Register Boniface VIII, the indiction of Holy Year 1300, Rome, a struggle for control of resources, a new European architecture of Gothic structures, great advances were made in chemistry.

3138, divided by 3.14 = 999, the number of the beast turned upside down. When the European population was gripped by mass hysteria and fear that the world was coming to an end. They didn't grow crops that year for fear of the coming Armageddon. A French mathematician became the Pope. The ancient city of Pyros burned to the ground.

3138, divided by 2.236 + 3.0 = 5.236 = 599 + 20 x 3.0 = 659 + 7.0 = 666. At the time people in middle age Europe were in a blind panic because to them 666 was the scary number of the Beast and they thought the Anti-Christ was going to appear. This so called number of the Beast was covered earlier. Whereby 360 degrees, divided by 54 degrees = 6.66 x 100 = 666 and it was determined the value 666 actually does have significance in terms of the Golden Ratio and no ill purpose is attached to it.

3138, divided by Pi + 5 = 8.14 = 385 B.C., the decline of Ancient Egypt, about the time Ancient World philosophy ceased to be an influence - Persian domination, the Greco - Roman period. This date matches with the available sketchy historical account of that time. Alexander the Great was born some time around 356 B.C. which would have put him in his prime in year 389 B.C. +/- at age 33 less a month when he died of mysterious causes.

3138, divided by the reciprocal of 0.9999999 = 1.0000001 = 3138.000314 - 3138 = 0.000314 x 10,000 = 3.14, Pi, that fundamental mathematical element of a cycle.

3138, divided by 1000 x 1.618 = 1618 - (10 x 5) = (- 50.00 - 8 = - 58.00) = 1618 - 58 = 1560, divided by 1000 = 1.560 / 3138, divided by 1.560 = 2012. It takes negative signs in the process to arrive at this year, however what this means exactly is undetermined as yet.

Perhaps the code in the Great Pyramid truly does relate to historical dates of importance. Therefore it can be viewed as a time machine as well as a marvel of engineering, a time capsule and astronomical instrument. At this stage it appears as though the key to viewing history past and future in terms of the Gregorian Calendar is locked up within the value 3138 and how it relates to the golden numbers such as 1.0, 3 / 5 and their square roots, 7, the solar number, 8, the number of infinity, Pi and Phi and all the other numbers. In this case 2 = 1 + 1, or 1 plus the mirror image of itself, 4 = 1+1+1+1, or 2 + 2 = 4 x 1, 6 = 2 x 3 and 9 = 3 x 3 etc. In the case where 3138, divided by 1000 = 3.138, rounded off a value of 3.14 is provided, Pi, the number that indicates cycles. Furthermore 3 + 1 + 3 = 7, the solar number and 8 = the number of infinity and 8 - 7 = 1.0. The number of all. It is quite possible future dates of historical events and birth dates are being predicted for Gregorian Calendar years in terms of Phi on the Golden Spiral of time and life. The above are just a handful of them and this aspect of the study is in its infancy at this stage, yet, to a certain degree it appears quite reasonable as if the way the numbers interact with each other somehow tell a story in mathematical terms, on the other hand it might be dismissed as nonsensical coincidence and conjecture by some. Try this method on your own or a loved one's birth date and see what develops. Acquire a list of historical events and people and see how the numbers correspond with the events. There will be no disappointments as it is quite fascinating. None the less, the true meanings of all numbers are a code in themselves that possess many wonders and explanations when their mystic or metaphysical meanings are interpreted, therefore the above proposal including how the Ancient Zodiac works makes sense, but because so much information has been lost over the ages after civilization divorced itself from the natural domain, we have been left behind in a shroud of darkness and ignorance about the true workings of anything to do with the Cosmos. Because of this I have come to believe that modern day astrology and numerology is out of touch with reality in their belief systems. For instance, a thought to share, it is possible the Gregorian year of a persons birth date added to 3138 might provide a persons celestial or spiritual life path number once the numbers are totaled in terms of numerology and perhaps the Gregorian birth date furnishes the terrestrial life path number. It occurs that any of the above derived numbers including the present year value added to 3138 might have meaning of significance when their numerological values are determined. It is at least, a consideration worth looking into. This area of the study will be addressed and expanded upon in chapter 8 near the end of the story titled, the Ancient Zodiac and the 12 planets of our solar system.

Construction of the Great Pyramid ?

Yet another unsolved mystery that presents major challenges. Theories abound on this issue to the point of hallucinations. Anything from a 10:1 ramp to its summit and spherical rocks being rolled upgrade to the top of the slope, then chiseled into cubes that fit for construction purposes and that it took 20 years or more to complete while thousands of slaves labored on it in the hot sun. The first question that comes up is, how was it that the base area was made so precisely level? A general thought on this is, there are remnants of walls around the structure and the area was flooded with water from the Nile and the simple principle of gravity at various levels of the water assisted the effort until it was excavated and leveled precisely to the desired grade elevation. This speaks of a pumping system and the question is, how was it powered? In this case an area of approximately 14.6 acres including the wall enclosure is considered and the volume of material removed is unknown but it was probably quite substantial. One might make the assumption the site was leveled by the sweat of the brow but types of machines not unlike a back hoe or bull dozer along with the equivalent of haul trucks that might have been pulled by camels and elephants or more advanced equipment shouldn't be ruled out. To ensure that the foundation preparation for a structure of this massive size and weight was just right any soft ground areas would have to be dug out, disposed of and backfilled to grade in compacted lifts of broken rock and gravel type material. It would have been well to have water available as its influence would help consolidate the fill material in the sub grade of the foundation area. With this said, it is possible this group used survey equipment more advanced than ours. I really can't help myself when it comes to the particulars of this debate because during my career I have seen many types of engineered developments take place. I know what works and doesn't work with regards to construction techniques. In the case of the vastness of the Great Pyramid construction and all its specialized details I look at the way we get work done in this era and come to the conclusion that the ancient ways must have been far superior to ours and this is the reverse of present day thinking on the matter because, very often the truth is stranger than fiction.

Some thoughts have been generated in determining what the dimensions of the stone blocks were actually based on. Socrates, the early Greek scholar reported that a standard interior block was approximately 2.5 x 10 x 10 feet but this doesn't tell us why it was these dimensions. Therefore the following input based on 3 places of decimal is made: Since the height of the capstone is 1.732 and the entire design of the structure is based on the Phi Ratio it works out that the slope value of the capstone is 2.203, therefore, 1.732 x 1+Phi squared = 4.534 x 2.203 = 9.988, divided by 4 = 2.497. This appears to be reasonable enough because the block dimensions must be based on values that relate to the size needs of the structure and its Phi based geometry. In 1837 a discovery by Howard Vyse showed that the side casing blocks at the base measured 5 x 8 x 12 feet with the slope in evidence on the 12 foot side. A way to explain these dimensions in terms of the Phi Ratio and the capstone in this sequence are as follows: 5 = 2 x 2.497 = 4.994, 8 = 2 x 2.497 = 4.994 + 3 = 7.994, 4 x 3 = 12. It has been determined there are probably two casing stones along the base with lengths of 10.24 ft. perhaps at the severely damaged corners, plus (61.8) 12 ft. lengths. Therefore 2 x 10.24 + 61.8 x 12 = 762.08 ft., divided by 1.732 = 440 Royal Cubits...see Royal Cubit Theorem. In addition, 12 - 10.24 = 1.76, divided by 2.236 = the square root of Phi, 0.786. From this it can be seen how the 2 x 2.497 = 4.994 foot dimension of the interior standard block fits into the scheme. Judging by the photos it appears as though this dimension of interior blocks was used for the most part in the horizontal and vertical planes. The height 484.68 − 1.732 = 482.948, divided by 4.994 = 96.706 and 0.706, divided by 96 = 0.007. If these values are to be relied upon there would be no room for mortar of any type between the blocks. Shortly the answer to this question will be addressed in a timely manner when another fabulous application of ancient technology is noted. It works out the same way in the horizontal plane when the 9.988 dimension is used. One of the most enlightening features is, the slope distance 616.53, divided by 4.99 = 123.6, divided by 100 = 1.236, and with no coincidence this is the square root of 5 - 1, another Bingo! The conclusion is, that with regards to the specific block dimensions and the type of rock material used, they must play a specific role in the energy generation network within the structure when it comes to the solar, piezoelectric and electromagnetic factors which are the result of heat, tremendous weight stresses and the basis of its design. It is believed the interior blocks are granite, which has a high content of silica and the limestone casing acted as an insulator. With regards to the alleged concave nature of the faces, theories suggest that this was done to increase the stability of the structure's mantle and this seems reasonable enough but the complexity of such an exercise would seem an intimidating challenge somewhat beyond our capacities. Besides that, since the faces might have been made to be concave, not unlike today's satellite dishes the thought occurs there could have been interstellar communication with other locations in the galaxy or beyond and around the planet at that time in history. The dimensions of the concavity are unknown at this time, however, further along an unexpected discovery appears to shed some light on this intriguing matter. We are unable to determine the particulars of the concave feature and the true reason for it due to the erosion factor, therefore no bingo points are scored on this one. The question is, what was the depth dimension used of what might have been a parabolic face and what mathematical significance was it based on. Such are the on going unsolved mysteries that relate to the true workings of the Great Pyramid.

We are told it took 2.3 million blocks to build the Great Pyramid and to this day it is truly puzzling how this daunting task was accomplished. As mentioned before the main problem with solving this riddle has been the thoughts on it have been falsely based and the opinion has prevailed that anything constructed so long ago must have employed primitive means in its workings. In fact this is not the case at all, the capabilities of the ancient designers and builders has been sorely underestimated. It must have been the other way around because whoever built pyramids in those times must have had a technology considerably more advanced than ours is today for reasons that are obvious. To begin with the simple truth is, we cannot duplicate the feat of constructing a pyramid of such size and quality as the Great Pyramid that contains the coded account of the Solar System, Milky Way and beyond, so it's back to the drawing board for those who have misjudged the abilities of the pyramid builders during this ancient time. It shouldn't be believed for one moment that any time was wasted by constructing a ramp to the summit in any way, shape or form and negating the effort and time schedule more so by removing it. It would have been a minimum effort for an optimal result, which is what we strive for in this day and age. In addition there never was, is not now, or ever will be such a thing as anti-gravity except in science fiction.

Please see the drawing titled: Great Pyramid Airlift Assembly that shows 4 sections of the structure that relate to a 2:1 slope crossing its vertical face in a zigzag manner. We are urged to remember the slopes of the Grand Gallery, Ascending and Descending Passages which are the same and what it means. The sections are determined where the slope intersects with the vertical at center. Angle 63 degrees, 26 minutes, 06 seconds, reappears, not once, but 6 times as if it is telling of something important, the main feature being, this arrangement would provide measurement control from center during construction as each level of blocks was placed. It would certainly be needed to place the blocks accurately for the concave faces of the structure. No doubt the distances from center lined up with points on the four cardinal directions and the four corners at the various levels were all predetermined. From the perspective gained from my practices it would take precision survey control in the form of application of the three point problem, or precise intersections of the cardinal direction lines to keep track of center and this could only be done by employing advanced survey methods and equipment, and/or perhaps a very advanced type of GPS system. Note has been taken that the upper fourth volume of section 4 divided by the over all volume x 100 = Pi - 3 = 0.1428 and the reciprocal value of this number is 7.0, the solar number.

In short, there is no other choice but to consider an air lift concept from the quarry to the construction site of the blocks that were jockeyed into position with the use of levers and so forth to a zero tolerance and this will be discussed in the following. Look on the web sites under the topic of air machines in ancient Egypt and it will be seen there are references to such, carved in stone some 300 miles east of Cairo from that time era, showing winged airplanes, helicopters and what looks like a submarine that could also easily pass for a zeppelin. If the textile industry in ancient Egypt could produce sails made of linen then certainly a balloon like or a dome form treated with tree resins supported by laminated papyrus reeds or such could have been produced. If there was such a thing as world travel at that time, and strong indications are that there were, imported balsa wood from South America for the frame could have been used and hard woods or very thick, strong rope could have been employed for the lower frame. It is very possible a lever crane of some sort was used on the lower levels of the structure. When Herodotus visited the site in 450 B.C. he was told immense machines were used during construction but that was over 3000 years after it had been erected so no one believed it and it was pushed to the side and forgotten. The Greeks are given credit for being the first in history to have used such a machine but we know they cultivated much technical information from old Egypt. The hydrogen filled Hindenburg logged 191,600 miles, the equivalent of eight voyages around the globe while transporting passengers and freight during its first year of service in 1936 until its disastrous ending occurred at the New Jersey airport in the United States. It had a length of 805 feet and a diameter of 135 feet and had a gross lift capacity of 242 tons, more than the weight of any block in the Great Pyramid where the heaviest ones, the casing stones, are approximately 30 tons when cut to the desired dimensions and an interior block weighs an average of 21 tons. These weights are based on one cubic foot of both solid igneous granite and sedimentary limestone long weighing 167.53 pounds, according to what can be deduced from today's scientific input which provides a weight of 168 pounds per cubic foot, rounded off. It can be analyzed this way, the volume of an interior block is 2.497 x 9.988 x 9.988 = 249.101 cu. ft. Therefore, 249.101 x 167.53 = *(1) 41732 cu. ft., divided by 2000 = *(2) 20.866 tons and there certainly are some familiar looking number values in these quotients which relate to the Royal Cubit Theorem. *(1) 41732 = 40 + 1.732 and *(2) 20.866 = 20 + 0.866, 1.732, divided by 2. This expresses the intricate way of how the dimensions of the blocks correspond with the structural needs and matrix of the energy grid system within this unique structure and it looks as though a double bingo attached to the Royal Cubit Theorem is at hand. It would seem a smaller scale hydrogen gas, as opposed to a hot air enclosure would have been used for transporting the blocks then, unless many blocks or entire sections at a time were mortared together on a nearby perfectly level area then air lifted to the various elevations and predetermined locations in the structure. A volume such as Noah's Arc outlined on page 146 further along serves as a workable example of a large volume configuration. A hydrogen filled enclosure with a length of 520 feet, a width of 84 feet and a height of 52 feet would have a lifting capacity of 45 to 50 tons. If more lift was required a larger one, perhaps twice this size could have been fabricated. The dimensions are based on the Phi Proportion and come from that era, therefore the possibility of this should be considered. Perhaps it is these dimensions the scribes used in the scriptures when they invented a myth and fairy tale fear factor story about it while the human masses were drowning in ignorance, or does the story come from the Out of Atlantis Theme. Another thought that comes to mind since this topic is touched on. The Tower of Babel story in the scriptures might be about an attempt by some group to construct a rising spiral installation with a base of 300 feet or so that would out class the Great Pyramid but it failed, not because the people involved with the project ended up speaking different languages, it was because they didn't know how to speak the right one, the one known as the Language of the Living Universe.

Hydrogen has a lifting capacity that is approximately equal to its volume and even in our time hydrogen and helium filled zeppelins have been used to transport people and freight, and it has been determined without question, our technology is not as advanced as the one that designed and built the pyramids in ancient times. We shouldn't assume for a moment that they didn't know the most plentiful and lightest element in the Universe was hydrogen and that they knew all about electrolysis in order to extract it from salty sea water using zinc and copper anodes in a controlled environment. Such a hydrogen filled craft was probably maneuvered by all direction propellers not unlike the way it is done in our time and this proposal makes absolute sense because the room at the top of the pyramid becomes diminished and inaccessible, therefore, what better, more simpler way would there be to build such a structure which might have taken one year, as in 365.24 days, and a minimum of manpower to complete. No doubt the work force on site was quite sizeable when it is considered there were other pyramids, temples and shrines being constructed on the Giza Plateau at the time. Earlier in this century archeologists located some of the encampments used by the builders approximately one mile west of the Great Pyramid. According to the findings they lived a simple but good standard of life and ate well. There is no evidence that suggests that they were enslaved by the ruling parties there. It is believed they had status in the community and might have carried on with the work for a number of generations while raising their families and training apprentices. It is believed it was this epoch era when the fundamentals of Free Masonry took root and further along it will be seen why. Most likely the overseers of the various projects and the entire work force were supremely educated and highly skilled in a grand science that was perhaps imported from Atlantis, the existence of which as yet, hasn't been proven beyond doubt.

As it stands the debate persists as to how the enormous blocks of stone were manufactured to such fine precision and what type of mortar was used to bond them together. The following explanation is offered: To begin with the work was far beyond the level of anything to do with copper chisels, ramps, ropes, pulleys and slave labor, therefore, the deduction made is stone fragments were vacuum scooped up to a tolerance of 3/4 inch minus into the holds of as many as 3,600 enormous hydrogen filled dirigibles from the quarry sites for the granite and limestone needs. That number fits as there are 3,600 seconds in an hour. Operations began by engulfing the areas with ignited hydrogen at 1,000 degrees Fahrenheit whereby the water byproduct cooled the rock and shattered it. The material was transported to the site foundries where it was super-heated to a molten state of 3,000 degrees Fahrenheit by way of huge hydrogen combustion furnaces. Then it was poured into correctly sized sand molds after which the blocks were air lifted into position in the structure according to the layout lines that provided alignment with the constellations of Sirius and Orion. This took place just as the cooling time was coming to an end when it became glass-like. Note: Heat fusion not mortar was used to bond the blocks together which were far more durable than today's concrete. The sand mold castings were made of Nile River clay. The full details will become available in the follow-up. The above description as to how the Great Pyramid at Giza including the enormous statues, obelisks and the Sphinx were constructed during that unrecorded period in history provides an explanation on how such wonders came into being. These projects were taken very seriously by the god-like beings who managed them. A 24 and 7 schedule was the approach and they were completed very quickly, not several years as put forward by today's researchers. Melted wax transformed into a candle or any other form supports an age old saying, there is nothing new under the Sun. Meaningful words, no doubt copied by the scribes from an ancient text source.

More interesting details are available with regards to the code about volumes and weights that the Great Pyramid can tell us about. As follows, our favorite values Phi and Ancient Pi come into the picture as a result of the computation process:
762.08 squared x 484.68, divided by 3 = 93,828,543.07 cu. ft., the volume of the Great Pyramid x 167.53 pounds = 15,720,000,000 cu. ft., (Ancient Pi, divided by 2, x 10,000,000,000) divided by 2000 = 7,860,000 tons, divided by 10,000,000 = 0.786 = square root of Phi.

The significance of the value 0.786 when it comes to volume and weight is, it can be seen that when divided by 10 = 0.0786 x 2 = 0.1572 x 10 = 1.572 = Ancient Pi, divided by 2. It appears then, 0.786 was a basic unit of weight like the pound having ten units that relates nicely to one half Ancient Pi because 15,720,000,000, divided by 2 = 7,860,000,000, divided by 0.786 = 10,000,000,000, and since granite and limestone are virtually the same weight per cubic foot it all makes sense. The volume of the Great Pyramid squared = 8.80 E 15 virtually, divided by 100 trillion = 88, the number of constellations x 10 = 880, twice its base length in Royal Cubits. Further along in the study it will be seen how the dimensions of the Great Pyramid correlate with those of Earth and the Sun. The results of this part of the inquiry are quite decent and some additional bingo's were to be had along the way. Each one is a piece of the puzzle and once we have enough of them organized properly the picture will tell us much more.

It can be seen in the study, when the right approach is taken, that which is being knowledgeable in the workings of the Golden Section to reach into the ancient past, a highly evolved and natural Universal System of time, weights and measures becomes known. Aside from that the drawing titled, The Great Pyramid Speaks, provides an update of the accomplishments acquired from the above survey of the heavens. It might be said, since the inner solar system is defined as being the Sun, Mercury, Venus and Earth, including their moons etc, what has been determined by this survey includes what might be known as the "Inner Universe", however, an enormous leap has been made where the matrix of the Cosmos has been determined and much of the rest of the story can be based on the simple principle of projection. At this stage it is more or less theorized that the distances to planets in other solar systems are in proportion with our sun once their diameters are determined, and that each of these systems have (12 planets), and that is in conjunction with the Royal Cubit Theorem. Unfortunately, we have been missing the boat on it until "Living with Geometry" made its presence known.

In closing with this fascinating section, the notion might occur that other great pyramids were constructed on earth by unknown intelligences during a time in the very remote past when previous Zodiac cycles were commencing and that is how the seeds of life were helped to germinate on this planet. Indeed, how possibly man, the prodigious child of God came into being on earth. The question of who built them is entirely up to speculation, our history books cannot even tell us just who built the most recent ones, going back 5,000 years. Since then those pyramids, if there ever were any, have turned to dust, and this will eventually be the state of the Great Pyramid. It is certainly food for thought after studying the energy processing that take place inside the Great Pyramid interior chambers further back in the study. For now, all we have to work with is the remnants of those megalithic structures left behind from an ancient time as we try to fit the pieces of the puzzle together in order to understand how this science can work for us today.

Back Inside the Great Pyramid:

Dimensions of the Sarcophagus Converted as Shown:

Length - 7.47 ft. divided by 1.732 = 4.313 Royal Cubits
Width - 3.20 ft. divided by 1.732 = 1.848 Royal Cubits

Height - 3.40 ft. divided by 1.732 = 1.963 Royal Cubits

It is seen there is room in the sarcophagus for a human body. The first thing of interest noticed is the conversion to Royal Cubits value for the length which is very close to the square of 3, divided by 4. If this was meant to be the case the length would measure at 7.50 feet. These measurements were recorded by someone who was competent with recording measurements I assume and erosion wouldn't be a factor in an enclosed area but rock shrinkage as stated earlier, can be attributed to air and moisture loss over a lengthy period of time and it has been proven that dehydration takes place in a structure having these ratios. The factor is 7.47 divided by 7.50 = 0.996% and a mental note on this is made. By adding close to three hundredths of a foot to the length, width and height then dividing each by 1.732, the length would be 4.33, the width would be 1.861 and the height 1.98 but the lid went missing, probably due to vandalism, therefore the true height is unknown. The wall thickness of the Sarcophagus is close to 6 inches, or 0.5 ft. leaving the interior length at 6.5 feet and an interior width of 2.223 feet x 12 = 26.7 inches. Since the average height of a man at that time was 5.6 feet there was more than enough room for a human body in the enclosure. We are told the inside height from the base to the top is 2.86 feet, therefore the base thickness would be 3.873 – 2.86 = 1.013 ft. With no information on the lid which might have been broken up and carried off by the early looters, or stored away by the builders its thickness is up for grabs, however, if it had a depth of 0.367, (reciprocal value of the capstone base 2.723) which seems reasonable + 3.873 = 4.24 / reciprocal = 0.236 or, divided by 1.732 = sq. rt. 6, and the height not including the lid 3.873, divided by 1.732 in terms of Royal Cubits = 2.236, the square root of 5. I believe that is the way the ancient designers would have it and created it to suit the Phi proportion and for the Sarcophagus to function properly. For the width it appears 1.861 is reasonable because this value squared = 2 x 1.732 = 3.464 = a double Royal Cubit.

Adjusted Dimensions of the Sarcophagus and Fully Converted:

Length ~ 7.50, divided by 1.732 = 4.33, divided by 0.0866 = *86.60
Width ~ 3.223, divided by 1.732 = 1.861, divided by 0.0866 = 37.22
Height ~ 3.873, divided by 1.732 = 2.236, divided by 0.0866 = 44.72

Each dimension has a Golden Ratio significance and it can be seen how the length relates to the sun as a 100 thousandths part of its radius.

Length ~ * 4.33 = square root of 3 divided by 4 x 10
Width ~ 1.861 squared = 2 x 1.732 = 3.464 = a double Royal Cubit
Height ~ 2.236 = square root of 5 *

That 0.0866 pyramid inch seems to work out just fine and the Royal Cubit Theorem is getting warmed up again. The length to height and the width to height are not in the Phi Proportion but when the dimensions in royal cubits are divided by Phi some interesting quotients materialize:

4.330 divided by Phi = 7.00
1.861 divided by Phi = 3.00
2.236 divided by Phi = 3.618

and the volume is 18 cubic cubits, 1 + 8 = 9, the full number. One can only imagine how the proportions of the Sarcophagus and its dimensions lend themselves to energy conduction and transmission of some sort and have something to do with transmutation and possibly teleportation of a living thing within it but at this stage it isn't understood. For now it remains as one of those million dollar questions.

Below is shown how the correct dimensions of the Great Pyramid correlate with earth's dimensions with regards to the square root of 3, divided by 40:

Earth Equatorial Radius = 3963.636 miles / re: the revised height dimension
3963.636 x 5,280 = 20,927,998 ft., divided by 484.68 = 43178 divided by 1,000,000 = .043178, or rounded off to 0.043. The square root of 3, divided by 20 = 0.0866, divided by 40 = 0.043, the ratio to which the height of the Great Pyramid is to the equatorial radius.

Earth Polar Radius = 3950.0 miles / re: the revised height dimension

3,950.0 x 5,280 = 20,856,000 ft., divided by 484.68 = 43030.45, divided by 1,000,000 = 0.04303453, or rounded off to 0.043. The square root of 3, divided by 20 = 0.0866, divided by 40 = 0.0433, the ratio to which the height of the Great Pyramid is to the distance between the center of the earth to the poles.

Earth Equatorial Circumference 3963.636 x 6.283 = 24,903.525 / re: the revised base perimeter

24,903.525 x 5280 = 13165812, divided by 3048.32 = 4319, divided by 100,000 = 0.04319, or rounded off to 0.043. The square root of 3, divided by 20 = 0.0866, divided by 40 = 0.0433, the ratio to which the base of the Great Pyramid is to the earth's equatorial circumference. Without question then, the dimensions of the Great Pyramid correspond with the earth dimensions to within three places of decimal of the square root of 3, divided by 40.

What's more: SOL = speed of light

The square root of the correct speed of light, 186,216.56 miles per second = 431.528, divided by the sq. rt. 3 = 249.150, divided by 2 x 2, divided by 1000 = 0.004, reciprocal, (250) / 249.150, divided by 250 = 0.997, close to that ratio value mentioned above. Also, SOL divided by 100,000 = 1.862 squared = 3.467, divided into a double cubit which is 2 x 1.732 = 3.464 = 0.999 = 1.0 for all intensive purposes. Modern Pi - 2.0, divided by 2 = 0.571, cubed x 1,000,000 = SOL x 0.999. Without a drawing of it envision a right angled triangle with an adjacent side of 3.0 and an opposite side of 1.732. The subtended angle becomes 30 degrees and the hypotenuse length = 3.464, the double cubit. The point of the discussion is, a remarkable relationship between the SOL, 3, the sq. rt. 3, the sq. rt. 3 x 2 becomes evident.

Earth's circumference, 24,903.525, divided by 110 1/3 Phi = 365.24 days in a year.

The 5 th. planet from the sun, Jupiter, the largest one in the solar system has a diameter of 86,753 miles, approximately 10 % of the sun's diameter. When divided by 2, the quotient is 433765... Interesting.

Other Royal Cubit Theorem Correlations

The Royal Cubit dimensions of the Ark of the Covenant and Noah's Ark established in chapter 6 are reviewed. Following are tables showing how the Royal Cubit Theorem equates with these dimensions.

Ark of the Covenant:

Length ~ 2.5 Royal Cubits x 1.732 = 4.33, divided by 0.0866 = 50

Width / Height ~ 1.545 Royal Cubits x 1.732 = 2.676, divided by 0.0866 = 30.90
The length, width and height are in the Phi Proportion. There is no correlation between the Sarcophagus and the Ark of the Covenant as thought by others.
Noah's Ark:

Length ~ 300 Royal Cubits x 1.732 = 519.6, divided by 0.0866 = 6000

Width ~ 48.54 Royal Cubits x 1.732 = 84.08, divided by 0.0866 = 970.80

Height ~ 30.00 Royal Cubits x 1.732 = 51.96, divided by 0.0866 = 600.00

Take note of the 4.330 value derived by multiplying the length of the Ark of the Covenant by the square root of 3. This is the square root of 3 divided by 4 x 10, and the pyramid inch value of 50 based on 0.0866 doesn't just seem to be, but is in fact, quite rational and it is seen that the width / height is precisely based on the Phi Ratio. Take note of the Royal Cubit length and height values of Noah's Ark which are based on the Royal Cubit times the square root of 3 divided by the pyramid inch of 0.0866 and the values of 600 and 6,000 are also more than rational, and of course the width to the length is in perfect agreement with the Phi Ratio.

The proofs of the Royal Cubit Theorem are before us and an enormous leap of understanding has taken place from what the Great Pyramid is telling us about the workings of life, our solar system, the Milky Way and the Universe itself. It is all based on an ancient measuring system that makes total sense and in the process, enlightenment is ours, and before long we will be visiting the constellations in depth. The worlds of mathematics, physics and astronomy have been transformed before our eyes and after centuries of being stuck in the dark, making rough guesses the green light is on for a "GO" in the process. It appears as though the revised, or what can now be called the true dimensions of the Great Pyramid correlate with the earth dimensions as well as the sun, the planets and our galaxy in such a way that the theme is totally to do with the Phi Ratio, the square root of 3, divided 20 = 0.0866, the square root of 3, divided by 4 = 0.4330, including the square root of 5 and this leads to more than some vague conclusions on the matter. Therefore, it would seem there is a 1:4330 ratio between the Great Pyramid, earth and sun dimensions intended in its design proportions and not a 1:43200 ratio as so determined by others. The square root of 3 divided by 20 = 0.0866, a ten millionth part of the sun's diameter and the square root of 3 divided by 40 = 0.0433, a ten millionth part of the sun's radius. This is a pure mathematical value that expresses "the four cardinal directions of the globe to which the Great Pyramid is perfectly in alignment with and is the Key to understanding the mystery of the Ancient Measuring System. It has become more than apparent the original base length of the Great Pyramid was 440 x 1.732 = 762.08 feet and this has been proven beyond any doubt. It would seem the design intention of the Great Pyramid was to construct it to the proportions of earth, the sun, our galaxy in such a way it would be in place for a very long time and fortunately for us it has, because it is finally telling us what we need to know. There was some grand plan behind it all and the knowledge to do this seems quite beyond run of the mill astronomy based on measuring shadow lengths with reference to known vertical distances on the ground and so forth. One can only assume or might imagine the only other way this could have been carried out by the ancients was for them to have had a satellite system in place that was far superior to ours and how they performed the computations is anyone's guess. Another question is what type of surfacing was used for the design blueprints. Some might claim this is totally preposterous but a review of the drawing entitled the Giza Phi Connection shows they somehow had a thorough knowledge of global geography. Let us leave it at that for now, there are no words to describe their processes and the answers to these questions are beyond our understanding at this time. All we know is, they got the right answers and we cannot duplicate a grand work such as the Great Pyramid, however we now have a much better handle on the science behind it and what it can tell us. It will be time well spent in referring to the mystic meanings of numbers 3 and 4 in chapter 6. A thorough review of all of them including the Seven Hermetic Laws of Nature at any time is suggested, because they say it all and help provide the reader with the mind set and mode of ancient time perspectives.

It might be assumed the dimensions of the Great Pyramid were based on this ratio to at least three if not seven places of decimal by the wizards of ancient times with regards to the square root of 3 divided by 20 = 0.0866 and 4 = 0.4330127. However, it does appear as though using three places of decimal does the job. We aren't sure what circumference was used with regards to the base dimension of the Great Pyramid. There is an indication the earth circumference used must be somewhat larger than 24,903.525 miles. This circumference is the one at sea level it is assumed and not accounting for a mean height of the global land mass elevation or the limits of its atmosphere. Going back to where the true polar radius is subtracted from the true equatorial radius, 3963.636 - 3950 = 13.636 miles this might be the limit of the stratosphere, or ozone layer in which life exists. In adding 13.636363 + 3963.636363 = 3977.2727 the circumference becomes x's 2 Pi = 24,989.205 miles x 5280 = 131,943,002.4 feet, divided by 3048.32 = 43283.842. It does it seem reasonable the ancients would use the stratosphere circumference to determine the perimeter of the Great Pyramid after using the sea level circumference radius and polar radius to determine the required measuring standard dimension such as the square root of 3. Perhaps this should not be ruled out, therefore the following is put forward to help answer this question. It appears reasonable as stated before the ancients were working to three places of decimal.

In addition:

3977.2727 x 5280 = 20,999,999.86, virtually 21,000,000
divided by 484.68 = 43328,
divided by 1,000,000 = 0.0433

Using the Royal Cubit Theorem: 1,319,430,024 ft., divided by 1.732 = 76,179,562.59, divided by 0.0866 = 880,000,000 virtually..., in other words the earth circumference used which the perimeter of the Great Pyramid is based on is 88,000,000 Pyramid Inches and that is the circumference that includes the stratosphere, 24,989.205 miles. The base of the Great Pyramid is 440 Royal Cubits x 2 = 880, furthermore it can be seen that attractive number based on 762.08, divided by 0.0866 = 8800, mentioned before hand is multiplied by 10,000 = 88,000,000, therefore that part of the question is answered. The number 88 is special because it expresses the double infinity of the mind and the Universe and the addition of these two numbers reduces to 7, the solar number of completion and that is what, it appears, the Great Pyramid and the workings of the Universe are basically all about.

As we know a distance such as 1.3 hundredths of a foot makes a difference of 5.72 feet over a distance of 440 true Royal Cubits which have, with no question, a value of the square root of 3. With regards to estimate errors based on false measurements a tenth of a mile is 528 feet, and if a dimension is out by a quarter of an inch the difference would be 110 feet per mile, therefore precise distances to at least three places of decimal based on a reliable baseline reference in dealing with large distances would have to be known in order to evaluate the relative location and dimensions of any object elsewhere with precision. This was learned from my practices in the disciplines of survey/engineering. For so long the exact dimensions of the Great Pyramid have been unknown and the solutions have been left up to educated guesses and fudged approximations until now. In a certain reference another opinion of a value 7899.988 for the total polar distance is given. This value divided by 2 = 3949.994, merely 49.63 feet longer than the 3949.9 mile radius. This is very close to the earlier approximation of 3950.08 miles. Therefore it might be safe to say the polar radius equals 3950 miles. When the 3949.994 value is run through the formula the result is 0.04303038. Subtracting this from 0.04330127 = 0.00027089, not bad. Another mental note is recorded. It might be asked if this radius is closer to the true value and there is something more to be said for the proposed 484.68 height if this radius is near correct. It would seem so. The volume of earth is 260,000,000,000 cubic miles. When the 762.08 base and 484.68 height are converted to miles then employed to compute the volume of the pyramid, this value divided into the earth volume = 4.33 E 14. Therefore it might seem apparent, the concept of the Great Pyramid being in the ratio of the square root of 3 divided by 4 with Earth's dimensions has developed into the more advanced stages. The Royal Cubit Theorem has also been applied successfully to the King's Chamber and the Sarcophagus dimensions. A quantum leap has been taken as a result of this process and it needs time to develop, in other words a large volume of food for thought has been ingested and it needs time to be thoroughly digested.

What They Have Been Telling Us About the Great Pyramid…is " Incorrect "

The standard information given on the Great Pyramid tells us the length of the base is 9131 pyramid inches and that inch is very close to the imperial value. The base length used is 756.36 feet, divided by 9131 = 0.08283, and 9131 is divided by a cubit equaling 25 imperial inches rounded off. Therefore it is an approximation to begin with and I was never comfortable with it. This was the dimension used by dividing the polar radius of 3949.9 miles converted to feet into 10,000,000 equal parts. Hence, the logic is, 9131 divided by 25 = 365.24 days in a year. It also goes on to say the perimeter of the base, 4 x 9131 = 36524, divided by 100 = 365.24 days in a year. This works out very well relative to a number that makes it work, however the fact is, when the polar radius 3949.9 miles x 5280 feet is divided by 10,000,000 the quotient is 2.0855472 ft. x 12 = 25.026566 inches. Therefore, when 9131 is divided by 25.026566 the quotient is 364.852 and 365.24 x 25.026566 = 9140.7029, therefore it appears something is amiss in this theory and it is a fudge job to make it work. What this formula says is an unknown divided by 25 = 365.24 and this odd, mumbo jumbo, incomprehensible, meaningless number 9131 comes up and it has no relationship with the Phi Ratio whatsoever. The problem is, today's dimensions of the Great Pyramid have been employed and the answers work out because this cubit of 25 imperial inches was used to manipulate and gain a desired result whereas the Royal Cubit Theorem does no such thing. I have come to the conclusion that the 25 inch cubit is a bogus dimension in the workings of this matter and that the polar radius of 3950 miles less13.636 miles = 3936.363, which is the difference between earth's equatorial radius and polar radius, subtracted from the polar radius, converted to feet, then divided into 12 million equal parts which equals the square root of 3 provides the answers. In my opinion those who have been using these pyramid inches to deduct prophecies have been wasting their own time and everybody else's because this value is meaningless in terms of the Golden Ratio. All this changed when the workings of the Ancient Zodiac were brought to light and the year 3138 B.C. was determined to be when the Great Pyramid was constructed.

For clarity on this matter refer back to the reference generated on this topic back in chapter 3. What makes sense is 756.36 x 1.0075625 = 762.08, or 440 x 1.732 = 762.08 because it deals specifically with the erosion, shrinkage, settlement factors and whatever else transpired when foundation material was being carted off to other locations sometime in the distant past. This relates to the 1.732, divided by 1.719 = 1.0075625 observation previously mentioned. Another way of computing the number of days in a year using the Royal Cubit Theorem concept is as follows: Circumference 390.972 x 0.0866 = 33.8581, divided by 360 degrees = 0.0940504 + 24 hours = 24.0940504, divided into 8800 = 365.24, the number of days in a year. There are 360 degrees in a circle during which a 24 hour period of time elapses and one revolution of this circle = 0.09405 including the 24 hours. This is yet another proof of the Royal Cubit Theorem. These numbers, based on the Phi Ratio and the square root of 3 are telling the truth, therefore it might be asked, how is it that the number of days in a year can be computed several ways in this scheme based on the Phi Ratio using a base dimension of 762.08 feet which is relative to the square root of 3 x 440 cubits, or that conversely 762.08 feet divided by 1.732 = 440? The only answer is, with very little doubt; 440 Royal Cubits x the square root of 3 = 762.08 feet x 4 = 3048.32 feet, and a height of 484.68 feet, and side slope distance 616.527 are the original dimensions of the Great Pyramid and "A long standing riddle is solved". As stated the result of a vesica pisces using two intersecting circles with diameter of one unit is the square root of 3. Coincidence or not, when the statute mile, 5,280 feet is divided by 12 the quotient just happens to be 440 and when this is multiplied by 1.732 the result is 762.08. In terms of the true pyramid inch the following solution to the question of where 440 comes from is presented: 762.08, divided by 0.0866 = 8800, divided by 10 = 880, divided by 2 = 440. Another indisputable explanation on this point comes to light in Chapter 8 when it is known there are 88 constellations x 10 = 880, divided by 2 = 440 Royal Cubits in the base of the Great Pyramid.

Following is yet another outstanding way to compute the number of days in a year using the true base dimension of the Great Pyramid:

*Base area, 762.08 x 762.08 = 580765.926,
divided by 43,560 (sq. ft. / acre) = 13.33 x
the reciprocal of 0.999 = 1.001 = 13.34,
Square root 13.34 = 3.6524 x 100 = 365.24 days in a year...an easy "Bingo"...

It has gotten to the point where I won't even look at the reference on hand about the Great Pyramid anymore because it is useless, false and more than misleading. It is enough to make the trackless sand dunes in the Sahara Desert yawn and the Universal Consciousness is deaf to it. The only solution is to "bury this hogwash with a large load of camel dung" and forget it. If there is an audience to the above and following input there is an opportunity to emerge completely from the dark ages. What I have ascertained is there has been a certain level of well meaning input by Great Pyramid enthusiasts who have been challenging modern science with their findings but they never had the true dimensions. In fact the above and following input proves that neither party has had the answers all along and the Church has been totally vacant on the topic since day one, even though reference was made to the Great Pyramid in the scriptures, but it is lacking the details on it.

The call is made because the correct answers are coming in using the Phi Ratio approach and it is the correct dimensions that are making the difference. What it boils down to is the true dimensions of the Great Pyramid have been determined simply by using a Royal Cubit equaling the square root of 3 x 440 from which the base, height and slope are computed with regards to the Golden Ratio Proportion and these new dimensions show a direct and sensible ratio correlation with Earth's dimensions and the sun's relative location and size including the correct value for the Astronomical Unit, the speed of light and the planet distances from the sun and their dimensions, the details of our galaxy and the star constellation patterns etc., and it will all make sense to the physicists and astronomers after they study the contents of this inquiry. The arguments have been listed showing proofs and a strong case in support of the Royal Cubit Theorem has been presented. The challenge is placed in the hands of all Great Pyramid enthusiasts, scholars and the world scientific community and all others who relate to this topic. The Great Pyramid is truly, not unlike a book or a storehouse of sacred scientific knowledge made of stone that was, and is still is tuned in with the life energies that encompass Earth, the solar system and the workings of the Galaxy and Universe itself in an extraordinary manner we are unable to fully comprehend at this time, and for reasons yet to be defined its presence, even in its more or less dysfunctional state, has cast a spell that has captivated the imaginations of all who have attempted to fathom its mysteries. And now, with the Royal Cubit Theorem and true dimensions available there is new life in the study and an opportunity to get on track and delve much deeper into the subject for the answers we need. Please see the drawing entitled, Great Pyramid - True Dimensions, complete with the capstone detail. This one provided me with a sense of triumph and pleasure.

As stated, with reference to earth shifts and settlement, the erosion, shrinkage factors and physical molestations etc, the original dimensions of the Great Pyramid were altered but access to them has been reclaimed based on 440 x the square root of 3 along with the Geometry of the Golden Section and the dimensions have proved out. The equatorial circumference and polar radius of earth are known to three thousandths of a foot therefore the computations work out to precisely the square root of 3 divided by 40 etc., and the Royal Cubit Theorem proves to be quite correct. Part of the problem has been when attempts are made to find the information needed it gets confounded with different opinions on earth dimensions and the records of them get jumbled, lost or mixed up, and they lack absolute accuracy, however, because the original dimensions of the Great Pyramid were established the matter was resolved and so far extraordinary results from this proposition have come in and the above listed are of much significance to the inquiry. Therefore, the process is well underway.

We are working in the shadows of great masters that were active on this planet long before our time. With diligence and perseverance we can follow their footsteps and learn from the many examples they have left behind for us to study, and in the process we might get a better handle on our own present and future. The big questions we are left to speculate on are "How and Why" did they carry out their works in this manner? "What was their plan and purposes. Who were they"? " Why was the Great Pyramid constructed and how exactly did it work"? It is a grand concept that students and other interested parties at all levels should cover these topics right in the classroom, however this is not the case because our education system is limited and mystery schools do not really exist, yet the world itself is the greatest institution for learning once the head space and reference material is at hand, when opportunities are made to go for it. I have always maintained that a simple thought or concept has the potential that can lead to extraordinary levels of enlightenment and this has been proven to be the case.

After deducting the difference between the equatorial radius and polar radius and subtracting that value from the polar radius, then dividing by 12,000,000 I found that this precisely equated to the square root of 3 and at that very moment I knew that I was on to something but I had no concept at the time how big it was. When the simple multiplication of 440 times the square root of 3 equaling 762.08 took place along with a few more computations that lead to computing the height and slope based on the Phi Proportion and establishing the ratio and proportion of the square root of 3 divided by 4 relationship between the Great Pyramid, the sun and the earth dimensions, and confirming with the sun's relative distance from earth, its size and the speed of light, the true distances from the sun to the planets, then the planet dimensions were computed along with input on their ellipsoidal orbit pathways around the sun etc., then it was on into the Milky Way itself and beyond into the constellations. Then an understanding of the Ancient Zodiac and the time the Great Pyramid was built came to us. There was "Magic in the Air, Eureka "!, within my grasp was the means to work with the true dimensions of the Great Pyramid and the greatest piece of work in geometry I have ever seen in my life revealed itself to me! Finding this has something to do with my years of practice in survey / engineering work and my need to tinker about with numbers, shapes, forms and different concepts etc. Very often solutions to problems manifest themselves with mental focus and the willpower that seeks agreeable outcomes, and often enough there are flashes of intuition mixed in with it. This is like one of those well oiled mathematical machines I spoke of earlier, only this one is "Big Time", If it works, practice with it until it becomes part of you. In this case it has made a survey of the heavens possible. Once a person applies themselves diligently and long enough to a problem and is patient in the process the answers materialize. The best way to explain it is, I have had a passion for geometry since I can remember and this must have somehow tuned me in with the Universal Conscience during which time I could not avoid becoming a geometrician or geometer on my own terms coupled with will and interest in the subject. Perhaps this is why I developed such a passion for survey / engineering but this breakthrough with the true Great Pyramid dimensions and where it has led is like a crowning touch to my efforts surpassing my wildest wishes for what my intentions were in the study. For my troubles I might get an honorary degree above zero in mathematics. In fact, when I was developing as an engineering surveyor there were frequent thoughts, like premonitions on what type of a higher form of mathematics I might be studying some day. I knew there had to be more to the story than I working with and studying at the time and I felt that I was missing out on it. The question at the time was, where could I go to learn about something I didn't quite have a definition for. Without fully realizing it earlier in life, I can say now, it is the study and applications of the Golden Section Geometry that fills the bill for me but there is no such thing like that available in the education system. This study has everything I need in it at this time and anything more that develops in future. I also believe the many meditation periods spent in my 10 foot Phi based pyramid over the past time period have had a positive influence on my mind and body. It has become my fortress of solitude that I nicknamed Ramses IV for some reason and whenever I visit in it I sense the energy and somehow it has had a positive effect on me. My physical being and mind relaxes and my brain waves are somehow stimulated at the same time and the only thing I can say is the sessions I have had in it provided desirable results. I am always perfectly healthy, in a continual state of being grounded, and centered and have tons of energy. All this started years ago when I first experimented with a worn out razor blade inside a 12 inch pyramid to find out it if it would really work to sharpen it. Use of a structure such as this for meditation or greenhouse purposes is highly recommended to any parties who are interested, and this of course leads to the logical conclusion in echoing the theme, our public buildings and domiciles should be constructed in terms of the Phi Proportions. As it works out, and this is truly a coincidence, my location on the globe with regards to latitude is within minutes of the slope angle of the Great Pyramid and the longitude places it approximately 11,000 miles north west from Giza, which is in a 3 2/3 x Phi proportion of the earth's circumference but what this had to do with where my inquiry ended up I am not sure. There was always a fascination with phi based pyramids, especially with the Great One, and this investigation has provided me with the opportunity to find the answers I needed and the freedom to share what I have learned with the world. To describe it in a metaphorical sense, my journey involved passage along the Golden Spiral Road of Life where I found what I was looking for, when I got a glimpse of God and learned about the workings of life and the Universe by engaging with the fundamentals of the ancient sciences of a distant yet glorious yesteryear. In plain terms, what has eluded the experts in western culture for hundreds if not thousands of years has been clarified and re-established, based on observation and deductive reasoning, to borrow from the well known memoirs of Sherlock Holmes series of mysteries and their solutions. The results of this inquiry are not an every day occurrence, yet the solutions are based on simplicity itself. Luck is defined as coming upon something that is desirable by chance, while providence is defined as the result of divine guidance or care. I believe the latter has somehow been at play in this development, and something most desirable has been the result. Notions of extra sensory perception the psychic and second sight are put to the side but as mentioned a degree of intuitive reasoning coupled with the facts is not totally ruled out. As far as I am concerned, when certain symbols and number values become available and there is a real need or will to know the truth of the details, intuition, like a sensation cuts in like a spark of insight, coupled with the power of deductive reasoning that can lead either quickly, or eventually to solutions, then the pieces connect and fall into place. The contents of chapters 3 and 4 are not to be misconstrued with a new age attempt to explain the workings of the ancient sciences because they are based on simple truths and sound scientific principles on the matter and though the results appear rather astounding, they are left in the capable hands of the reader.

Thank you for your interest with the details and enduring the child's play number crunching that provided the answers needed. I trust the reader and other interested parties will give the Royal Cubit Theorem and true dimensions of the Great Pyramid a whirl and do their own thing with it. The object was to make a contribution to the well of knowledge that relates to the Great Pyramid by setting the record straight on the true earth dimensions, the Royal Cubit issue and reclaiming its original dimensions, for, to be sure these have led us on an incredible journey. The truth of it is, we have been stuck in the realms of rough approximations for a very long time without them. It was worth going after because the truth on these matters has finally come to light and I trust this concept will spawn another generation of researchers and the education system with the straight goods to carry on in study. Once the establishment gets a hold of this theorem and information there will be a critical analyses period to begin with, once the information is verified a number of heads will be rolling down the back streets, and they will literally go into orbit when it is seen that it is the truth and no doubt a lot of false information such as the 25 inch cubit and bogus pyramid inch will fall by the wayside as a result. And when the astrophysicists in the NASA Space Program get a hold of it they will put it to a good use. History and science textbooks will be rewritten and it will all be for a good cause because education in these matters will be on track with reality. Figuratively speaking, by knowing its true dimensions, and working with them as shown the Great Pyramid re-awakened and started to talk to us, and in time it will tell us much more about what we need to know, the solar system, the Milky Way and beyond will be more fully traversed and the stars will no longer be a limit, and the mind of man is sure to become supremely elevated in the process. We will know exactly what to do in our planetariums before long and if I am not mistaken, one of the most important purposes of the Great Pyramid was to educate us and that is why it was built to outlive time. In a sense the influence of its creators, whoever they were, are still with us. On the imagination level, though it might not seem believable by some or even to most, when the true dimensions of the Great Pyramid were determined and the verification checks that were run on them proved out, I sensed a heart beat deep within the massive life angle sloping walls of this ageless wonder, then I felt a tremor and heard a sigh of relief because, not unlike a living time capsule of coded messages, its spirit was finally free to tell its story. I have been thinking about it for so long now the Great Pyramid has become like a mysterious acquaintance that I befriended long ago, and now it has become not unlike my most trusted associate in the quest for real knowledge and the truth of it is, through due process an ancient memory of a grand science has been tapped into and there is no turning back now. What else can be said at this stage, but, ah yes, " The Great Pyramid has Spoken!" Once it started to talk to me I was compelled to see where it would go and there were no disappointments. In some circles I know of, there will be great joy over this exposition of enlightenment from which so much has been learned. Whatever the case might be, the scientists, philosophers, educators, artists, meta-physicists and students etc. of today now have something tangible to work with. The thought which occurs is, this information might have been available somewhere deep in the archives at the library in Alexandria before it was sacked and burned to the ground, or else it was in the possession of an elite group who kept the secrets of it to themselves for reasons unknown. The answer to this may never be known but we have the details now, and perhaps it is timely in our process that we do.

The United States of America, Great Pyramid and the Free Masons

On the over left side of the American dollar bill is seen a graphic of the Great Pyramid placed there by the Free Masons who were the founding fathers of this great nation. The front face is lighted while the west side is in darkness and at the time this currency was developing western civilization was venturing into the unknown and the general consensus amongst its leaders was, there was something very special about the Great Pyramid. Within the capstone, lighted all around is the " All Seeing Eye " and " In God We Trust " is written on the currency. In Latin above the pyramid is written ANNUT COEPTIS, meaning " God has favored our undertaking ", and beneath the pyramid, also in Latin is written "NOVUS ORDO SECCORUM", meaning " A New Order Has Begun. " On the base of the pyramid is year 1776 written in Roman numerals. The founders of this unique civilization and its high ideals gave the Great Pyramid the respect it commanded but the true understanding of its mathematics was on a par with the Newton following at that time, however, the unanimous decision between myself and the publisher was made to use this part of the American one dollar bill and the Washington Monument on the book cover. This suits me solidly as I have family relation ties in the United States and I accept their philosophy of democracy, which is virtually the same as Canada's. I haven't forgotten, our first Prime Minister was a Free Mason. Forever, it seems I have kept an American one dollar bill stashed away in the back of my wallet and it appears there was an underlying reason for this as well as my taking up the study of Free Masonry, the oldest world wide order of the brother hood of man to which I have a most powerful affinity with. Perhaps then, "there truly are no coincidences in life", as my dear old Friend would say during my study period in the Craft. These features in the United States Capital City, Washington D.C. are part of an attempt and tribute of respect to import influences from an ancient time on earth into the New World. Now truly, A New Order of Understanding Has Begun, with reference to this Ancient Wonder, because its secrets have at long last been decoded. It is time to step out of from the shadows and fear of the unknown and into the light and freedom that the truth and enlightenment offers in order "to be" at one, with the grand purpose of life and the workings of the Living Universe…the infinity sign on the book cover speaks for itself.

To carry on with the Ancient Royal Cubit Theorem Applications and Research:

Stonehenge: A timely visit to Stonehenge in the process tells us the radius of the Sarson Circle is approximately 108 feet. Following are some deductions generated on this delightful topic, based on the Royal Cubit Theorem:

108 divided by 1.732 = 62.355658 (hmmm, you look kind of familiar)
The 108 feet is given as an approximate diameter, therefore the end quotient is also an approximation but there is a number in the Great Pyramid true dimensions discussion further back that is close to this value. It is close enough for an investigation. Therefore, another hunch or intuitive thought will be played out. To begin with the correct diameter of this circle is needed. As I recall 8800 divided by 100 were the adjacent and opposite sides of a right angled triangle which produced a hypotenuse value, divided by 2 that became a radius of 62.2254 which when multiplied by 2Pi provided a circumference of 390.962.

The square root of (88.00 squared x 2) = 124.45079,
divided by 2 = radius 62.2254 x 2Pi = circumference 390.962.
Then there was the hypotenuse length of that right angle triangle from the Royal Cubit Theorem drawing with an adjacent side of 1.0 and opposite side of Phi squared = 1.07044, or the sq. rt. Of 3, divided by 1+Phi. Hence 390.962, divided by 1.07044 = 365.24 days in a year and the true diameter of Sarson Circle at Stonehenge becomes known, 62.2254 x 1.732 = 107.774 ft., Wow!

Here is another familiar looking number. Refer back to chapter 3 where it was determined the dimension of the Great Pyramid was originally 1077.74 feet. As it turns out the diameter of the Sarson Circle is by no coincidence precisely 10%, or one tenth of this value, almost too fantastic to believe but there it is before us.

* Since this is the case, 107.774, divided by 2 = the radius 53.887 x 2 Pi = 338.572, divided by 0.618 x 1.5 = 365.24 days in a year.

Another way to compute this value using the Royal Cubit Theorem is: 124.45079 x 0.0866 x 10 = 107.774. With this dimension evaluated the circumference of the circle is: 107.774 divided by 2 = 53.88719 x 2Pi = 338.58279, divided by the 30 upright lintel support blocks around the perimeter = 11.286093 and this arc subtended by the radius is exactly a 12 degree portion of the circle. Also, 338.58279, divided by 360 = 0.090355 x 24 = 24.09040355, divided into 8800 = 365.24 days in a year. It is also timely to mention in the process that when the distance from the Great Pyramid to Stonehenge becomes known another revelation is at hand.

Distance: 2,232.97 miles x 5,280 = 11,790,081.6 feet,
divided by 1.732 = 6,807,206.467 Royal Cubits,
divided by 0.0866 = 78,605,155.5 (squared) = 6.18 E 15,
divided by 10,000,000,000,000,000 = *Phi.

There doesn't appear to be any coincidence that the location of Stonehenge is linked to the Great Pyramid in terms of the Golden Section. The archaeologists have found evidence of a large human population in the area which indicates that the site might have been a very important center in that ancient time for reasons unknown today. The above values are familiar to us by now because they are seen in the previous computations for the Great Pyramid true dimensions write-up. In this we have the basis for yet another proof of the Royal Cubit Theorem at a distant ancient site location and it shows that the measurement system employed for Stonehenge and the Great

Pyramid were the same, through the square root of 3 relationship expressed in the Royal Cubit Theorem and it also indicates that these structures were part of a grand ancient plan that we don't fully understand. We know that both of these installations were used to keep accurate track of the time and that their precise latitudes and longitudes were known in ancient times. An approximation of the lintel support block widths by others is estimated to be 3.2 m. divided by 0.3048 = 10.50 ft. Then it is determined 11.286 - 10.50 = 0.786 which is the square root of Phi. Therefore, it is a very decent estimate because it jives with the Golden Ratio. Stonehenge is determined to be about the same vintage as the Great Pyramid and was developed in phases over future years. Here again is an ancient stone structure that has been effected by the erosion factor which made it difficult for the researchers to determine the true diameter of this circle. Only seventeen of the lintel support blocks are remaining and those that are left are in poor condition due to erosion. Most interesting is the fact that the dimensions used for Stonehenge are mathematically related to those of the true dimensions of the Great Pyramid. Another comparison is: the Sarson Circle circumference 338.58279 divided into the base perimeter of the Great Pyramid 3048.32 = 9.00 precisely. The diameter 107.774 x 1.732 = 186.66 Royal Cubits, another nice looking number. When half this value is multiplied by 2 Pi, then divided into the base perimeter of 440 x 4 = 1760 the quotient = 3.0, and the updated base dimension of the Great Pyramid perimeter, 3048.32 divided by 107.774 = 28.284, divided by 20 = sq. rt. 2. Then, 762.08 divided by 107.774 = 7.071, divided by 10 = 0.7071, divided into 1.0 = the square root of 2. Also of significance, when the radius of the Sarson circle is divided into half the diagonal dimension of the Great Pyramid and the quotient is divided by the square root of 5 the result is Phi, divided by 2 + 1.0, divided by 10 = Phi divided by 2. In terms of the numbers: 538.872, divided by 58.887 = 9.151, divided by 2.236 = 4.09 − 1.0 = 3.09, divided by 10 = 0.309 x 2 = 0.618 = Phi. The hope is this type of input will serve as an assist to the dedicated researchers in order that they gain a better understanding of its true purpose and workings. There is no reason for Stonehenge not to be included as an ancient wonder of the world, as it has also captured the imaginations of so many minds around the world. Nothing will be gained by singling out Newton as a misguided genius, it is time to move ahead, and I dare say, as a result of this inquiry, we are already well on the way. It will be proven there are other ancient sites across the ocean to the west of Giza which were based on this ancient measuring system.

Teotihuacan, Mexico

Another interesting visit to the City of the Gods at Teotihuacan, Mexico in a distant land from Giza on the west side of the Atlantic Ocean provides observations and deductions which become a revelation. There at a distant site from Egypt on the east side of the Atlantic Ocean is the Pyramid of the Sun which the archaeologists estimate to have a base dimension of 738.24 feet but the erosion factor was never considered where heavy rainfall in the area has had a large effect on a stone structure such as this. The object is to determine the original dimensions by approximating the erosion factor which would be higher than 1.0076 the one used in the case of the Great Pyramid research. To make the point and taking an intuitive shortcut 738.24 x 1.01593% = 750 feet. To help understand why it is called the Pyramid of the Sun the following computation is put forward: Its height 750 feet, divided by *5,280 = 0.142, divided by 2 = 0.0710 x 1+Phi = 0.1149 the slope distance in miles, divided by the AU 91,796,000 x 100,000,000 = 0.125, reciprocal = 8.0 the number of Infinity. The same number that describes the never ending expanse of the Universe is provided when the perimeter of the Pyramid of the Sun 750 x 4 = 3,000, divided by that of the Great Pyramid

3048.32 = 0.984, square root of which = 0.992, reciprocal = 1.008 – 1.0 = 0.008 x 1000 = *8.0. There are no coincidences it would appear.

There is a substantial difference of 11.8 feet between the actual base length of this pyramid and its existing dimension compared to the one of 5.72 feet at the Great Pyramid. Everything falls into place once it is understood Egypt had a much dryer climate and the erosion factor was less. So far the results of this study are proving to be conclusive.

To the north west of this site is the Pyramid of the Moon which we are told has a base length of 698.87 feet x 1.01593% using the above mentioned erosion factor = 710 feet its original dimension. A remarkable feature provided by this pyramid is when the distance from earth to the moon is multiplied by 2/3 1+Phi, divided by 710 = 365.24 days in a year. Another interesting feature to take note of is the square root of 1460 the sum of 750 + 710, the base lengths of the Sun and Moon Pyramids, divided by 100 = 0.382, the square root of which = Phi, reciprocal = 1+Phi. The orbit circumference of earth around the sun is Phi x 1,000,000,000 = 618,000,000 miles and the orbit perimeter of the moon around earth is 1+Phi x 1,000,000 = 1,618,000 miles. Also, its perimeter 4 x 710 = 2840, divided by that of the Great Pyramid = 0.931661, square root of = 0.965225, reciprocal = 1.036 – 1.0 = 0.036 x 100,000 = 3600 seconds in an earth hour. It is noted that the ground cover at these sites could only be eroded stone which has become sand.

Teotihuacan is approximately 9,000 miles of arc distance on the earth's surface from the Great Pyramid and is directly on the Ley Lines described in the drawing titled Giza Phi Connection. I don't know why this site is not mentioned on that drawing when it should be. I think this is because records get lost and misplaced, jumbled or the connection was not made before. The distance divided into earth's circumference provides a value very close to 4.5 x Phi. This ancient site was a big as Rome and it once bustled with life housing more than 250,000 thousand inhabitants when trade, commerce and religious ceremony was brisk. No doubt, in an earlier time there was some sort of communication between Giza and Teotihuacan based on a science we don't understand. A mica enriched powder was found at the top of the Pyramid of the Sun and that is a strong indication of this deduction. There is no mention of anything like Sarcophagus's in these pyramids but there might have been at one time. Eventually the site was abandoned for reasons not clear to us by historical accounts. I believe there is a misconception the site is only around 1500 years old and if the archeological digging went deeper and the appraisals were fine tuned it would be found to be the same vintage as the Great Pyramid at Giza.

Another stop of interest along the way is at the ancient capital of Xian in China where the climate has a history of being more or less moderate with rainy summers. The Archeologists report there are 90 to 100 ancient pyramids in the area including one called the Great White Pyramid. The basic measurements reveal the base length to be roughly 1500 feet with a height of approximately 1000 feet. First of all the climate wasn't as dry as Egypt and not as humid as in Mexico where those other ancient pyramids are situated. An estimate on the erosion factor after running some numbers through comes in at 0.984148646%, reciprocal = 1.016106667%. Therefore 1500 x 1.016106667% = *1524.16 feet, divided by 1.732 = 880 Royal Cubits and Bingo! The Great White Pyramid was exactly twice the size of the Great Pyramid and we are left to wonder if it once had a cap stone and there are chambers and passageways within it. Also it is just slightly larger than the Pyramid of the Sun in Teotihuacan.. We might ask then, what is the real story on this topic, are there other ancient sites on this planet that are as sophisticated as the one at Giza? Unfortunately the Chinese authorities will not allow foreigners to enter their country to have a good look around at these sites and this is rather a shame. Ancient and gigantic pyramids have been recently discovered in the Ukraine and others have been found on the other continents throughout the world. The problem is, insufficient data on their dimensions hampers the research, therefore the Royal Cubit Theorem cannot be applied to them but the odds are, according to the above findings, it will suit them very well. Hopefully, in time this information will become available. We absolutely need to know more about the science behind these undertakings.

The above is more than very interesting, it is rather astounding and fascinating at the same time. As stated, either the ancients were from other worlds or they were from a time zone of great sophistication with a people on earth we don't know about. Whatever the case might be there is a strong indication that world travel was a reality in a remote time in world history and the parties or intelligences who constructed the works at Giza, Stone Henge, the City of the Gods, China and many other places in the ancient world were indeed using the same measuring system. I hereby make a motion that all other ancient sites in the world be investigated using the Royal Cubit Theorem along with intelligent evaluation of the area erosion factors for each site because, as we know, the lack of this consideration has been hindering our progress for far too long.

Getting back to Stonehenge for the moment, it is one of approximately 180 sacred sites considered to be situated along the Ley Lines in the British Isles. There are such sites all through the rest of Europe and around the world, many of which have not been discovered yet or are inaccessible beneath the ocean's depths, hidden in the jungles, or are obliterated due to land movement, or urban expansion etc, but in the British Isles there seems to be a rather large concentration of them and it isn't known why. It can only be speculated it might have had something to do with Atlantis. I have always wondered what influence the energies, if indeed there are any, from these sites might have had, or are presently having on the people who live there and on those who's ancestry is from there but have relocated to other parts of the world over the past, more recent centuries. It must occur that perhaps by coincidence or by some other means the sun rises and sets on the British Common Wealth twenty four and seven. From this tiny area of the earth its leaders and people developed and manifested a vast empire and its language is spoken in every corner of the globe. Evidently, to say the least, that is the way it has turned out and it is a mystery how this came about. Though there could be something to such a notion, we don't really understand it, making it is difficult to prove and it has nothing to do with race, creed or religion. Never the less, the possibility is there and this thought has often lingered in my mind for some reason because my heritage is Welsh and Irish going back a couple of centuries. I am a North American citizen it might be said, with a blood tie to peoples of the British Isles but my real heritage, like everyone else in the world, could only somehow belong to the workings of the stars in the Cosmos. This appears as an invitation to say we are not from here and our origins are from elsewhere in the Universe. This is not necessarily impossible to prove.

The ancient world has not only left its mark , it is also still a powerful influence on us today, and it has a force of its own to contend with in this time period. The Great Pyramid, Stonehenge and many other ancient sites from that era are still with us. This is why efforts should be made to more fully understand them and carry on with the research into their many mysteries. The benefits are ours if we do so. What has been determined by the Royal Cubit Theorem and the true dimensions of the Great Pyramid is a prime example of such. If Pythagorus, Archimedes or Euclid, to name a few, had come up with it 2500 years ago, or if it had never been lost in the first place it would have done us all a very large favor, and for certain personages such as Galileo, Bernoulli, Einstein and Tesla, who are just a few names on a long list of notable scientists, I am quite sure they would have made good use of this Theorem. All be it said, I believe it has a future now.

Back to the Gag, Grind and Groan in the Present Time Zone

Work is in progress on the follow up to Living with Geometry, but first I need input from the readers so that I know we are on the same page with regards to coming to an understanding the Language of the Universe and that we are in agreement with what it has done for us. Alas, the above enticing subject areas and intriguing adventures of the mind in chapters 3 and 4 will be put to the side for now as the humdrum, partial tale that our historians have been telling us carries on. Eventually, we are told, the extra palm was dropped off the body parts version of the Royal Cubit, or it could be said it was shortened to something close to 18 imperial inches, the forearm to finger distance, and worked with to that standard in a later period. This could have been about the time Egypt in all its splendor and sophisticated applications in the ancient sciences was going into decline after the real Royal Cubit and its sacred qualities were lost. It was no coincidence that in time a double cubit or 2 x 18 inches = 36 inches became a yard in the imperial system, one third of which is one foot of course, however there are differences of opinion on this. Some believe the yard is derived from the length of a mans stride or pace while another postulate is it is the girth of a mans waist while others claim the yard was invented by Henry the 1 st. of England as being the distance from the tip of his nose to the end of his thumb. This is what we were told in grade school and since then I have determined the education system never was the most reliable source of information. A double cubit equaling a yard makes the most sense and in the case of the metre being a bogus value our heritage and link to a natural measurement system of units based on earth dimensions from an ancient and glorious time in world history is lost by using the metric system. I treat this system the same way I deal with those bogus pyramid inches by converting them to feet. And here again, I make the claim that the foot, yard and mile existed without being specifically named as such but never the less were part of the ancient measuring system a very long time ago, otherwise it would not have been possible to make the above and following agreeable computations. As claimed previously, the square root of 3, or the true royal cubit, or 20 x 0.0866 true pyramid inches, one ten millionth part of the sun's diameter = 1.732 are indirectly based on one foot. Now we are stuck with 1.0 yard = 0.9144 metres or 1.0 metre = 3.2808398 feet using the conversion factor of 0.3048. I can handle it, my best friend at work is a conversion calculator but I don't really appreciate it, though I know my complaints fall on deaf ears. It became interesting and somewhat challenging when a project to work on was originally surveyed in chains and links, then the conversion from this measurement system had to be equated to feet in the imperial system then to metres in the metric system. While working in the imperial system it was common practice to convert from inches to tenths of a foot and vice versa using the 0.08333 factor and it was all just numbers to me anyway, but I truly do resent the metric system. It is a measuring system that doesn't make a lot of sense and it didn't have to happen that way. Very often we get plans that show the site details in the metric system but the building dimensions are yet in the imperial system because the Canadian sawmills didn't convert to metric due to the trade system with the USA which remained in the imperial system.

For interest, the history behind the Gunter's chain which equals 66 feet or 22 yards was the measuring device used by the surveyors in England after the disastrous fire of London took place. The buildings were too close together and that is why the fire damage was excessive. From that time until today, at least in British Commonwealth countries and in the United States, the road allowances are a minimum of one Gunter's chain, 66 feet or 20 metres now, which equals 65.617 feet. All buildings now must be set to fire code setbacks by-law from road allowances and lot lines between individual buildings on their assigned lots. In the cities where there is zero building setback the fire code is beefed up but there are serious problems whenever there is a fire. Whatever the units might be, in survey work, to measure is to know descriptively and proportionately, and its proper practices often make the difference between chaos and order. When I look back on all the years I used a 100 foot measuring chain for survey work I didn't realize that by dividing this value by 0.0866 = 1154.7344 x 100 it yielded the diameter of the Milky Way and before that, those who used the 66 foot chain could have divided that value by the square root of 3 to come up with Phi Squared, yet the answers to these important questions were never very far away.

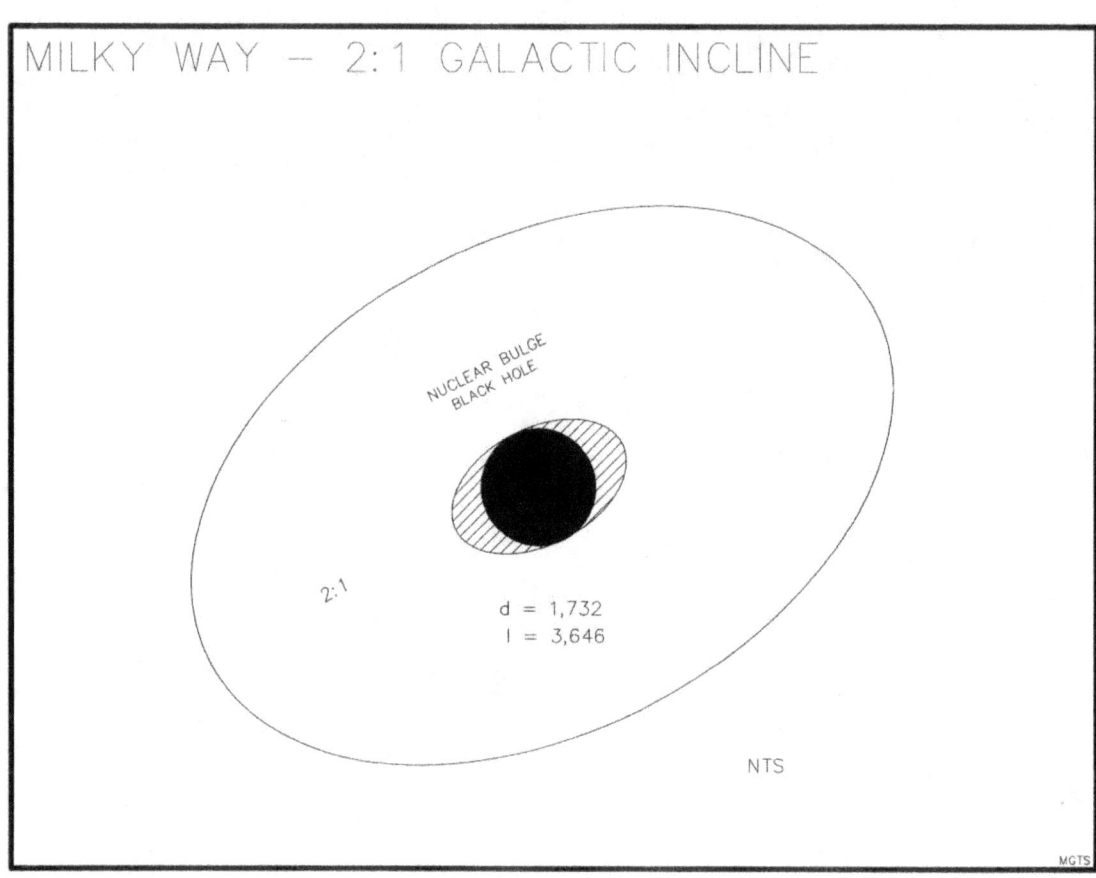

MILKY WAY — 2:1 GALACTIC INCLINE

NUCLEAR BULGE
BLACK HOLE

2:1

d = 1,732
l = 3,646

NTS

MGTS

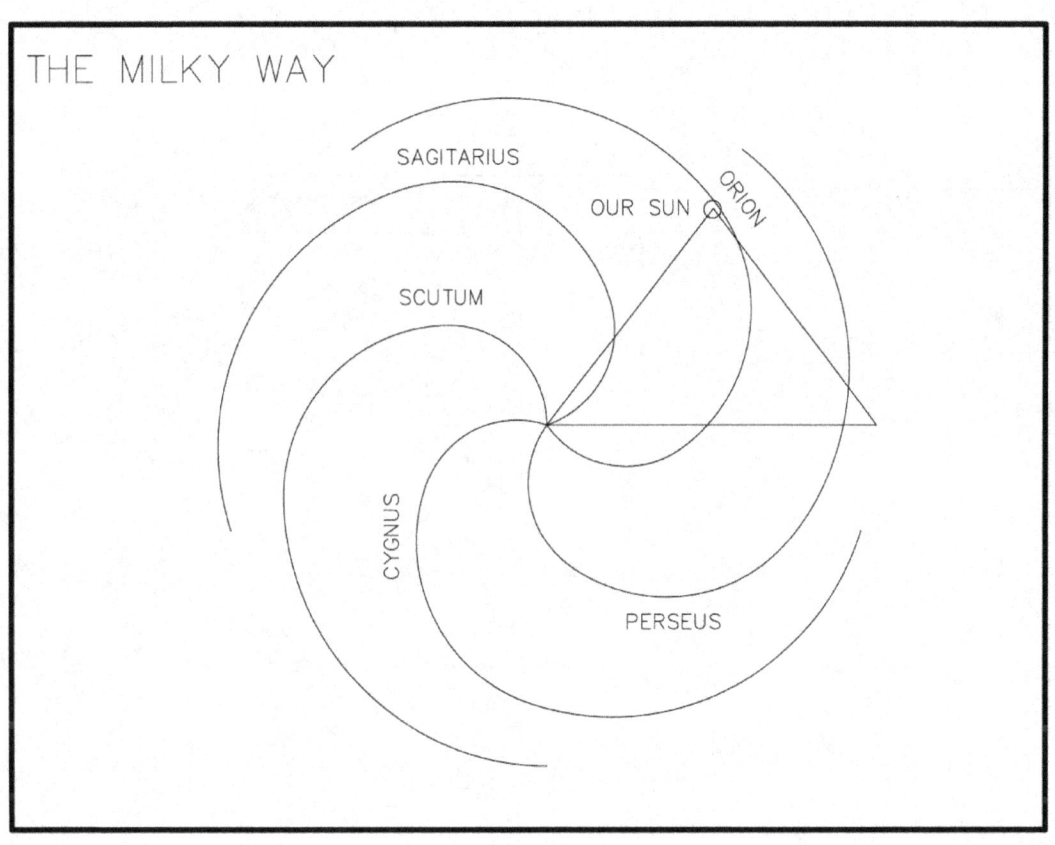

THE MILKY WAY

SAGITARIUS

OUR SUN   ORION

SCUTUM

CYGNUS

PERSEUS

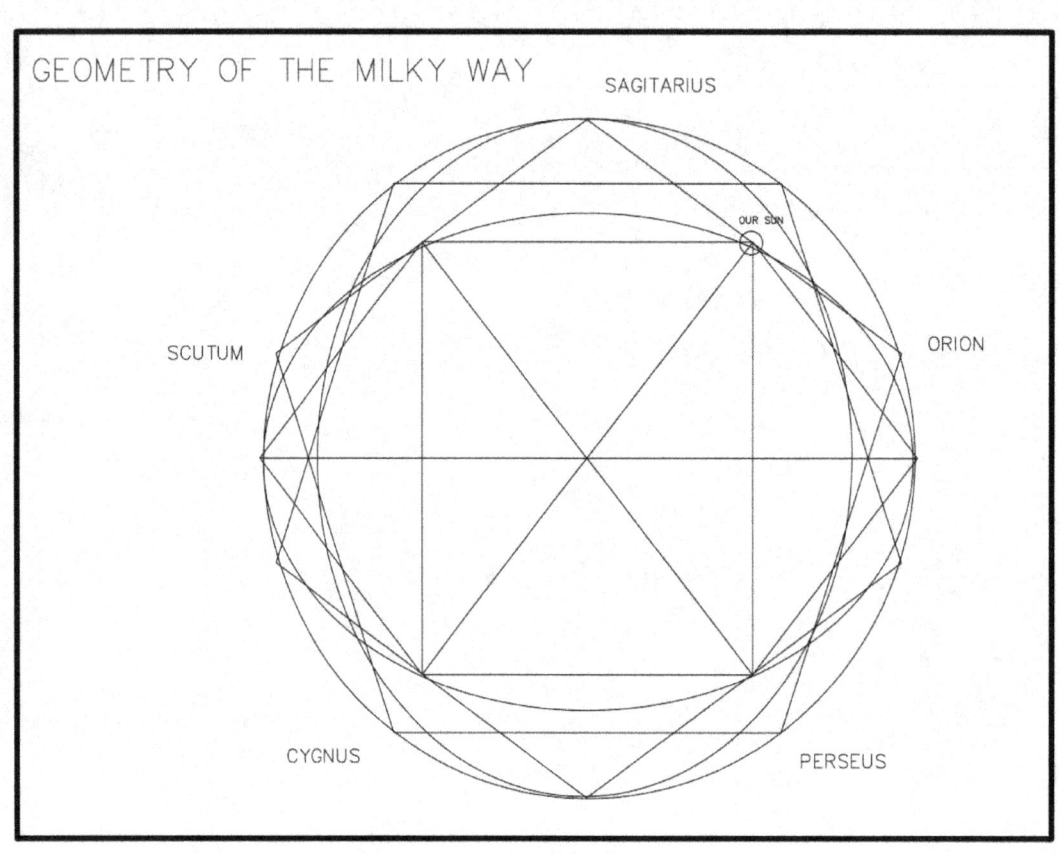

GEOMETRY OF THE MILKY WAY

SAGITARIUS

OUR SUN

SCUTUM

ORION

CYGNUS

PERSEUS

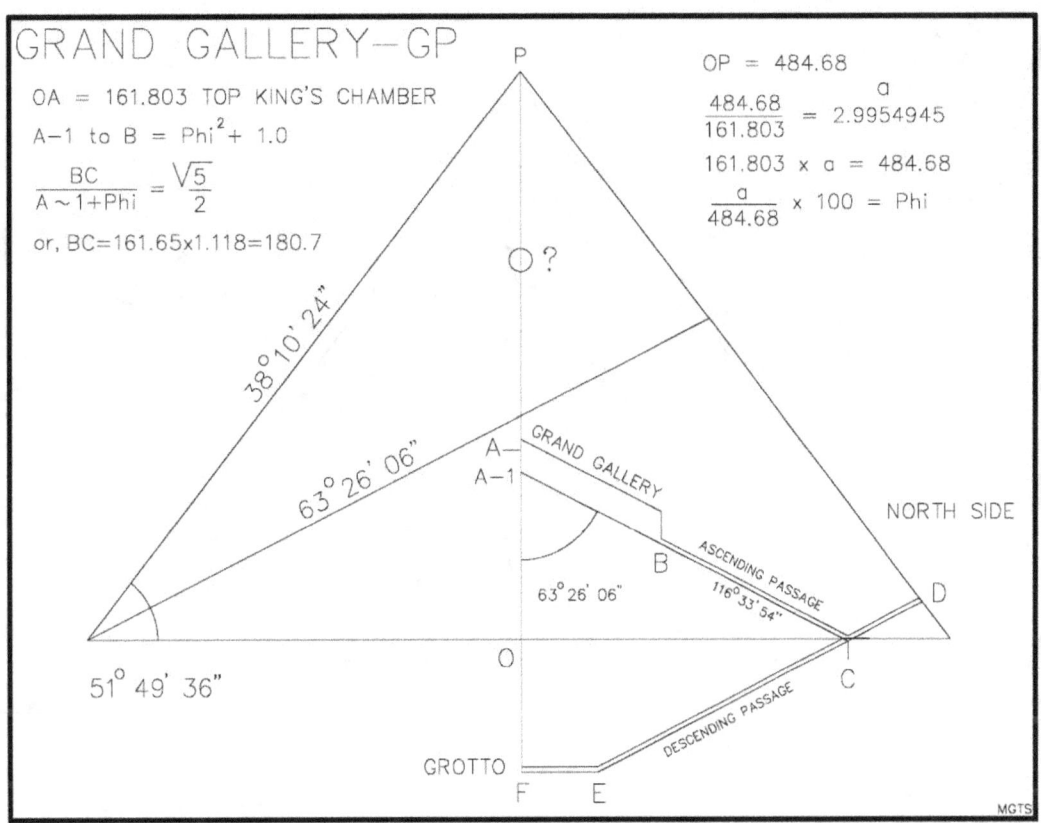

GRAND GALLERY—GP

OA = 161.803 TOP KING'S CHAMBER

A−1 to B = Phi^2+ 1.0

$$\frac{BC}{A \sim 1+Phi} = \frac{\sqrt{5}}{2}$$

or, BC=161.65x1.118=180.7

OP = 484.68

$$\frac{484.68}{161.803} = 2.9954945$$

161.803 x a = 484.68

$$\frac{a}{484.68} \times 100 = Phi$$

P

O ?

38° 10' 24"

63° 26' 06"

51° 49' 36"

A—
A−1   GRAND GALLERY

B

63° 26' 06"

ASCENDING PASSAGE

116° 33' 54"

NORTH SIDE

D

O

C

DESCENDING PASSAGE

GROTTO

F   E

MGTS

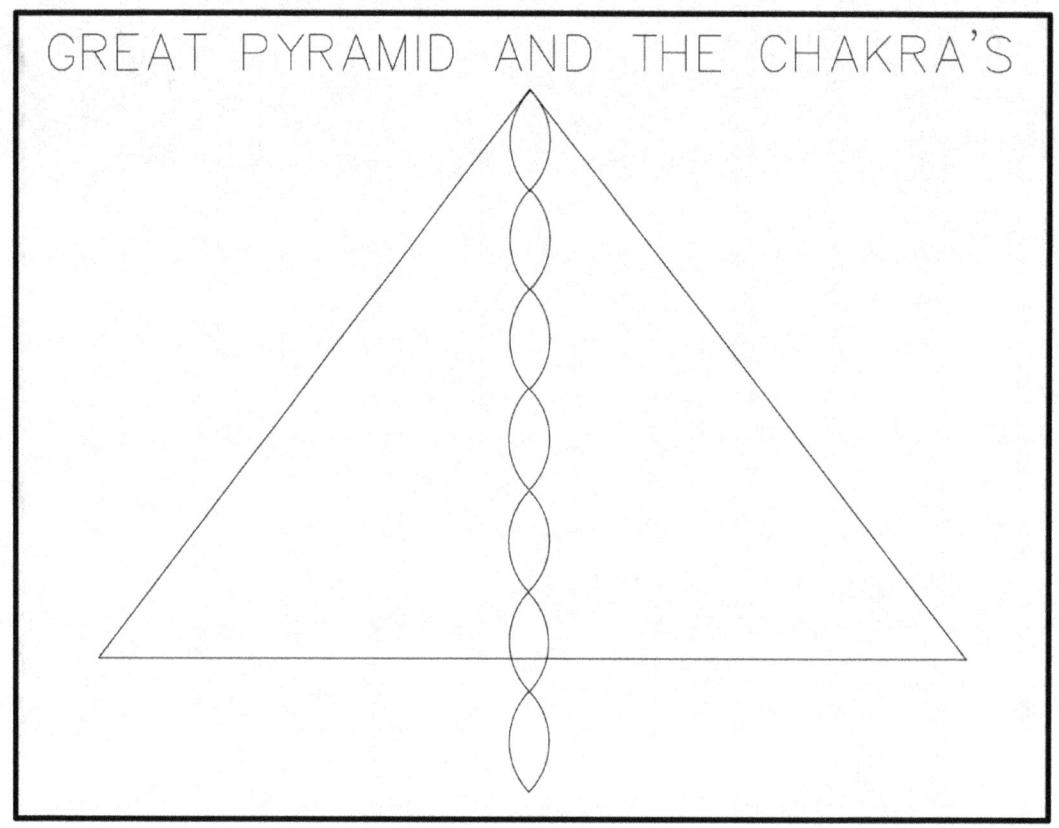

GREAT PYRAMID AND THE CHAKRA'S

MILKY WAY PYRAMID CRYSTAL

GLUE TAB (typ)

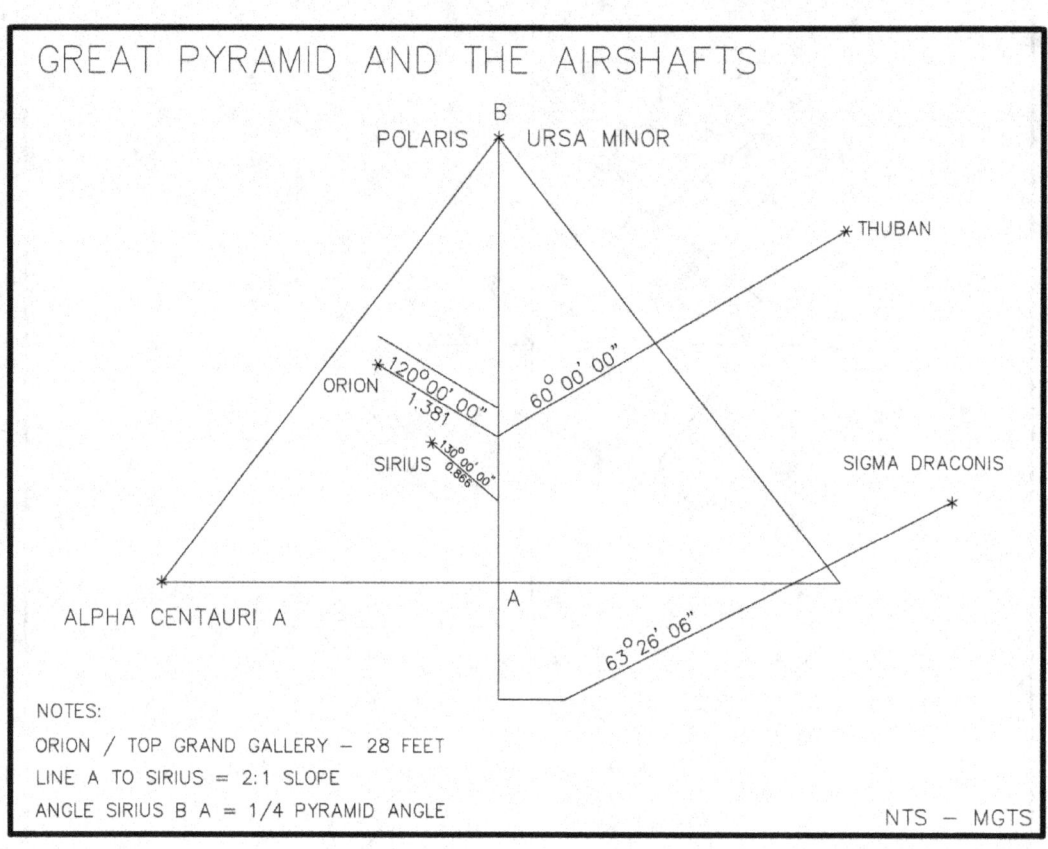

GREAT PYRAMID AND THE AIRSHAFTS

B
POLARIS ✶ URSA MINOR

✶ THUBAN

ORION ✶ 120° 00' 00"
1.381

60° 00' 00"

SIRIUS ✶ 130° 00' 00"
0.866

SIGMA DRACONIS
✶

ALPHA CENTAURI A
✶

A

63° 26' 06"

NOTES:

ORION / TOP GRAND GALLERY — 28 FEET

LINE A TO SIRIUS = 2:1 SLOPE

ANGLE SIRIUS B A = 1/4 PYRAMID ANGLE

NTS — MGTS

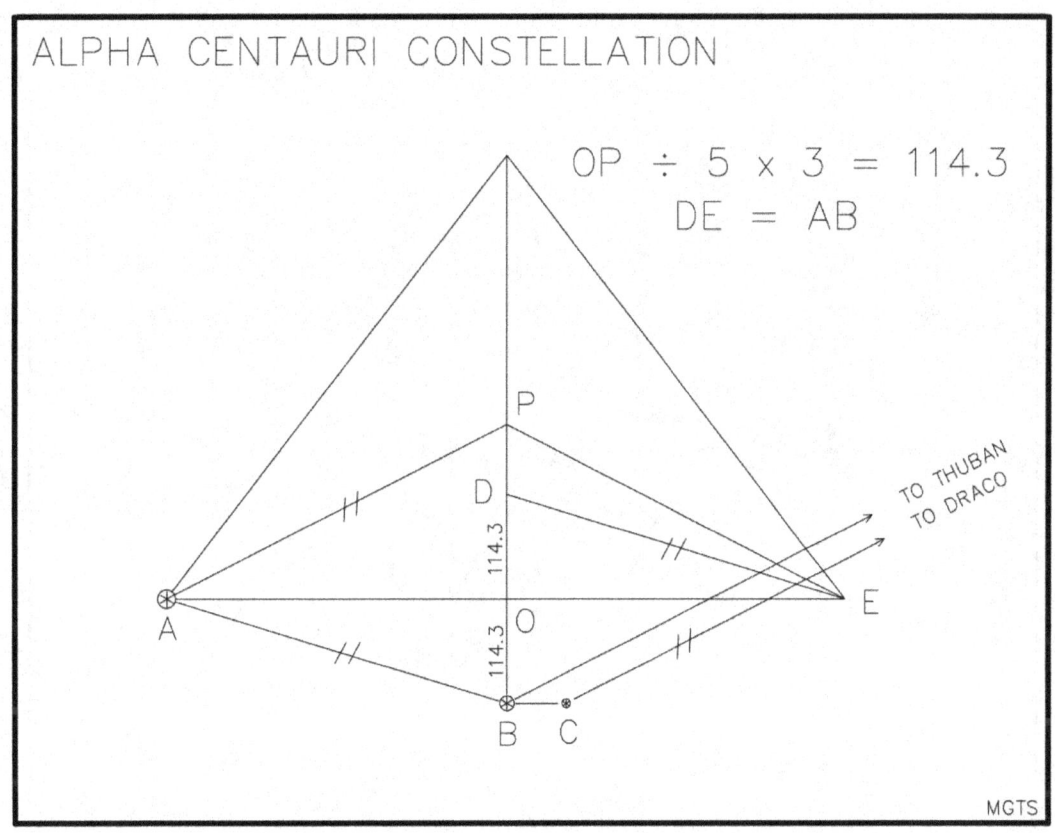

ALPHA CENTAURI CONSTELLATION

OP ÷ 5 × 3 = 114.3
DE = AB

TO THUBAN
TO DRACO

MGTS

CONSTELLATION OF ORION

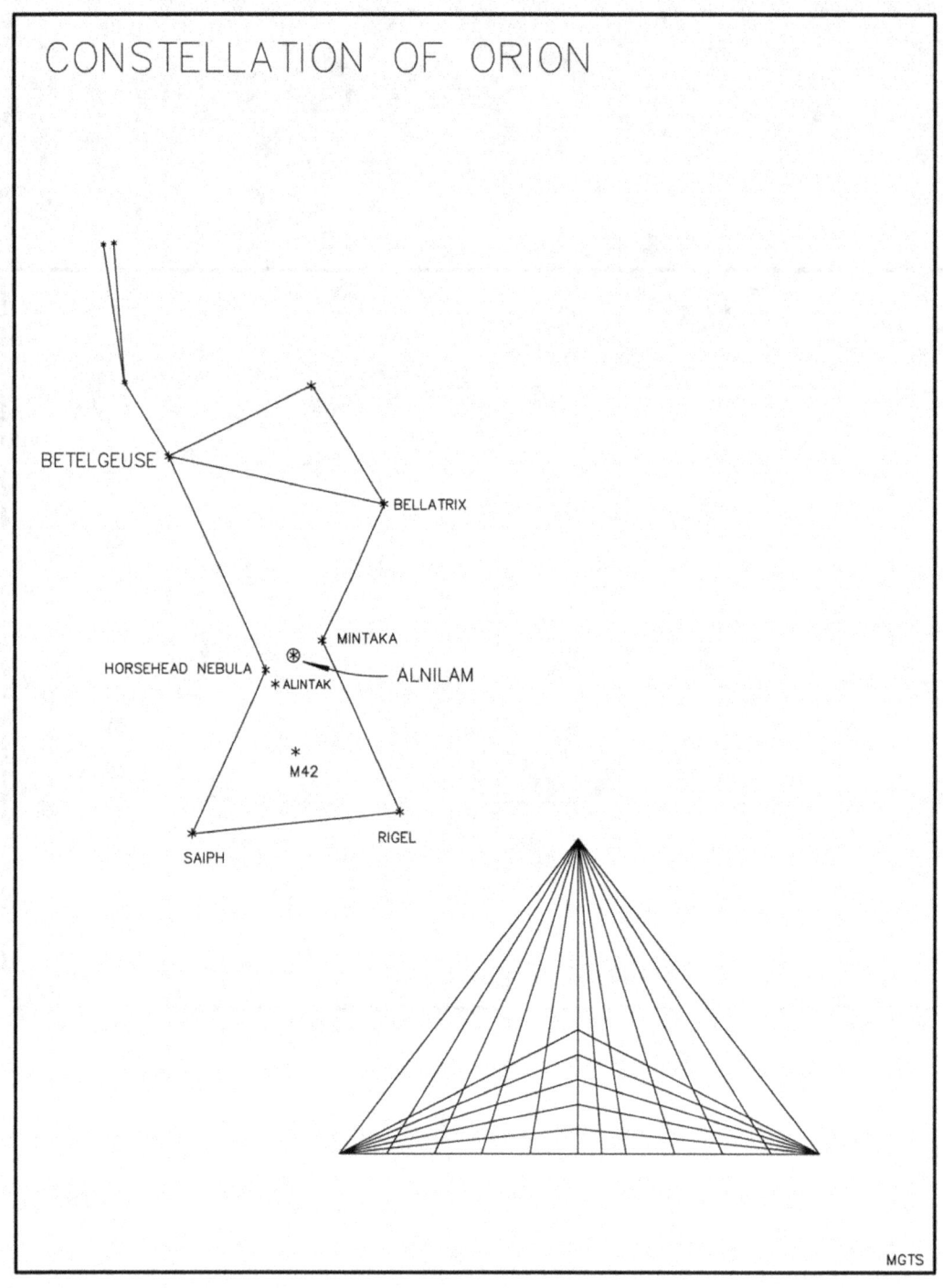

BETELGEUSE

BELLATRIX

MINTAKA

HORSEHEAD NEBULA

ALINTAK

ALNILAM

M42

SAIPH

RIGEL

MGTS

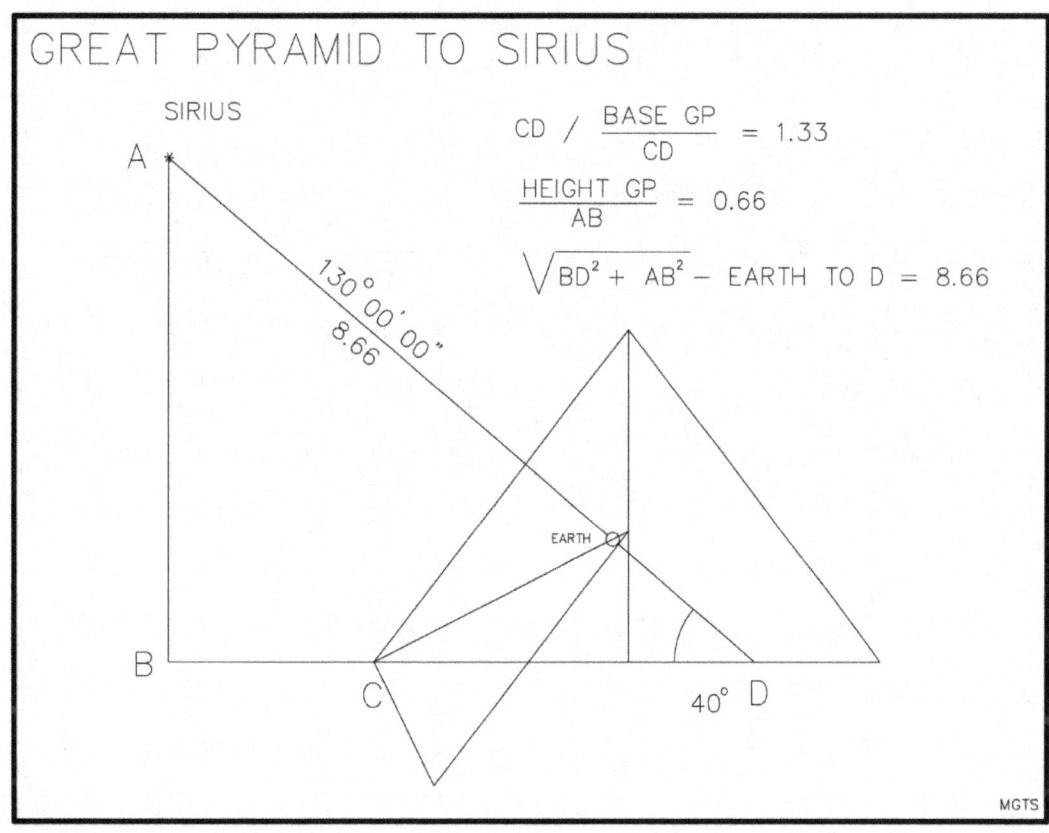

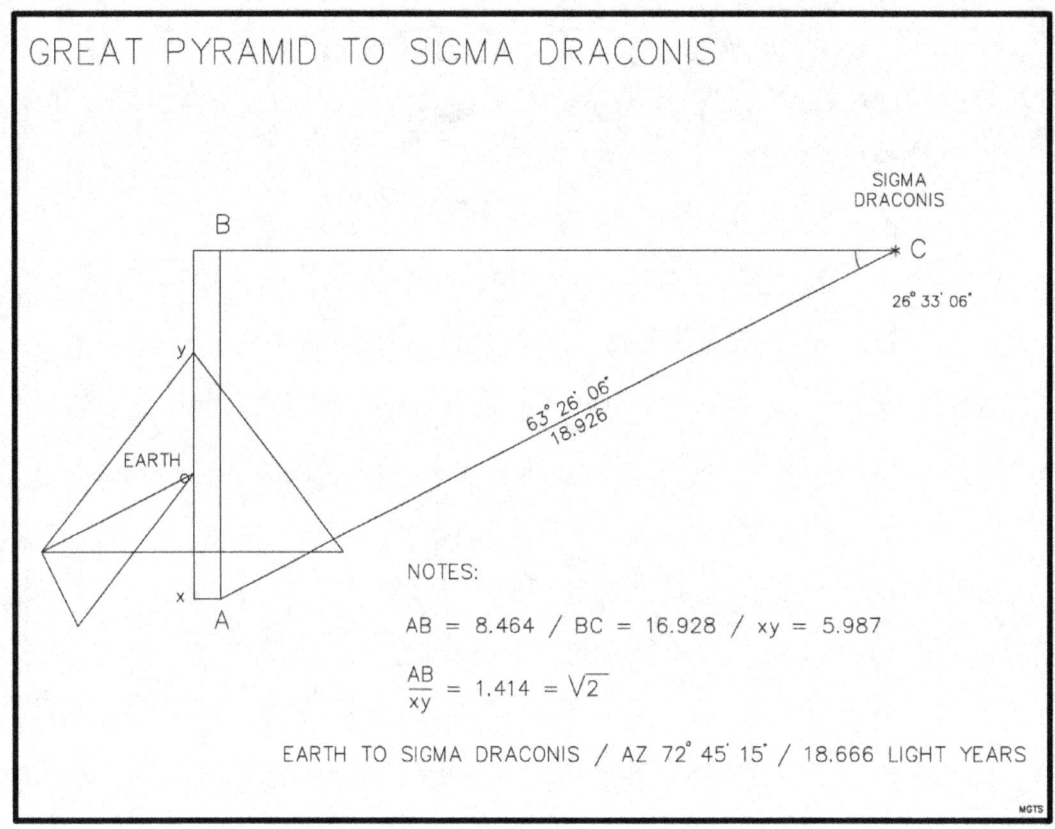

GREAT PYRAMID TO SIGMA DRACONIS

SIGMA
DRACONIS

B   * C

26° 33' 06"

y

63° 26' 06"
18.926

EARTH

NOTES:

x

A   AB = 8.464 / BC = 16.928 / xy = 5.987

$$\frac{AB}{xy} = 1.414 = \sqrt{2}$$

EARTH TO SIGMA DRACONIS / AZ 72° 45' 15" / 18.666 LIGHT YEARS

MGTS

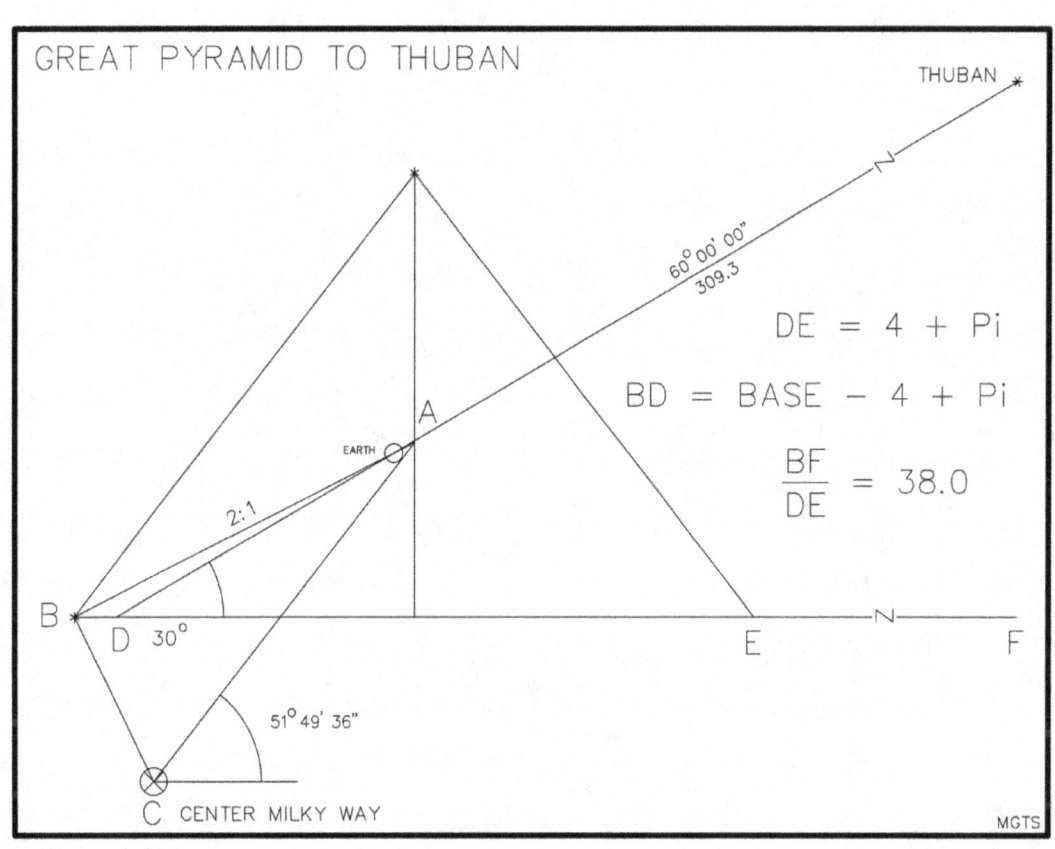

GREAT PYRAMID TO THUBAN

THUBAN

60° 00' 00"
309.3

DE = 4 + Pi

BD = BASE − 4 + Pi

$$\frac{BF}{DE} = 38.0$$

A

EARTH

2:1

B

D   30°

E

N

F

51° 49' 36"

C   CENTER MILKY WAY

MGTS

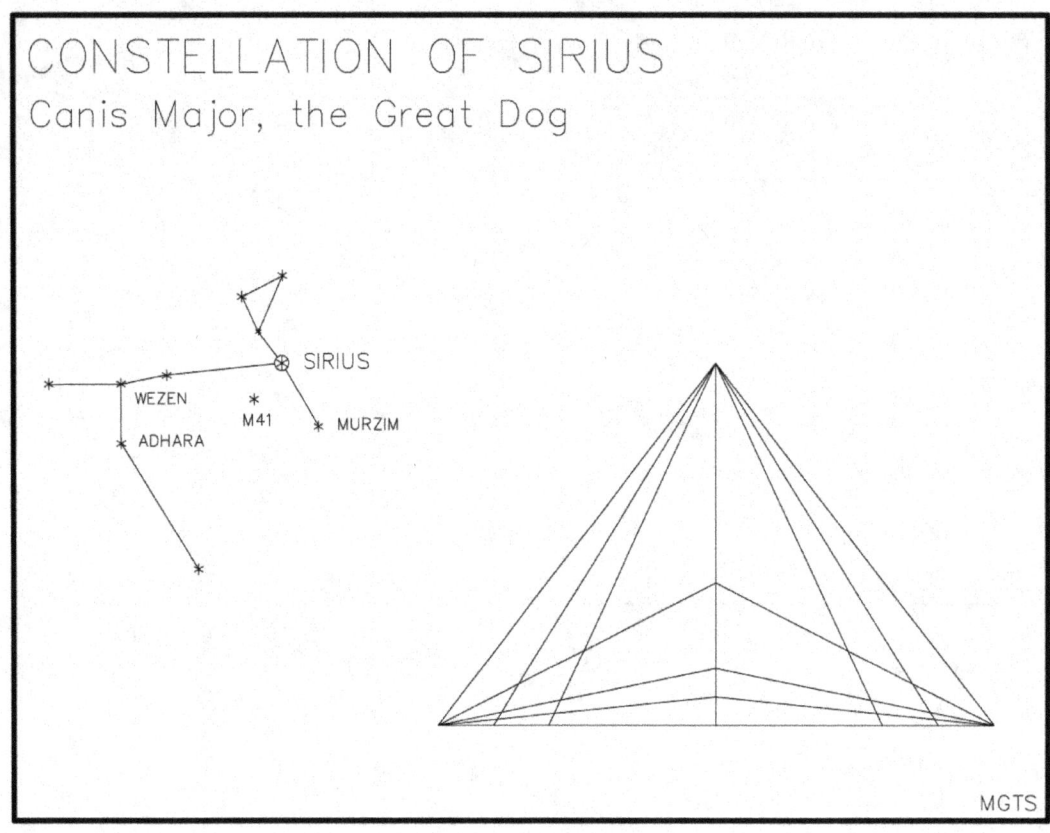

CONSTELLATION OF SIRIUS
Canis Major, the Great Dog

SIRIUS

WEZEN

M41

MURZIM

ADHARA

MGTS

CONSTELLATION OF ORION

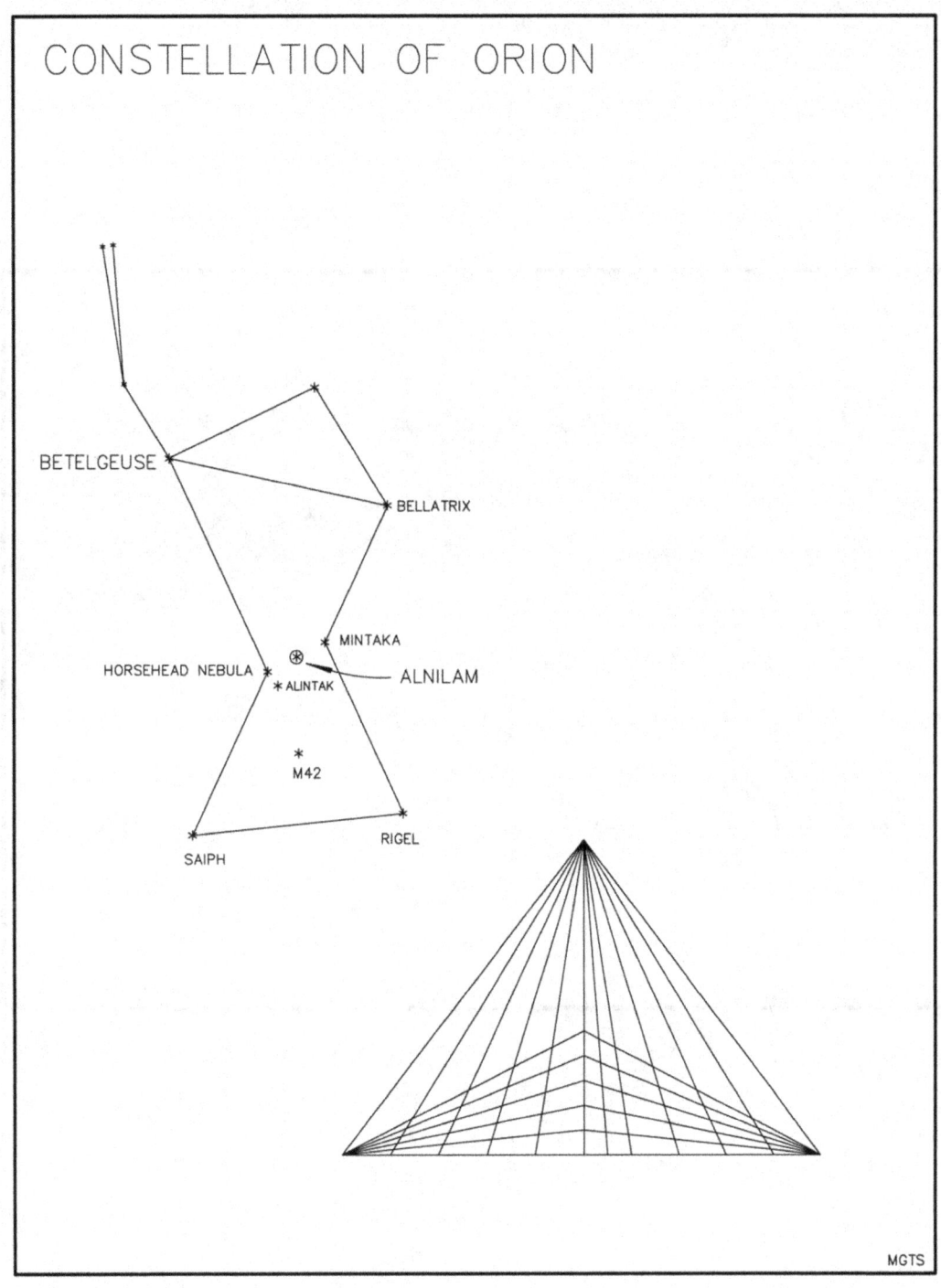

BETELGEUSE

BELLATRIX

MINTAKA

HORSEHEAD NEBULA

ALINTAK

ALNILAM

M42

SAIPH

RIGEL

MGTS

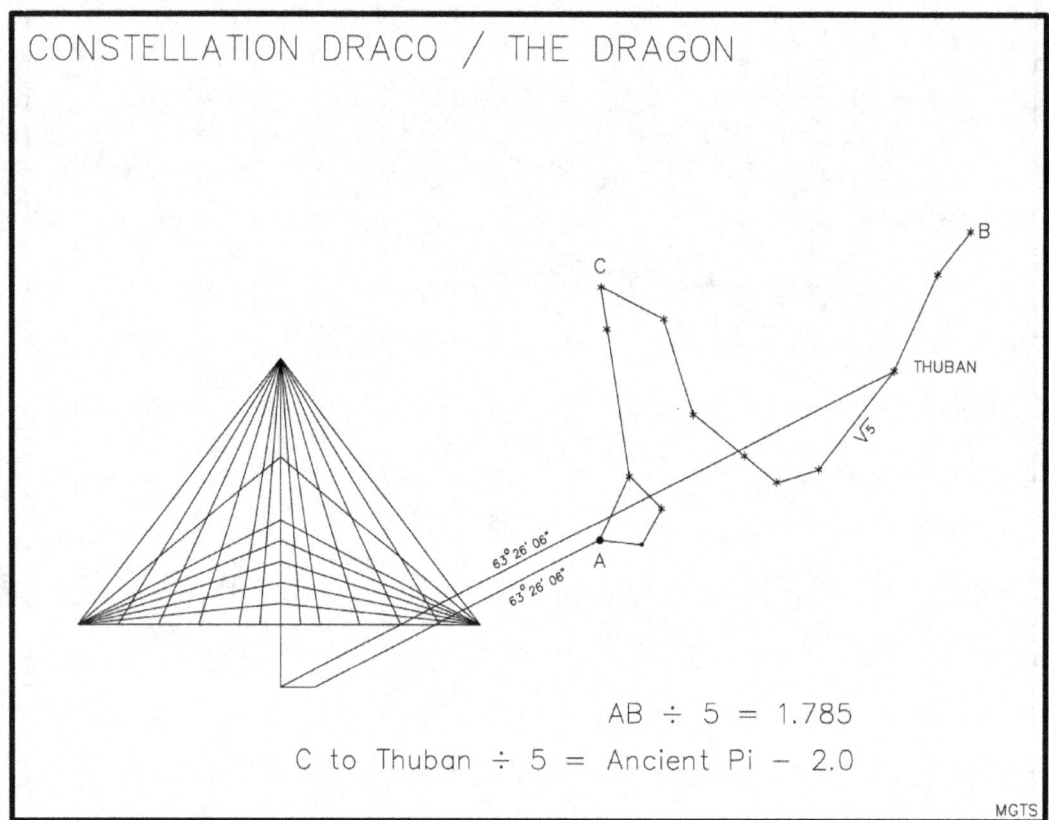

CONSTELLATION DRACO / THE DRAGON

AB ÷ 5 = 1.785

C to Thuban ÷ 5 = Ancient Pi − 2.0

MGTS

CASSIOPEIA'S CHAIR

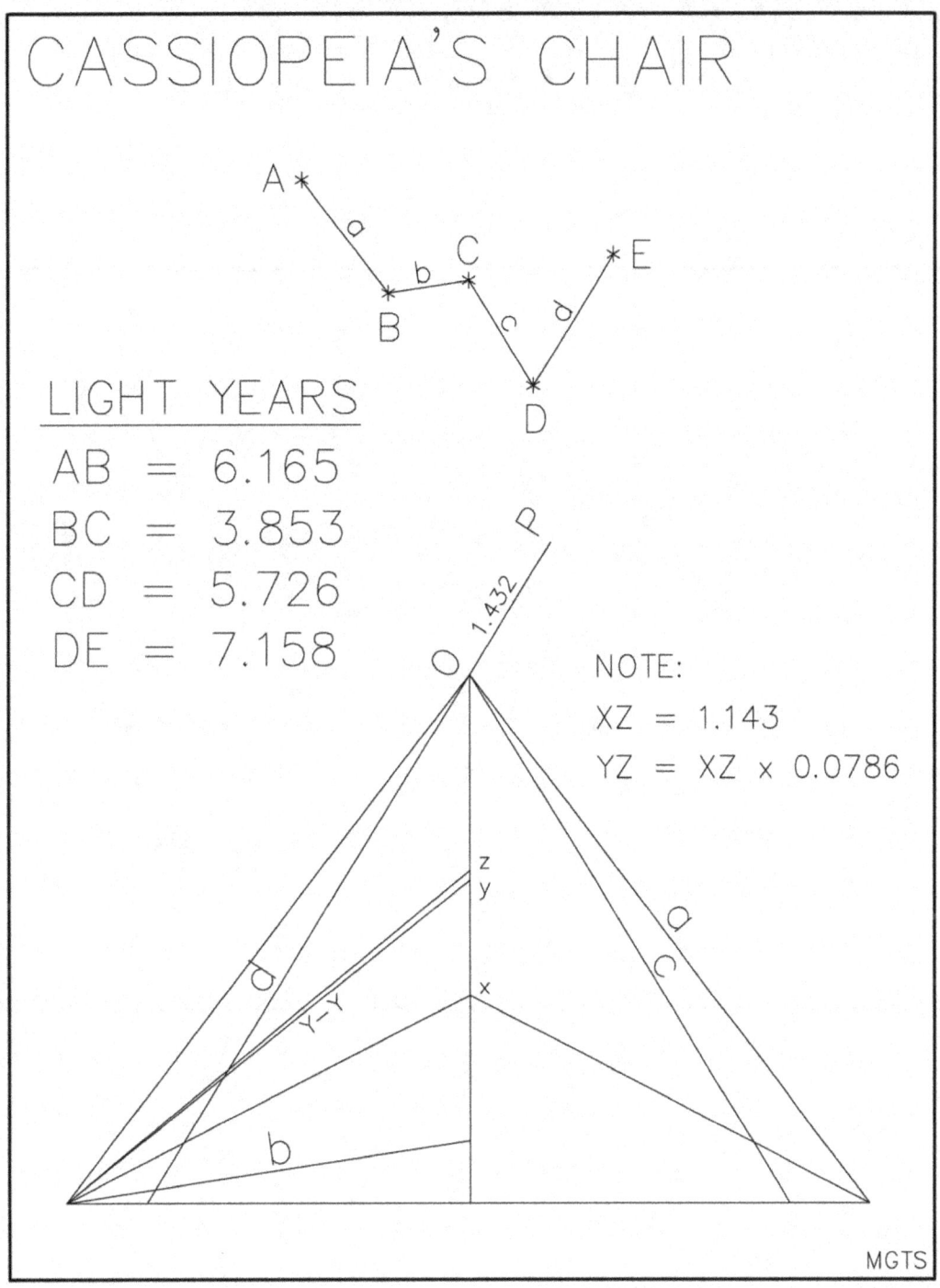

LIGHT YEARS

AB = 6.165

BC = 3.853

CD = 5.726

DE = 7.158

NOTE:

XZ = 1.143

YZ = XZ × 0.0786

MGTS

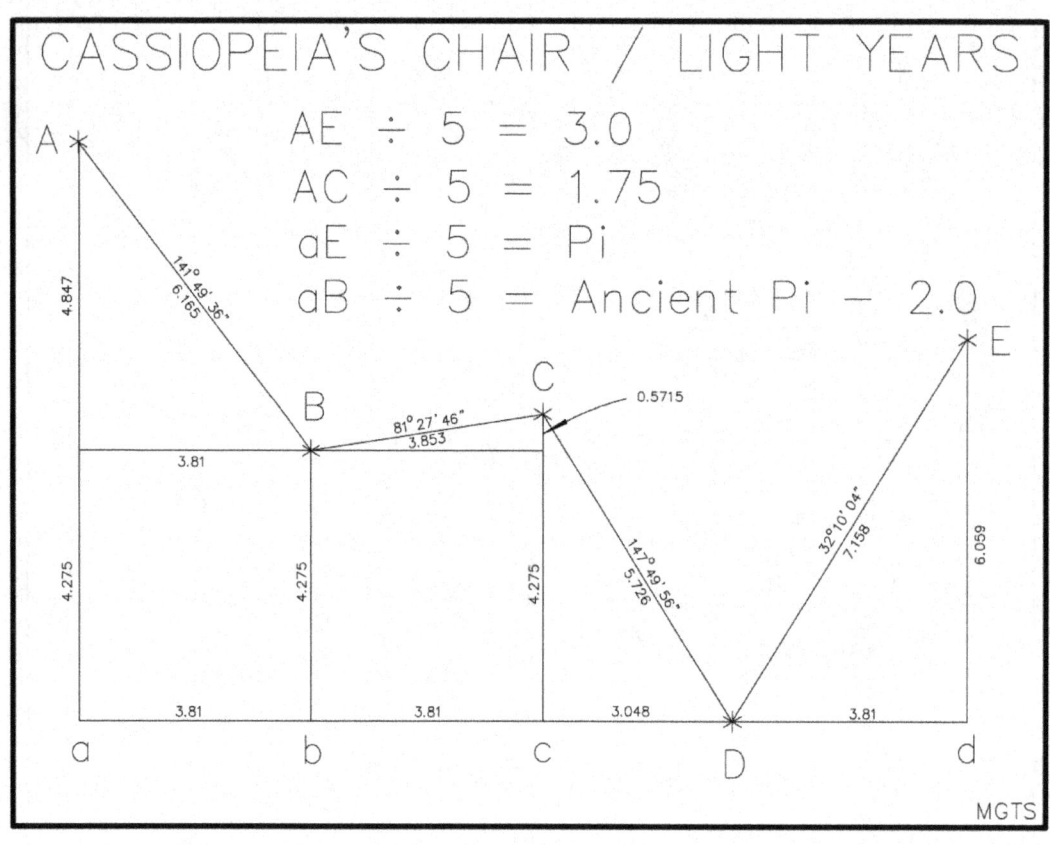

CASSIOPEIA'S CHAIR / LIGHT YEARS

AE ÷ 5 = 3.0
AC ÷ 5 = 1.75
aE ÷ 5 = Pi
aB ÷ 5 = Ancient Pi − 2.0

MGTS

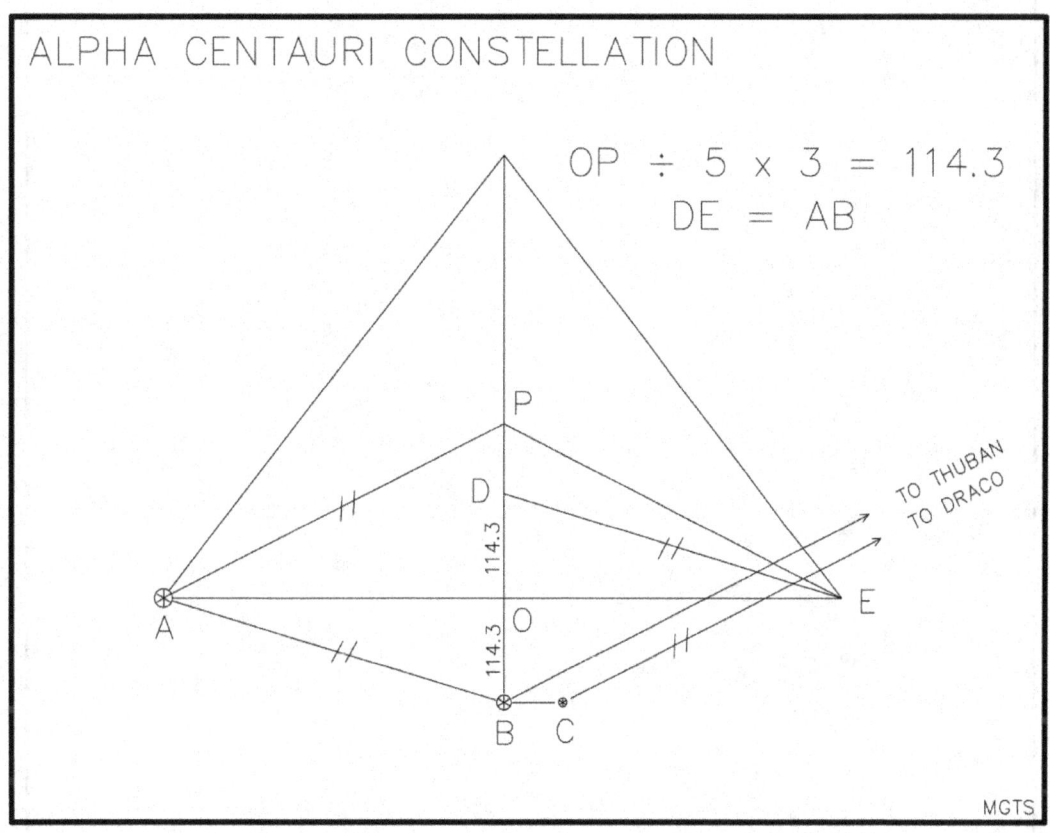

ALPHA CENTAURI CONSTELLATION

OP ÷ 5 x 3 = 114.3
DE = AB

CONSTELLATION OF ARIES

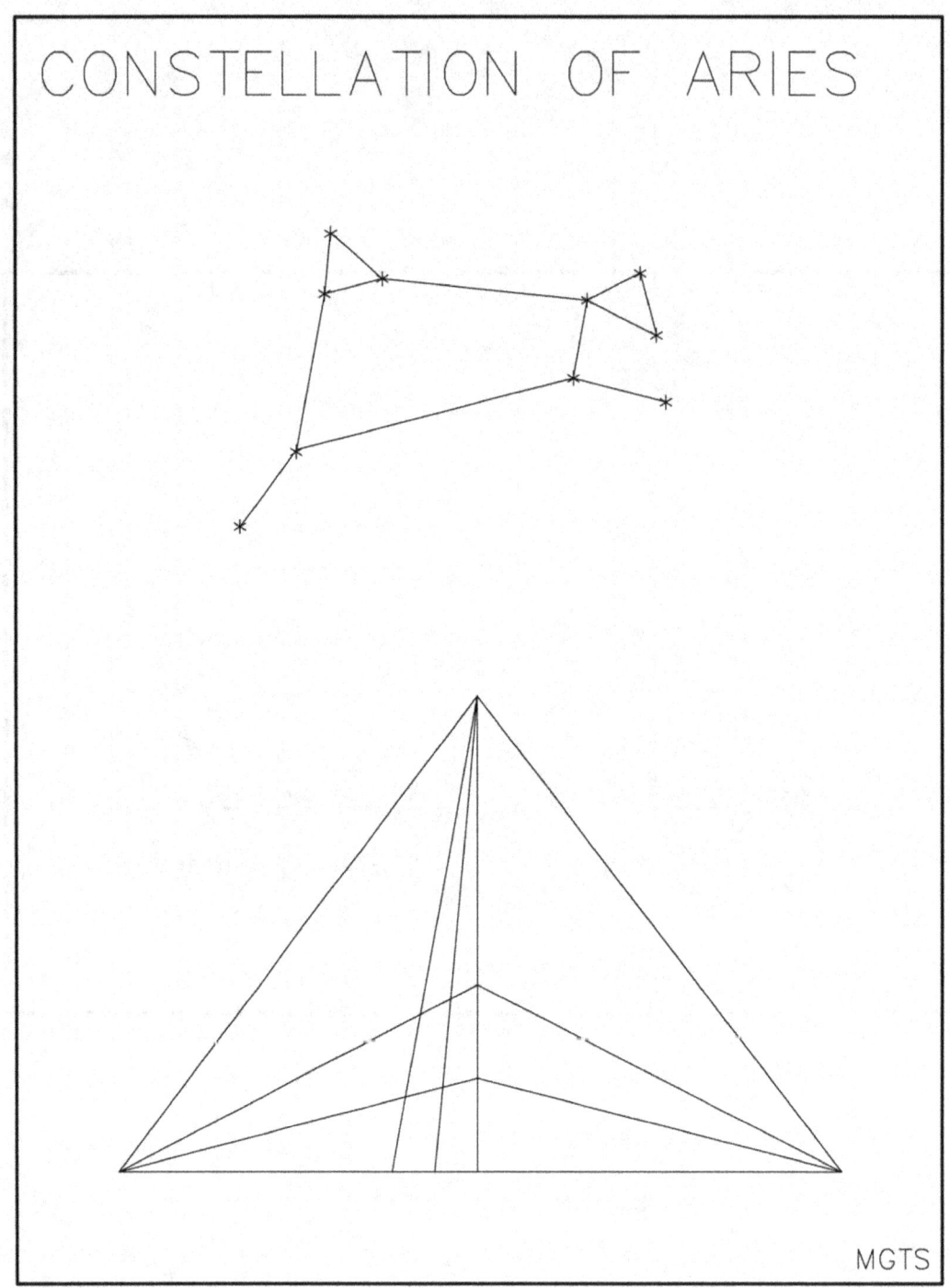

URSA MINOR

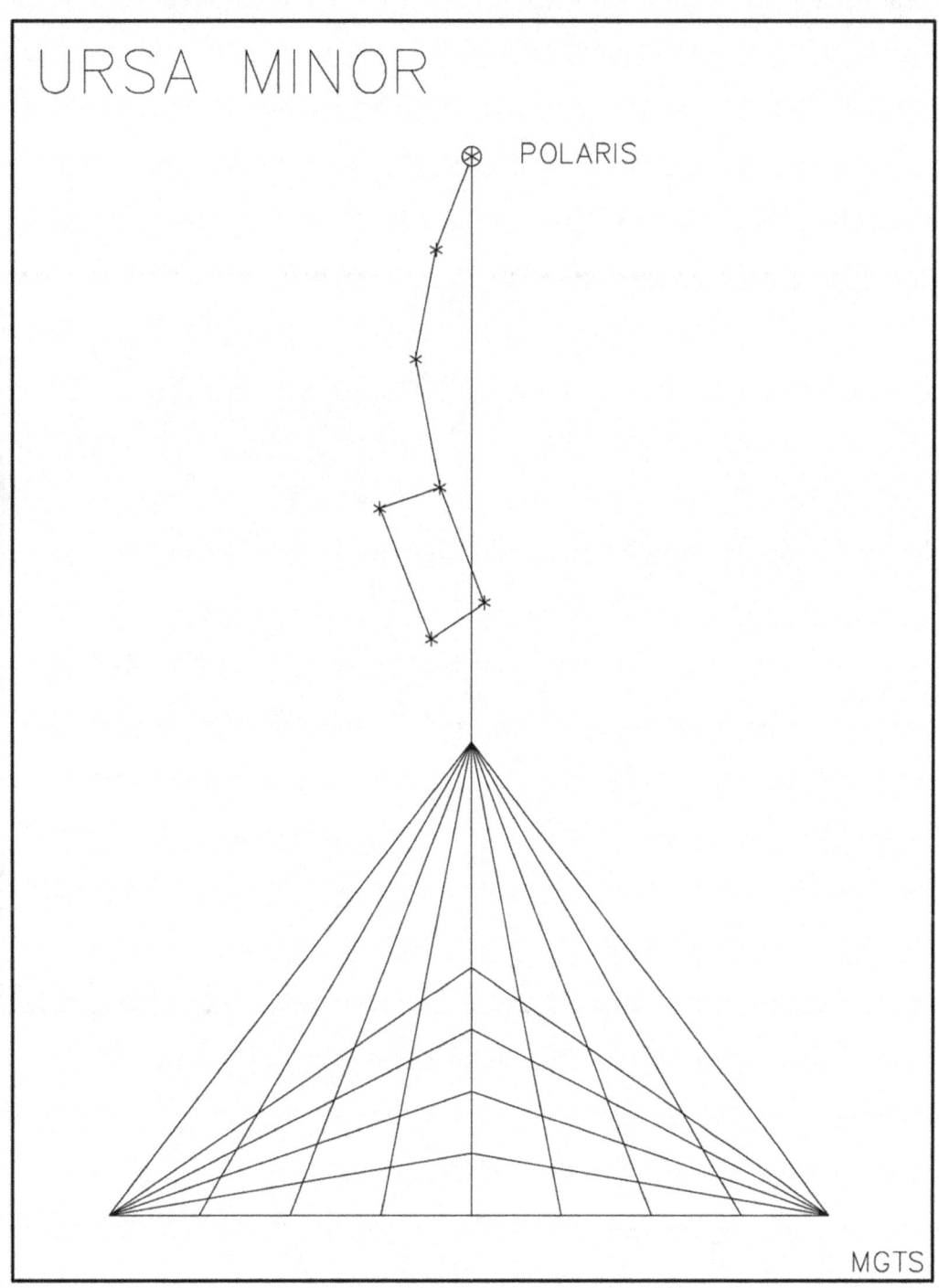

POLARIS

MGTS

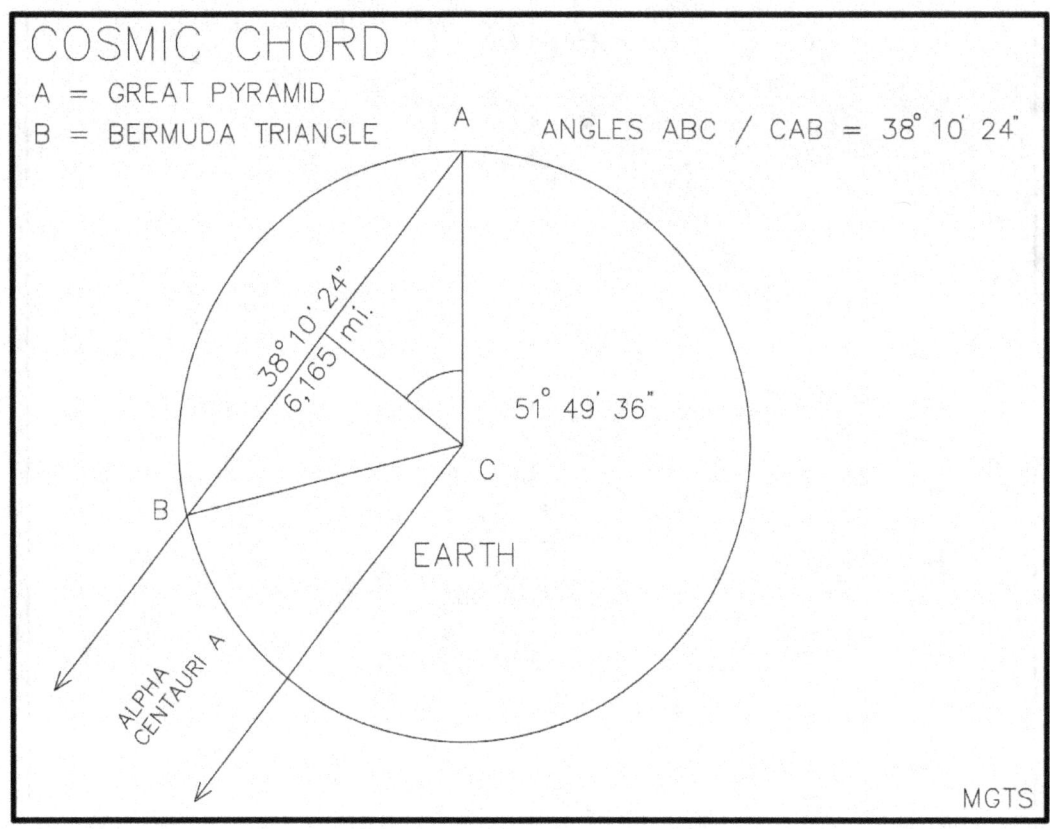

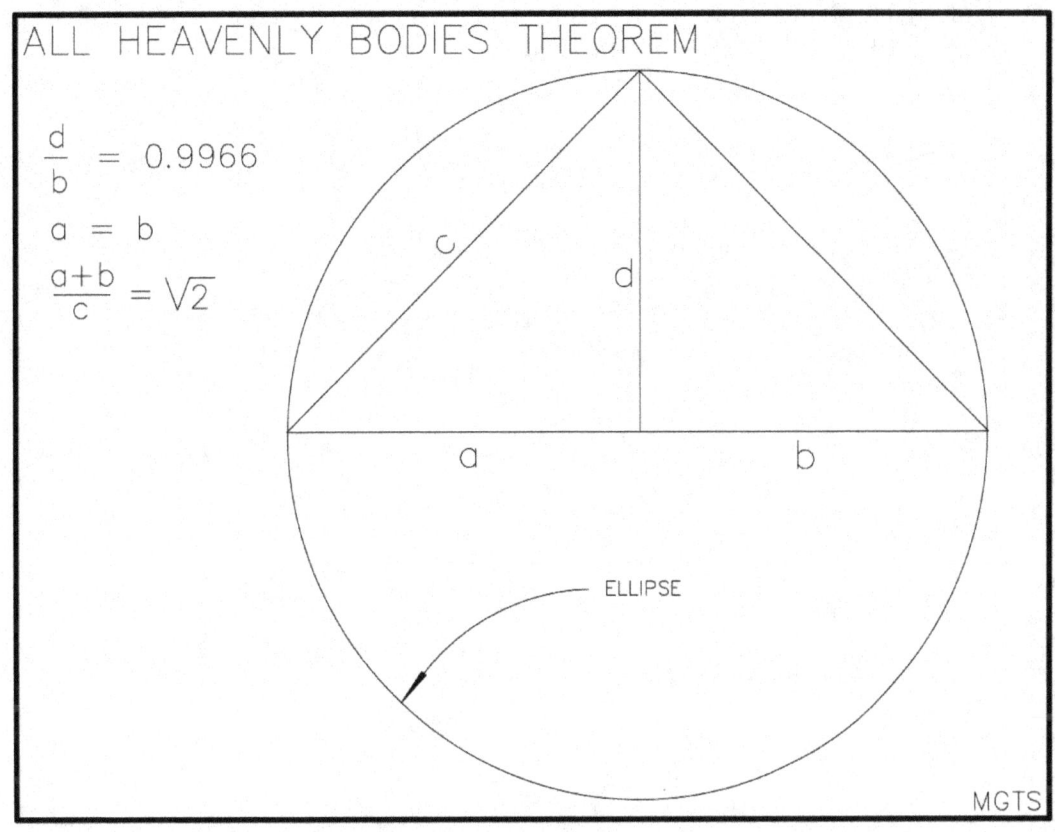

ALL HEAVENLY BODIES THEOREM

$$\frac{d}{b} = 0.9966$$

$$a = b$$

$$\frac{a+b}{c} = \sqrt{2}$$

c

d

a

b

ELLIPSE

MGTS

GREAT LIGHT YEAR PYRAMID

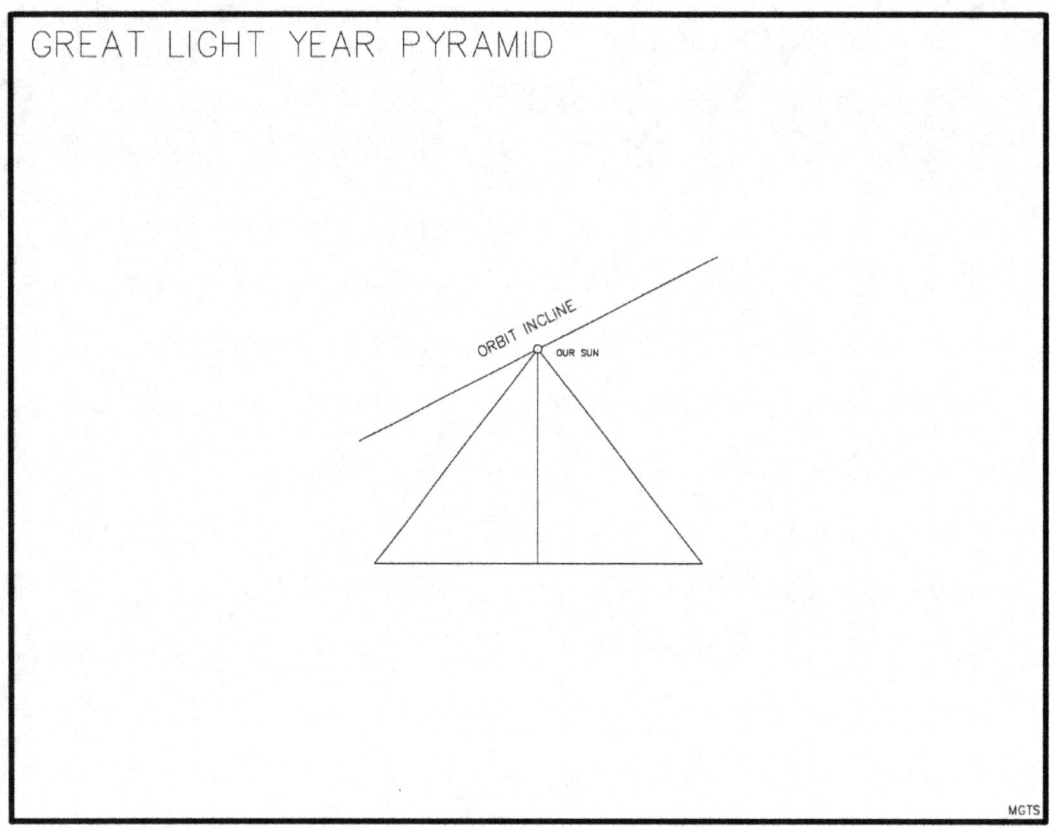

MGTS

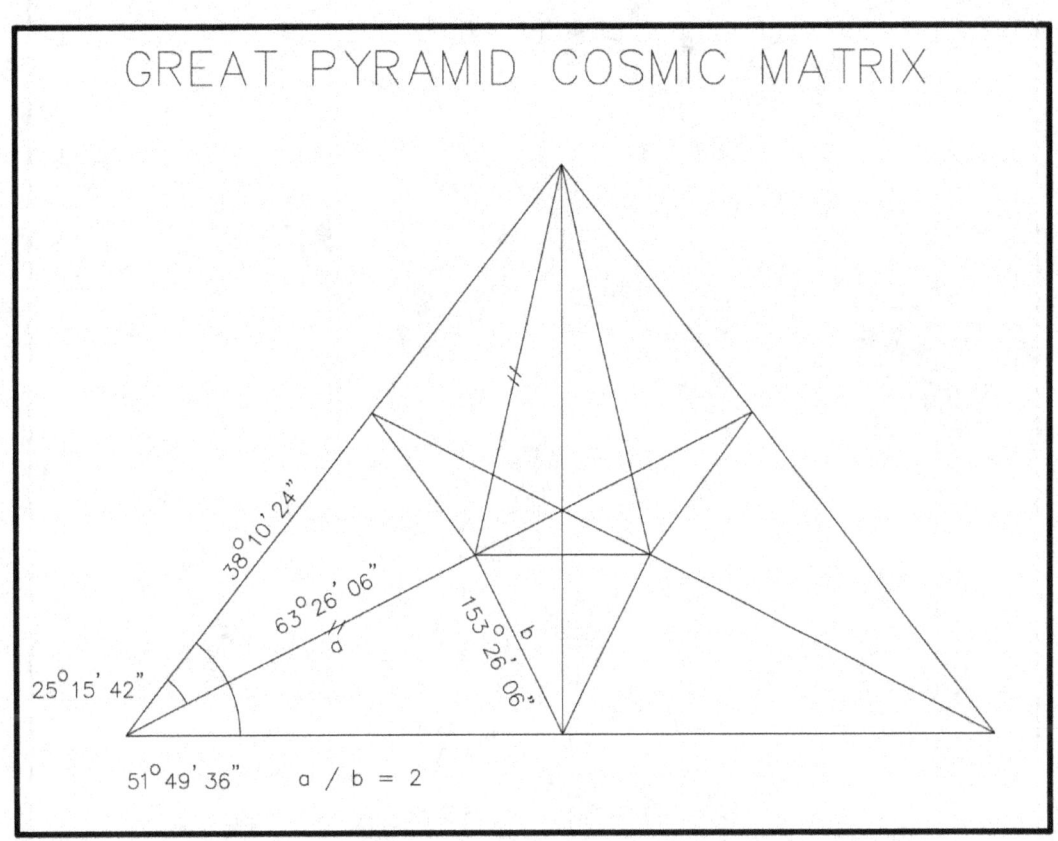

GREAT PYRAMID COSMIC MATRIX

AGE OF AQUARIUS 2012 to 7162

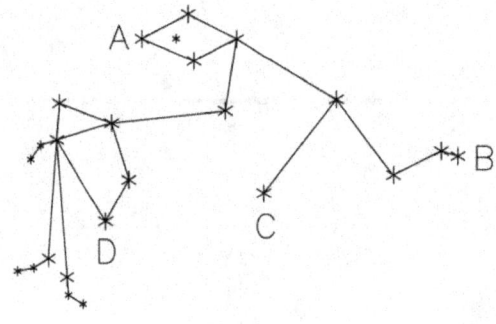

AB = 2.885 ÷ 5 = 0.577
AC = 1.665 ÷ 5 = 0.333
DB = 3.08 ÷ 5 = 0.616

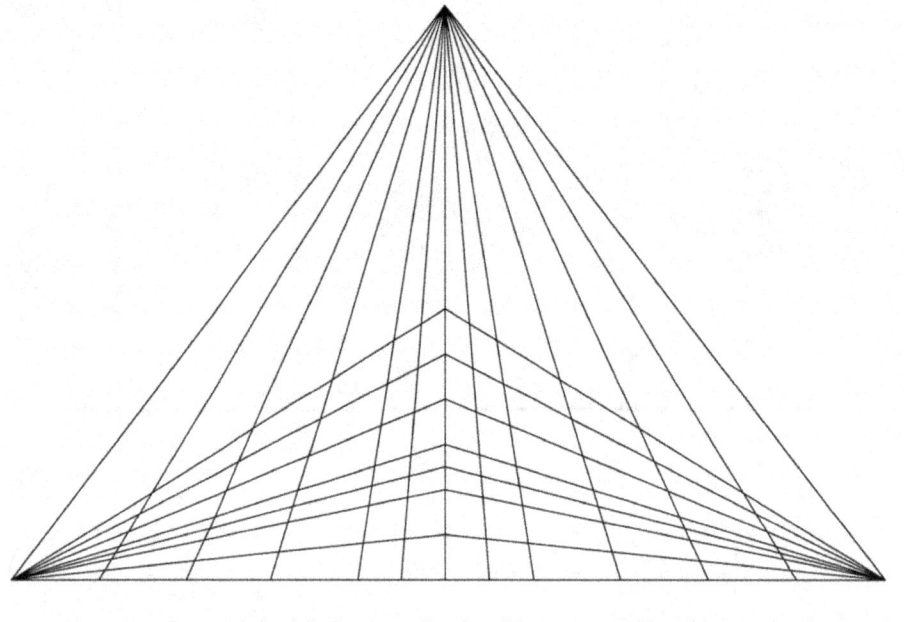

...DANCING WITH THE STARS...

MGTS

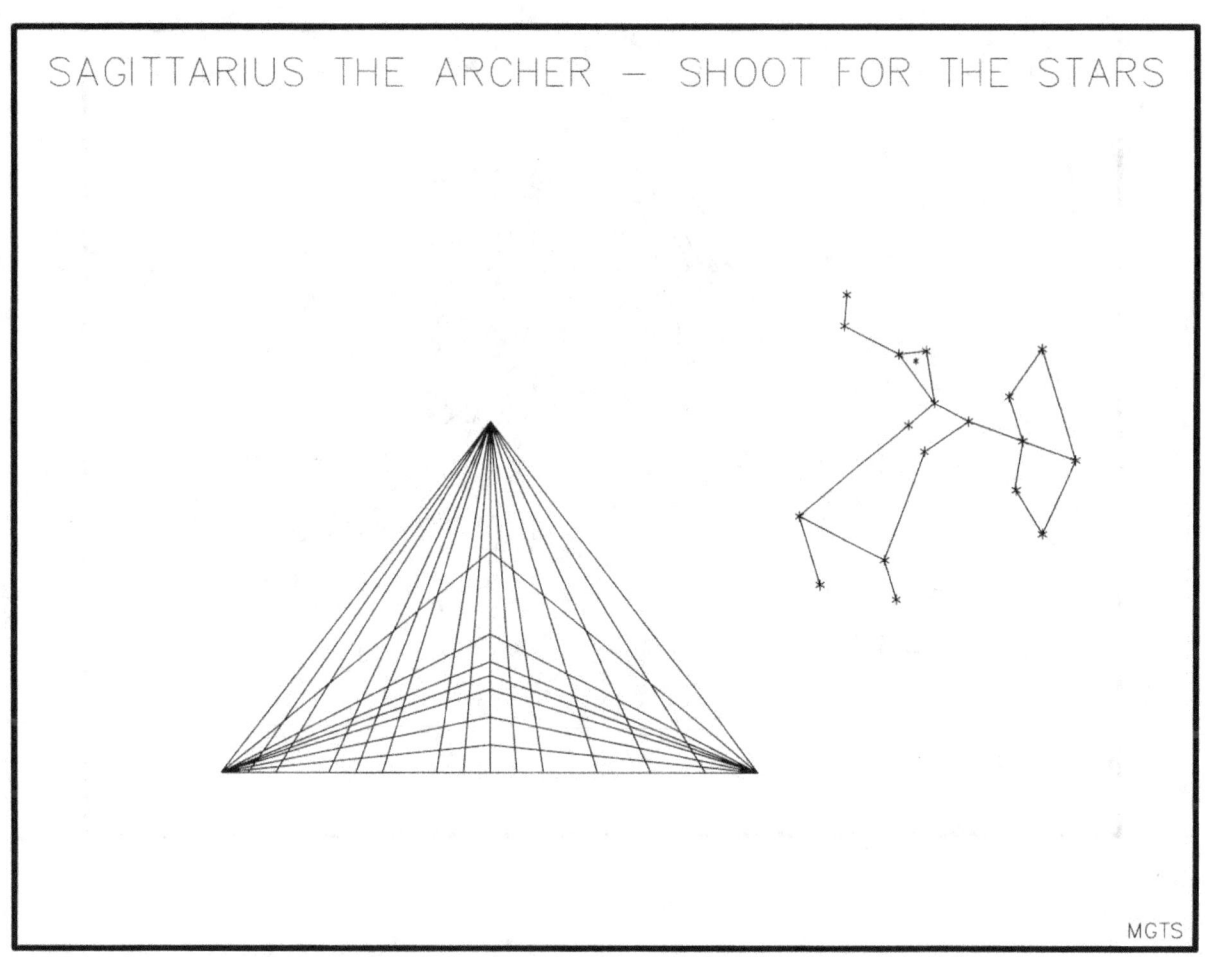

MGTS

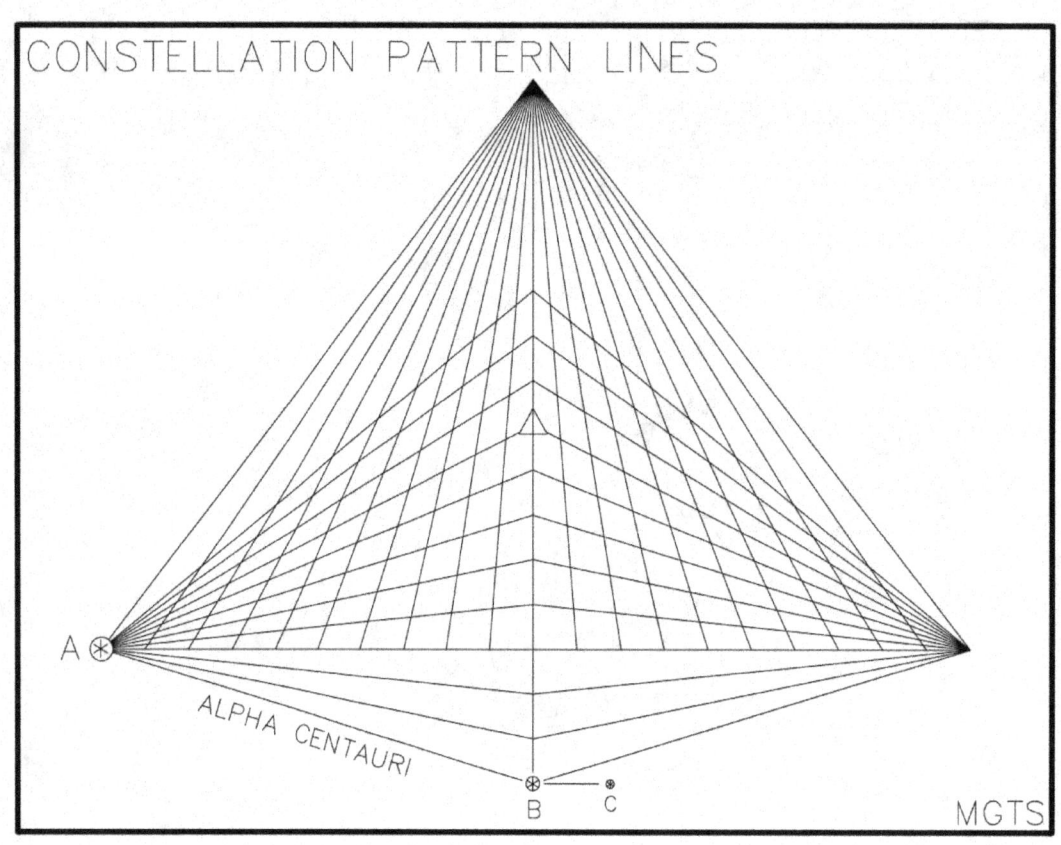

TAURUS

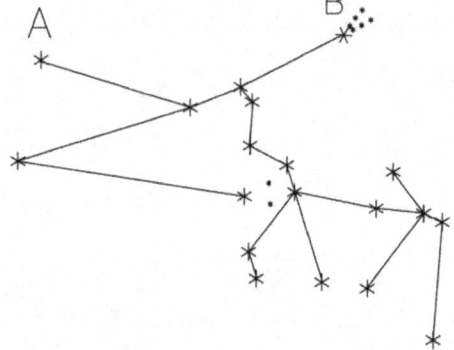

$$AB \div 5 = (5 \div 10) + (\sqrt{5} - 2.0 \div 10)$$

THE LANGUAGE OF THE LIVING UNIVERSE

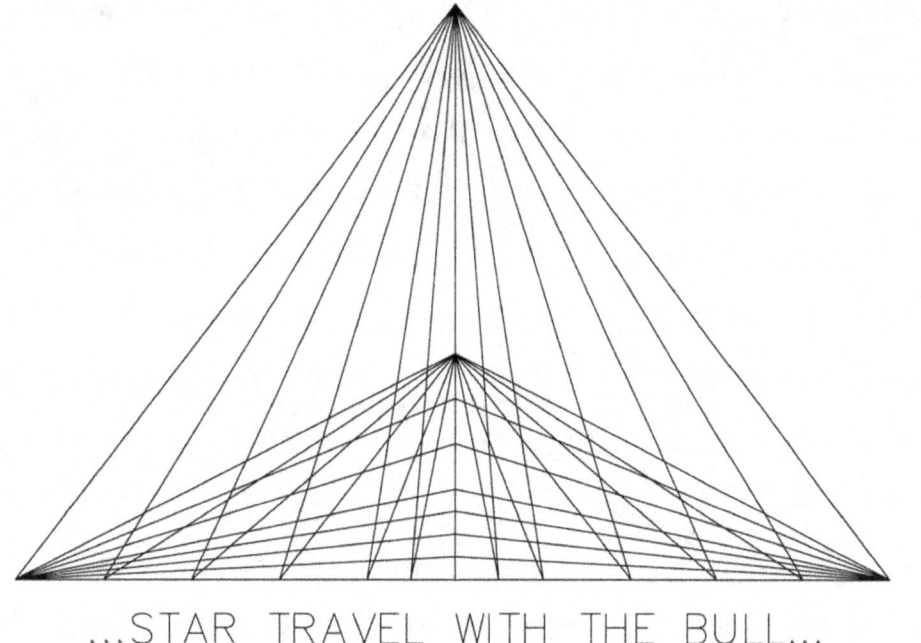

...STAR TRAVEL WITH THE BULL...

MGTS

GEMINI

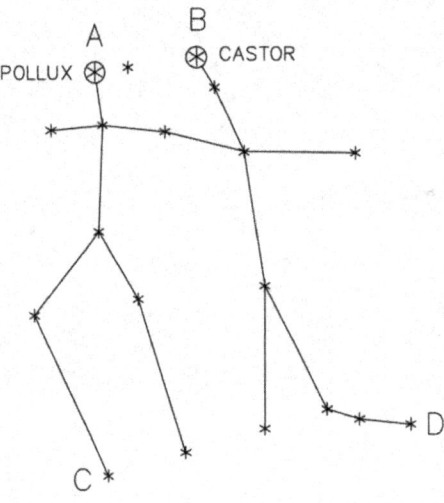

POLLUX ⊕ A *   B ⊗ CASTOR

$$AB \div 5 = \frac{1}{0.2} = 5$$

$$AC \div 5 = \frac{1}{0.75} = 1.333$$

$$BD \div 5 = 0.8 \times 10 = 8, \text{ THE NUMBER OF INFINITY}$$

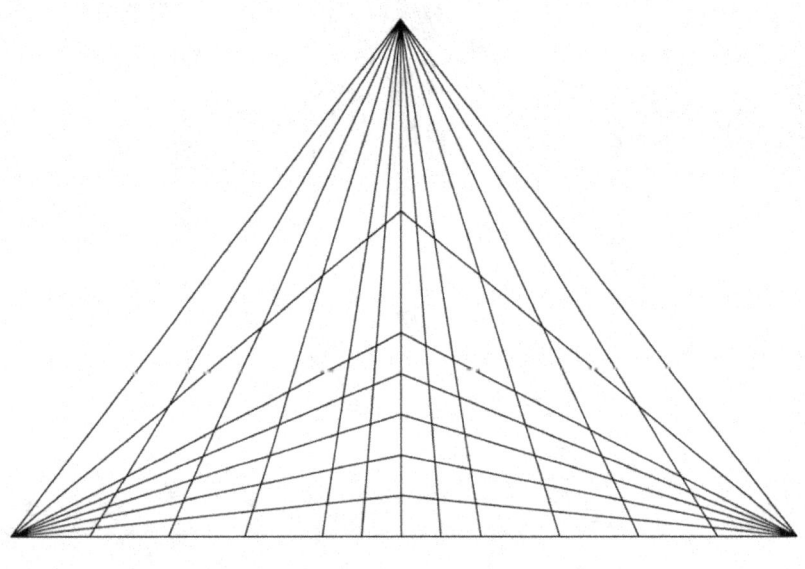

GEMINI EPITOMIZES THE DUAL NATURE OF THE UNIVERSE

MGTS

LEO

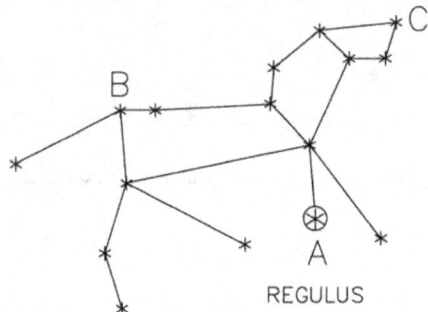

REGULUS

$AB \div 5 = Phi^2$

$AC \div 5 = \dfrac{1}{\sqrt{Phi} + 2}$

$BC \div 5 = 0.5$

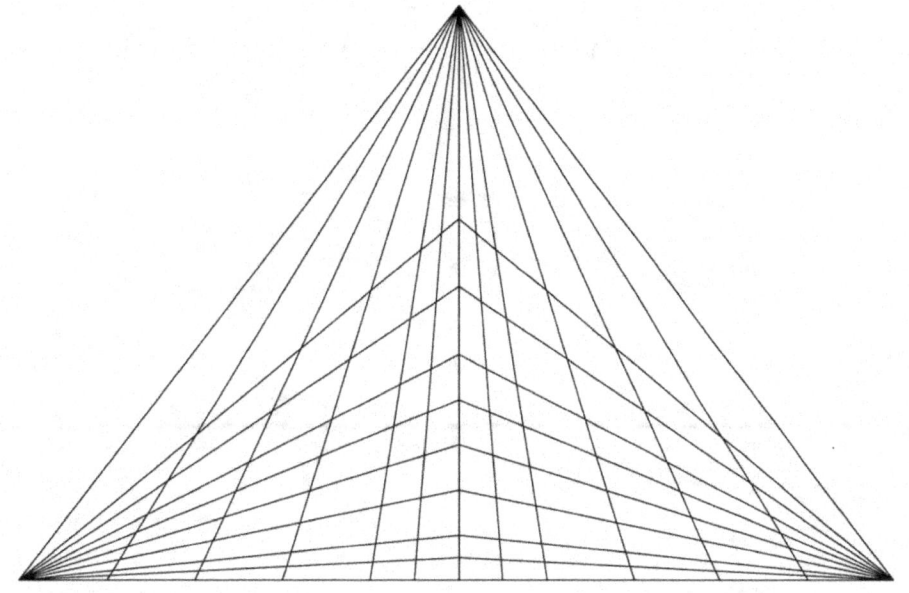

LEO the SPHINX / PROTECTOR of the SOLAR SYSTEM

MGTS

THE SPHINX

PHI BASED OBELISK

$$\frac{AB}{CD} = 1.25$$

$$\frac{1.0}{AB} = \text{Ancient Pi}$$

$\frac{Phi}{2}$

1+Phi

C   D

A   B

MGTS

OBELISK of TUTHMOSIS

Square Root of 3 Scheme

Height = 75.0 ft.

Base = 6.0 ft.

$$\frac{AB}{CD} = \sqrt{3}$$

$$\frac{x}{75.0} = \text{Ancient Pi} - 3.0$$

The Obelisk is a mystery...

telling of astronomical values...

see The Tower/Rider—Waite Tarot Cards

MGTS

WASHINGTON MONUMENT

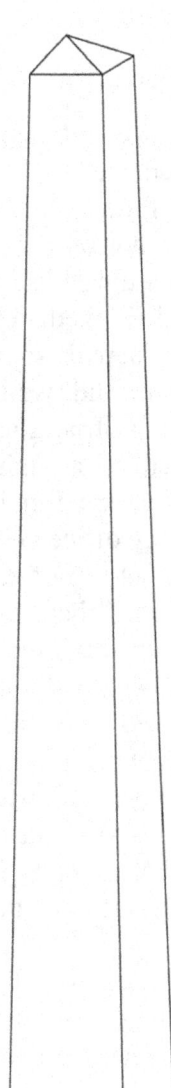

DIMENSIONS:

BASE = 55.07 ft.

MAIN BODY H = 500 ft.

PYRAMIDIAN BASE = 34.49 ft.

PYRAMIDIAN H = 55.56 ft.

...IS NOT A Phi RATIO OBELISK,
BECAUSE THE GOLDEN SECTION
WAS NOT UNDERSTOOD AT THE
TIME OF ITS CONSTRUCTION...

MGTS

CHAPTER 5

What is Freemasonry ?

In my process at a time earlier in life, at age 39, I came to know of yet another form of geometry which provided me with much motivation and interest. It became as important to me as anything else in life because it was as if my spirit had taken form and had somewhat of a definition attached to it. It started when I took note of a certain symbol on a nicely made plaque consisting of the compasses and square on a wall upstairs in the old building I had my office in from where I was managing a consulting practice. Of course I related to it immediately and I asked an elderly gentleman who was present about it and he said it had to do with some very important aspects of life. I said that G in the middle must stand for geometry. He said, well, you're close but there is a little more to it than that. I came to know he was a Freemason and at the time I knew absolutely nothing about this organization. Was it to be suspected of anything anti social or dangerous? Really, I didn't have a clue about it. As time went on we would end up talking about every thing under the moon and the stars during his frequent visits to my office and I came to realize he was a very wise man. As time went on I did some research on the subject of Freemasonry. I had been studying comparative religions for years by then and had concluded there was nothing in that area that made any real sense to me. Somehow a book on the subject was found under some old newspapers in my office and when I discovered what Freemasonry was basically about and how old it was I became more curious and enthusiastic about it. When I expressed an interest in Freemasonry to my new acquaintance I was invited to a meeting the very next day where five other Freemasons were in attendance and I was bombarded with questions about who I was and what I was all about and did I believe in the Supreme Being or not. My answer to this was yes, but privately I had no clear definition for exactly what it meant at the time and I kept that to myself. My new friend vouched for me as I had no family members who were Freemasons and from that day forward I was accepted into the Brotherhood.

I felt it was the correct thing for me to do though I didn't fully understand it all but I liked what I knew about it. I needed some answers and Freemasonry seemed to make a lot of sense as far as I could see at the time. I was told I could back out of it any time if I so desired. Therefore, of my own free will and accord I ventured into the study of Ancient and Accepted Freemasonry. In my regular work I had learned that to measure was to know and define but I needed more. After all I was on a quest for truth and understanding in life. How best could I contribute to society and as a member of the human race? What position could I take in life aside from just being in it? As it turned out, by coincidence, the east side of the old building upstairs where I was renting my office space was a Masonic Lodge by the name of Rainbow Lodge # 180. Here I was working in a room adjacent to a Lodge and the Freemasons were my landlords. After covering the Entered Apprentice and Fellow Craft work which took approximately four years I received full light for my masters degree in a distant Lodge where I learned of the title, Grand Architect which also means Supreme Being. It was then I concluded the Supreme Being must be a builder in esoteric terms. I will always be reminded that on my left was a Muslim holding his Koran and a Catholic to my right with the Holy Bible in hand during the examination session for the Master Mason's Degree. A copy of the Old Testament was handed to me as I reflected on my thoughts and views about religion which I kept to myself. This didn't matter to Freemasonry as long as I acknowledged the existence of the Supreme Being, which, by then I certainly did.

From this study I had food for thought to last a lifetime and developed a constructive and philosophical frame of mind that has never let me down in the 20 years I have been a member. Modern Freemasonry came into being by an order of the King of Scotland in the late 1700's. There are more lies about it than it has secrets but that is the price paid when indignation is heaped upon a treasure trove of wisdom by the less informed who are bent on destruction of anything of value in life. In fact it is the oldest and largest worldwide fraternity entrusted to the Brotherhood of Man predating all known religions by thousands of years. I embraced Freemasonry because it had withstood the test of time, in the same sense Golden Section Geometry has. To me it was totally trustworthy and is a living link to an ancient period in history that had great appeal to me. I became aware in my conscious thoughts there had been a world of great wonders on this planet long before our time and I took on the challenge to find out about it and to my way of thinking, I felt that membership in Free Masonry might help lead the way.

Freemasonry is not a religion, cult or secret society. It accepts all members irregardless of race, creed or religion and with no exceptions its lodges are found in every part of the world. From this study one learns to build or rebuild his own life and those of others using the elements and esoteric applications of right angles, horizontals and perpendiculars based on an ancient philosophy that developed thousands of years ago. The study provides a man with reason, substance, truth and motivation in life which indeed, provides that all important crossroad between science and spirituality. My mentor, the aforementioned elderly gentleman once said to me, "There are no coincidences in life". It might be known it took only a handshake between two members of high acclaim from opposite sides of the globe in modern times to mark the beginning of the end of the Cold War. Perhaps it is of no coincidence the first president of the United States and the first Prime Minister of Canada were Freemasons.

My own thoughts led me to conclude there is no such thing as perfection in life and Man never was, is now or ever will be, but he will always "be free to be", or "not to be". We are challenged by corruption and negative forces on all fronts and at all levels, yet if a man would learn to trust in his higher being and cultivate that union with due care and attention the worthlessness in life is expelled. Once a man learns to love and laugh at himself and transcend life's imperfections through useful daily endeavor, controlled emotion, humor and constructive thought then providence, peace and harmony prevail and a lifetime becomes a sweet blink of the eye everlasting in eternal bliss.

A Freemason does not compete with or criticize religions, or expound from the rooftops upon what is right or wrong, though he might constructively and thoughtfully express an opinion or share viewpoints on life and its imperfections based on his own experience and training of thought. My mentor once said to me, as I got deeper into the study, " Alcohol is the greatest entertainer but the poorest of masters ". Here was a truth, for its abuse has been known to topple giants. Then I learned that Dr. Bob was a Freemason, who's influence with Alcoholics Anonymous has served to mend many broken lives. My mentor mentioned that I was a Freemason now because I had been one in my heart all along. Then he added, I just know there is a book in you, and the time will come to share what you know with your brethren and the world. I found these to be curious but somewhat comforting statements because my mother and my tutor shared these very same thoughts with me about my somewhat latent writing skills. By this time I knew that a number of Free Masons were prolific writers and eloquent speakers. It was after I became a Freemason I discovered the author of the Sherlock Holmes stories, Conan Doyle, was also a Freemason and Rudyard Kipling was another. Then I discovered Pythagorus was a Freemason of his time in ancient Greece. With their influences on my mind it could be said I had been training to be a Freemason without fully realizing it since my youth. Then I remembered telling my tutor that what I had been doing for a living might lead to something special and I had found so much of what I had been looking for in Freemasonry. Here were some coincidences in life that might be called, simply, providence, for I could effectively think and work with geometry on the physical level and equate with life on the esoteric plane as well. I had come to the conclusion that the G in the Masonic Symbol could stand for both geometry and God because when I look at the unfolding arms of the Golden Spiral there is a strong resemblance to the Letter G when it is drawn clockwise. Hence, the geometry of the Grand Architect sits well with me and always will. After my mentor passed away I was destroyed because he was the only person I could classify as a trusted friend during the adult stages of my life. During our last conversation I mentioned that I was near ready to get started on that book and he said, " I am certainly looking forward to reading it, the world has been waiting for it a long time now ". Life moves on and adjustments are made, in many ways I privately say to myself, " Living with Geometry " is dedicated to my dear old Friend and my other Brethren around the world but to be fair about it every living sole on this planet is welcomed to know the truth of such matters about God, life and the Universe.

Indeed, many truths on the nature of life are learned from the study including the answer to an age old riddle; What are the heaviest, most destructive and dangerous elements in the Universe which have no mass or substance? The greatest challenge in life is its imperfections. A graduate and practitioner of this most ancient and accepted philosophy makes a steadfast pledge to be trustworthy and honorable in his heart, upright in stature, square in all dealings with passions subdued, on the level and within due bounds of all fellows and God. These simple guidelines for life are immortalized by the compasses and square with G in the middle, the symbol of Freemasonry, and on his ring because all men can have the will "to be".

The answer to the riddle, what are the heaviest, most destructive and dangerous aspects or conditions in life which have no mass or substance?

To be, or not to be…that is the question. A question all men might ask of themselves. Incidentally, William Shakespeare was a Freemason at a time in history when free speech had dire consequences. The answer to the question is rather, if a man chooses not to be… resulting from a negative mind set he will live with an uneasy conscience, dishonorable actions and less than honorable thoughts coupled with indignity, jealousy, contempt, prejudice and malice inclined toward his fellows and rejection of values such as the all seeing eye and existence the Supreme Being.

These are a product of the Lower Nature in Man, the main source of his difficulties and his greatest enemy. As it is known and most often is the case, the only thing in a mans way is himself. Though no man is perfect, he will come to know of life's terms and learn to deal with its differences and imperfections to rise above them. Primarily this is what Freemasonry is about.

One might only reflect on imperfections when the living evidence of the conquests made by that lower nature in man rejects those grand qualities of life. These are portrayed in the news and law enforcement files every waking moment, in our penal institutions and asylums, its reflection are seen in the human tragedies on Skid Road. One might examine the futility and agony of war. Man will run into trouble every time if he does not keep himself in check by staying within due bounds. As history tells us, he often chooses not to be, yet the choices in life are always two fold in nature. The opposing forces of the Universe are perpetually at work, Death vs. Life, Failure vs. Success, Destruction vs. Creativity, Hate vs. Love, War vs. Peace, Abuse vs. Abstinence, Chaos vs. Order, Oppression vs. Freedom etc. It can be concluded the only pathway to self mastery must lie within, for it is no where else found. The lower nature in man exists at all levels but so does the essence of that greater spirit of being from whence comes that most precious gift, the free will to reason, make those sagacious choices in life "to be" at one with the true spirit of God and the Living Universe.

Throughout the Dark Ages in western culture while severe socioeconomic deterioration was taking place after the decline of the Roman Empire the Church was leveling its indignity toward Freemasonry because to the establishment of that period in history it was unacceptable for any man to be free. As discussed the operations of opposing forces of the Universe are not unlike those of man's affairs and they are perpetually at work. In this case it was oppression vs. freedom. In an earlier time the Jews went through this with ancient Egypt for a time, or so it is said, and the Christians and Jews went through it with a corrupt Rome, then endured it internally until its religions became fractured into something like the shattered milk jug that couldn't be mended, yet Freemasonry was always there and eventually it will help pick up the pieces. There is no power on Earth that can dispute its truths and wisdom and its doorways are always open to the seeker.

Could there ever be a singular belief system that works for all Men, or a unification called the Brotherhood of the World? There exists a semblance of a fraternity between nations but it doesn't include all of them. Yes, of course, this would be the ideal, but we know perfection does not exist and imperfection is our greatest challenge because the opposing forces of the Universe are perpetually at work. Freemasonry is based on astounding yet simple truths. Its origins were founded so long ago today's members are left to dwell upon those elusive lost keys which are not really lost, even in today's clamor. Knowing from what era it comes from, and the evidence of good will toward the betterment of its fellows and world affairs found in the study of Freemasonry is conclusive enough to determine its origins vastly predate any known religion of this day. It's values are to be cherished and honored. To a Member of the oldest and largest worldwide fraternity dedicated to the Brotherhood of Man for untold millennium it provides motivation and drive for excellence in all he undertakes in life. Whether I found Freemasonry or it found me, I had no equivocations whatsoever in becoming a member of this ancient brotherhood of dedicated men from all walks of life from every culture and corner of the world.

Following this all too brief chapter are drawings titled, Geometry of the Masonic Symbol, Free Masonry and the Golden Section, The Great Pyramid and Free Masonry in which are seen the simple yet powerful principles of the horizontal, perpendicular and right angle. The square and the compasses are not only the tools for building in the material plane but are those which lead to an understanding with God, Life and the Universe on the esoteric level because in this symbol of the Universe are found the values for the square roots of 2, 3, 5, Ancient Pi and Phi, those which were discussed intensively throughout the previous chapters. Predating the drawing for Elements of the Golden Section, it says much more in that the *12 factor is seen in the divisions of the square on both sides and the eighth divisions of these from the point of the square, the number of infinity, are hidden by the arms of the compasses because this unapproachable value can be accessed only by the mind of an adept, and the Golden Spiral in the middle that closely resembles the letter G, does indeed provide the signature on life provided by the Grand Architect. In a graphic following, The Pledges of Free Masonry wherein the Ancient Pi circle circumference and letter G, most commonly seen in the symbol are made available. It is more than evident that the mysteries of Free Masonry relate directly to that grand science of an ancient and wondrous yesteryear, and it becomes apparent, this enlightened group were well acquainted with the Language of the Universe.

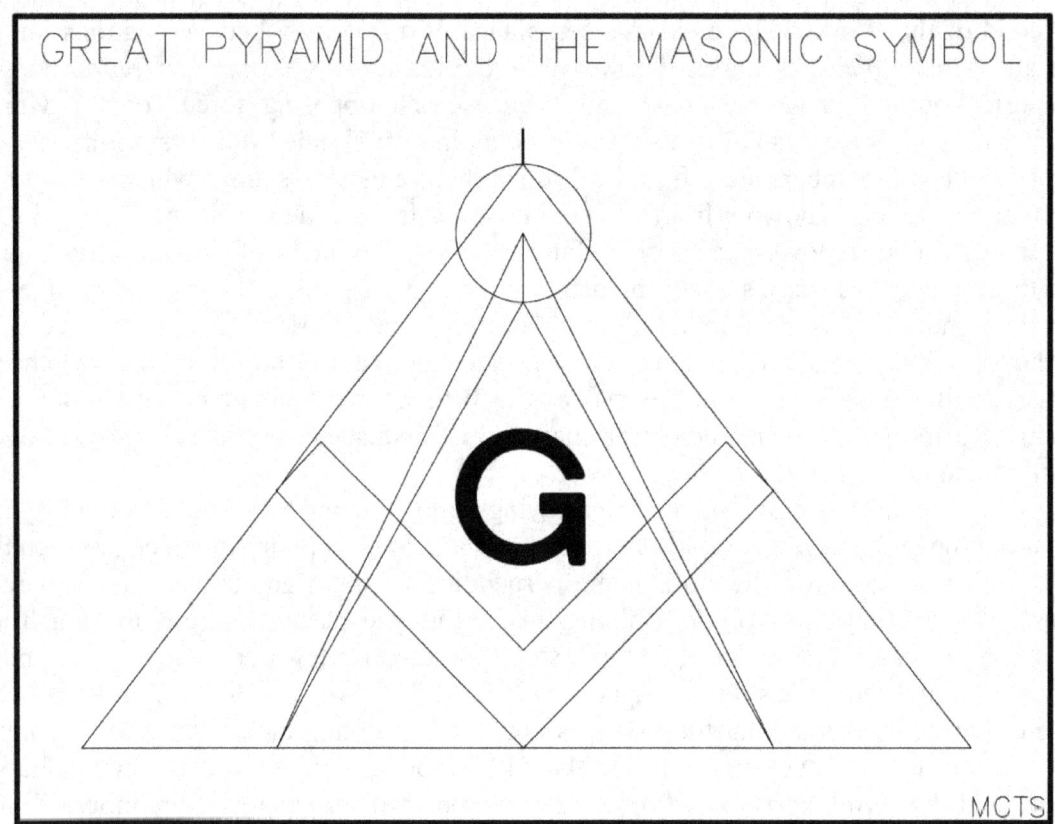

GREAT PYRAMID AND THE MASONIC SYMBOL

MCTS

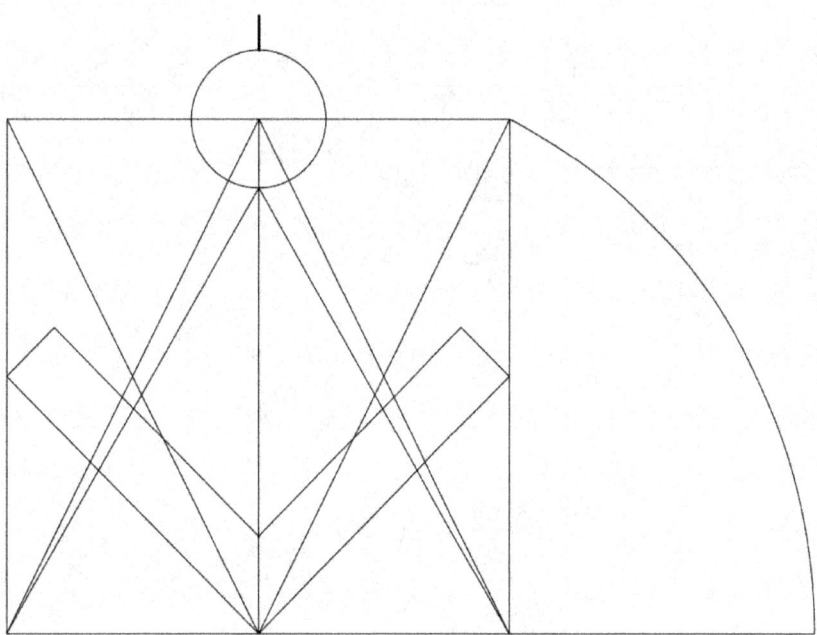

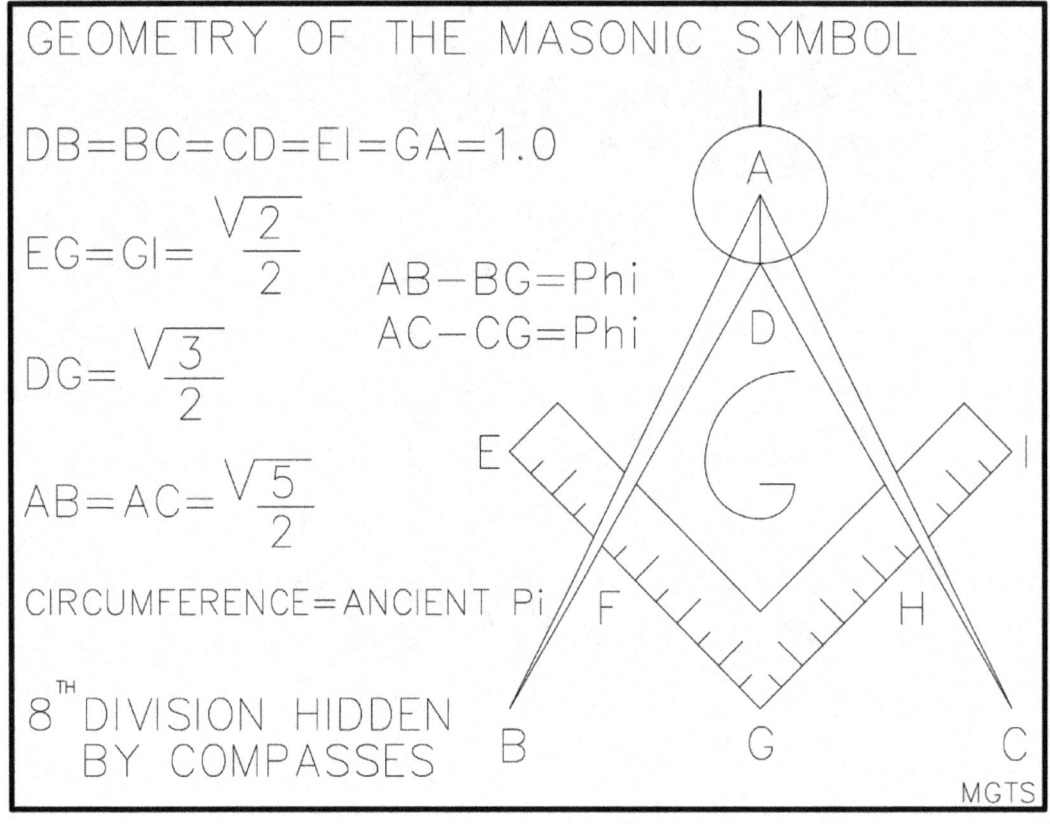

GEOMETRY OF THE MASONIC SYMBOL

DB=BC=CD=EI=GA=1.0

$EG=GI=\dfrac{\sqrt{2}}{2}$

$DG=\dfrac{\sqrt{3}}{2}$

$AB=AC=\dfrac{\sqrt{5}}{2}$

AB−BG=Phi
AC−CG=Phi

CIRCUMFERENCE=ANCIENT Pi

8TH DIVISION HIDDEN
BY COMPASSES

A
D
E
I
F
H
B
G
C

MGTS

THE PLEDGES OF FREE MASONRY

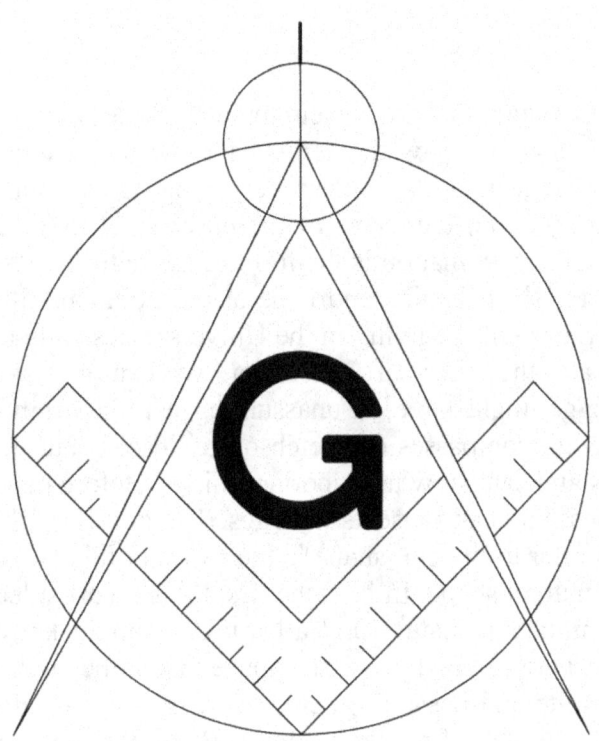

WITH STEADFAST COMMITMENT "TO BE":

UPRIGHT IN STATURE, SQUARE IN ALL DEALINGS,
PASSIONS SUBDUED...ON THE LEVEL AND WITHIN,
DUE BOUNDS OF ALL FELLOWS AND GOD.

WHILST IN PURSUIT OF THAT HIGHER,
MORE NOBLE AND GLORIUS PURPOSE.

bro. green

CHAPTER 6

Other Thoughts to Ponder Upon

Take a moment and return to drawing circles and expressing the finite making use of the infinite or vise / versa and how the terms of finite suit our needs as well as our fellow creatures on this planet. One day not so very long ago I happened to look at a common chicken egg with that inquiry about the shape of the universe at its infinite perimeter in mind and "Bingo !" The most reasonable explanation is the form of the outer limits of the Living Universe is a never ending Golden Spiral which begins and ends at the point of Infinity while the finite space inside an egg houses the embryo of the living in the Universe. Basically a side view of an egg shows one part of it is elliptical and the other part is circular. We can only assume that chickens and the other creatures who lay eggs might have an unassuming natural strain of genius in them as well. They can't scribe circles with compasses or sketch out ellipsoids but they are here with us in life, having their own purposes and can show us important things. Before our time dinosaurs were laying eggs. All of today's reptiles, birds and insects lay eggs. The question that comes to mind is, why is an egg the shape it is ? Similar to the egg shape is the oval but this is something we draw based on finite geometry, it is not a natural shape. In fact, the first thing an artist learns is, there are no straight lines or perfection of symmetry in nature and all artisans should know about the Golden Ratio. Simply stated eggs exist and are hatched in a finite space to suit the environment yet the geometry of the shell expresses the infinite and basically there is a reason for all geometric forms in nature. A truth is, the finite coexists with the infinite at all times. Being that the Universe is a living organism the very shape of an egg is what the procreation of life and how nature protects itself is all about. After coming to terms with the geometry of the Milky Way it would seem the limits of the Universe might be defined, or described as an ellipse of life energy with an infinite circumference and within it an equally infinite sphere which is in the phi proportion with the ellipse. It would seem then, this arrangement is the basis of the birth and continuation of life energy and because Phi and Pi are inseparable a fundamental arrangement first exists in terms of an ellipse having a long width axis of 2.0, a short width axis of 1.0 and a circumference of 5.0. In this sense, one becomes two, then three and this is the order of the cosmos. It is seen that the diagonal = Phi + 0.5, or 1.118 - 0.5 = Phi and this ratio and the potential for 1+Phi are within the configuration as outlined in the Golden Section drawing. Also, Phi + 1+Phi = the square root of 5. From there, according to the processes of the Golden Section in action the long width axis of the ellipse remains as 2.0 and the short width axis becomes Phi and the length of the egg become 1+Phi. This is seen in the egg shape derived from the dimensions of the Great Pyramid where the slope divided by ½ the base = 1+Phi, and the height = Phi. When the right half of the ellipsoid is erased and half the circle circumference is shown on the side opposite the half ellipsoid remaining the result is a perfect phi based egg shape with its small end to the left. When the reverse of the above is done a perfect phi based egg shape with its small end to the right appears and this speaks of the dual makeup and symmetry of the Universe discussed earlier. To be clear on this, the length of the egg = 1+Phi and its width = 2 x Phi. Therefore the geometry of an egg is the byproduct of the interaction between Phi and Pi in living things and this says it all in that the circular part is based on Pi x Phi, the length = 1+Phi, the width = 2 x Phi, and

234

the circumference = the square root of 5.0 + 3, or 1+Phi + Phi + 3.0. And it is also known that the square root of the sum of the squares of 1+Phi and Phi = the square root of 3, and furthermore 1.0 - Phi = Phi Squared. The circumferences of the egg is therefore, (2 x 1+Phi over 2) + (Phi x Pi). The circumference of the ellipse is, (2 x Phi) + (2 x 2) = square root of 5 + 3 and the arc = the square root of 5 + 3, divided by 2 = 1+ Phi over 2 = 1.309. In chapter 4 a mental note was made about the perimeter of six Phi Based light year pyramids that became evident in the geometry of the Milky Way, the same value is seen in the circumference of the above mentioned ellipse. This entire expression hinges on the Golden Section principle where one is to the whole, as the whole is to one. This profound statement of creation and the way life protects itself is captivated in an explanatory graphic titled, The Cosmic Egg and the conclusion is, this is the reason why an egg has such a unique shape. In the human female womb the basis of the geometry of an egg is seen and in the human skull cavity for the brain. Perhaps the term Egg Head is somehow related to this observation. The geometry of an egg is seen in plant seeds and in our early space capsules. The yoke in an egg is first spherical, suspended in fluid, then it becomes an embryo within an egg shell after fertilization with its feet pointed toward the small end and its body in the larger end that provides room for physical growth, hence the geometry of all eggs are in the Phi/Pi Proportion because they are part of the life support system during fertilization until birth. The geometry of an egg is the protective housing for a living thing while incubating and this is a clear example of how the living Universe operates. The reasoning behind the inquiry is, the geometry of an egg is not only a fascinating topic, it has the potential to play a valuable role in the architecture of the living , namely our species. The following theme reflects on constructive ways and means of accomplishing this while working with the grandness of nature's geometry, or better said, with the geometry of the living universe.

The next step was to get busy with the cad coordinate geometry program and draw an egg showing its semi circular section and the geometry of its elliptic part as precisely as possible. This was the more difficult part of it. The ellipse is a tough problem and as indicated nature handles them much more efficiently than we can. There is a marginal differences of around 2% in the ellipse section between the following drawing and the above because of this but in a practical sense the exercise will serve the purpose. Included there is a drawing titled, Ellipse Spline which uses Phi Ratios to define the elliptical curvature more accurately and this would be used when there is the need for a higher degree of precision. The word Spline is an engineering slang term which means to draw a line between points on a curve that has no specific radius point. Then please see the drawing titled Golden Egg Theorem. Also is another drawing that debates this alternate method of drawing an egg using radii.

Compasses and the straightedge are great and I will forever admire and be indebted to those historical figures such as Pythagorus, Euclid and Archimedes, to name just a few, who made their mark in the study, but today's technological advantages are hard to beat, though there is room for improvement. There is no more hen scratching on paper or chalk squeaking on the blackboard, but it is kind of fun to make sketches in the sand down on the beach. Another wonderful thing about present day technology is we can access the work of these historical giants through the Net. We can read about their lives and study their great accomplishments which laid the foundations for the technology of present time. Initially the problem and questions were what ratios could be used with regards to length and width to draw an egg? The above mentioned ellipse and circle seemed to be the right way to go. In chapter 3 the discussion came up about the elliptical Phi Based egg shape within the long and short radii using the Great Pyramid dimensions. Once the Golden Section and Pi came into play, similar to the Sign of Infinity drawing, the Golden Egg Theorem was developed, or hatched so to speak. It consists of four compound curves whose radii are in relation to Phi and Pi. By doing so the elliptical portion of the egg is defined using circular curves with known radii and from this drawing a concept for construction in the Phi Ratio was conceived.

Golden Egg Theorem: An Exercise in Possibility Thinking…

Consider what can be done with a Phi based egg geometry structure as it relates to architecture…as opposed to our conventional rectangular and square structures, and / or as opposed to circular domes.

1.) The geometric aesthetics are pleasing to the eye and easier to live with and in …

2.) They would stand up better to hurricane force winds… save lives and costs…

3.) They would be more durable, safer and stable during an earthquake event…save lives and costs…

4.) There would be advantages in areas where heavy rain and snow prevail…save lives and costs…

5.) They are cheaper to heat, ventilate and maintain…

6.) In the case of a coliseum sports arena or train and airport facilities etc. the seating capacity and people traffic areas would be increased and there would be no loss to the main area.

7.) The cost of such structures are far cheaper than conventional ones.

8.) The insurance on such structures would be cheaper once their track record was proven out.

9.) In the case of a Phi based or Golden Section Ratio Egg Dome structure the interior environment is most likely healthier for human occupancy. We know electromagnet energy is summoned in a Phi based structure such as that of a Golden Section Ratio Pyramid. The previously mentioned advantages provide us with the motivation to proceed with the venture whether it be for individual smaller family dwellings, apartment blocks or public buildings, especially hospitals and including those previously mentioned. Without doubt this is a winning proposition. We are reminded of the Eskimo igloo and the North American Indigenous Peoples Wigwam. These ancient people had the right idea all along and we have been missing the boat.

Note: The Problem with the Circular Geodesic Dome Concept:

What good does it do anyone to dwell or spend time in a semi-sphere circular dome? It would be like staying under something like half a large basketball or beach ball and not do you all that much good unless the Phi Ratio is introduced. Sure it will be resistant to wind velocities, deal with rain/snow conditions nicely and light would be allowed in through fancy curved windows and such but what else? Consider a dome with an arc in the Phi proportion to its diameter as shown in the drawing titled Phi Dome. With this arrangement the benefits of the Phi Energy Complex would be appreciated by its inhabitants because Phi and Pi are given a chance to work together. As shown the Arc is subtracted from 1+Phi and the difference divided by 2.0 is added to the Arc at A and C. This difference amounts to merely 3 % of the total arc but it will make a world of difference in its operation. The simple equations Arc = 0.5 x Pi + sq. rt. 5 - 2, divided by 5 as shown and Arc over D = 1+Phi provide what is needed. Simply stated it is an option for a simple dome structure that is worth pursuing but it doesn't have to be limited to this when the Golden Section Geometry of the Egg is considered.

Before we move on to the Phi Based Egg concept please have a look at the drawing entitled Construct a Phi Dome and give it a try. The basics are there in the drawing but the recommended sequence for construction is a little different. Two drawings of the same scale will be required. A suggestion is to double the dimensions to begin with. This will make it easier to construct. Using bristle board paper, first cut out the base plate circle including the glue tabs so the perimeter circle piece doesn't get distorted when the arced piece is fits in to it. Second, cut out the perimeter piece and glue as shown. Third, glue the base plate circle inside the now circular perimeter piece. If the measurements are true these will fit together nicely. Fourth, as the glue dries in the previous steps cut out the dome piece then fit and glue these inside the circles as any dome would be formed. Overlay the glue tabs on the ones in the base circle. I sprayed mine with speckle paint first, then Gold, for the obvious reasons. I placed the Phi Dome alongside the Phi Based Pyramid and Teepee and they make a nice set. After the paint dries manipulate the dome with your fingers until it takes form. Of course when materials such as steel reinforced concrete or laminated wood beams are used for a life sized dome this would not have to be done, at this stage we are only using paper. This adjustment is quite possibly a step toward revolutionizing Geodesic Dome Construction making it more suitable in the sense of healthy energy generation for its inhabitants. We know by now using Pi without Phi has its limitations. Return to the picture of ice the age hunters housed inside a skin covered mammoth rib cage which would be loaded with the Phi / Pi interactive energies and we see it wasn't such a bad idea.

Proposed Structural Detail for a Phi Egg Dome

With reference to the Golden Egg theorem we see a plan view. The next step is to find a practical, simple yet workable and effective way of constructing a dome enclosure with structural component detail. At first it appears an impossible problem but once the basic components of the dome are studied an organized approach can be made. In Drawing A there is a typical East View that shows computed arc length values from A through to E. Take note of the side piece detail. This plays a roll in the dome construction later on. Then go to Drawing B and the top piece and ½ the perimeter for the structure are seen. From this detail the dome can be constructed using our favorite material, bristle board paper. Now that construction of a number of platonic solids has been mastered, dealing only with straight lines, it is time to move on and take up the challenge of constructing using arced lines.

Following are the steps and commentary -

1.) A working scale has been determined and you are aware of the proportions, distances and angular values. Based on what is given first draw the full egg including the line details then carefully cut it out using scissors. Make it 12 " long if you so wish. See Golden Egg Theorem and Dwg. A.

2.) Note: Using a razor knife for paper cutting is suggested. As shown in Dwg. B draw and cut out the ½ perimeter detail twice, Make the bends for the glue tabs, then glue the first half around half the perimeter taking care so that the upright perimeter piece is as close as possible to the edge of the egg shape cut out. To make it easier the other option is to fix the glue tabs on the underside along the side of the egg plate. Paste the other half of the perimeter piece to the first one as shown and complete this phase.

3.) The next step is to cut out the dome structure taking much care not to get any bends or kinks in it. As shown in Dwg. B the chord lengths AB, BC etc. are the arc lengths determined from the work on Dwg. A to suit the purpose. It is seen that positions A,C,G and H are that part of the structure which is circular. Once these are fixed into position the vertical curvature from C to F is remarkably close to what it should be for the desired shape. This is because the chord length equals the arc length. Attach A and F at either end of the egg plate first, then fix G and H into place. Once this is accomplished the other lateral pieces can be carefully tucked and curled into position on the inside of the perimeter enclosure with a dab of glue on them. The best thing to do with these pieces is cut them extra long to begin with because it is very difficult if not impossible to compute their lengths for an exact fitting. Once these pieces are gently positioned to suit snip them off at the base of the egg shape. Allow some time for the glue to set and we are almost there.

4.) Next cut out the two side pieces that need to fit on the sides and fix them in place as required. Have a look at the photos to see where they fit. If you want to add to the side detail cut two pieces parallel to the top of the egg and fit them in place too. Also strips parallel to the golden curves for the nautilus shell you drew can be cut out and fit into the form. Take a moment to see what has been accomplished. Once your masterpiece is spray painted with gold or colored white with gold speckles or the other way around including some gold glitter it is finished and ready to be added to your collection for the mantle as an ornament, an inspiration and a conversation piece. The basis for all living things in the Universe, an egg shape, has been created simply out of a piece of paper. With a little ingenuity a full egg model can be constructed to hang from the ceiling in your living quarters and those healthy vibrations of Phi energy will radiate from it. To do this a double ply egg plate needs to be cut out having a parallel width equal to the structural members then after the perimeter pieces are established the process for the dome and side pieces on both sides are assembled and fixed in place. The gap on the sides between the perimeter pieces can be filled with silicone or by gluing a strip of bristle board paper equal to that width around it. You could figure out a way to suspend a quartz crystal in the form and as an added challenge fit an ivy plant in it. Once it is suspended from the ceiling try sitting under it quiet and still with your eyes closed in a meditative mode to become enveloped in the energy of the life force charges emanating from it. You won't be disappointed. Though the form is only made of paper its Phi proportions still summon life enhancing energies.

Of course it would be great to construct one of these using 1/8 "silicon bronze, gold or silver rods etc. If a person enjoys welding and soldering and has a budget for materials, it's worth a try. Another material that could be used for an egg dome construction is strips of balsa wood. It is very pliable yet it has a high tensile strength. Who knows what ingenious spin off invention could come of this activity. There are such creations available in the market place today and business is brisk but the first ones you make with your own hands are more special.

Building materials for a Full Sized Phi based Egg Structure

There are a number of choices available for dome building materials. It would depend on the intended use. Plexi glass or real glass for terrarium or green house structures it might be supposed and a tarp arrangement for a sports coliseum or airport facility as is already being done in some centers. Styro foam, milled wood, fiberglass or concrete along with lots of steel reinforcing bar for a dwelling or public building. Whatever the case might be, as discussed Phi egg structures are the type of adaptation and innovation we might very well look forward to in future once we overcome old ways of thinking and perceptions on how to be more agreeable with nature. A structure such as this surely reconnects Pi and Phi. Presently there is a movement in industry that is going in that direction with reference to the geodesic near circular dome concept but this has limitations, yet one might think the egg structure would some day be the ultimate when all the advantages are considered. On a much smaller scale an entertaining thought for that well known recreation past time would be to construct phi based egg shaped tents for camping excursions. The base might be a made of fold out aluminum tubing, fiber glass or wood and the egg dome frame could be made of tough nylon or fiber glass strips and the tailor made canvas cover would be easily secured onto and over the structure. The assumption is, sleeping or just spending time in such an enclosure could only promote the good health and well being of its inhabitants. We have to admit though, as far as the tent concept goes which is a healthy place to spend time in, it is hard to beat a Phi based pyramid shape or even a Teepee constructed to those proportions, which is something else we have been missing out on. The options are there before us if we only take the time to think it through and experiment with what nature and the Golden Section has to offer. All we need to do is pool resources and get those creative juices flowing and go for it. One way or the other, we need to get out of those boxes we call homes and public buildings.

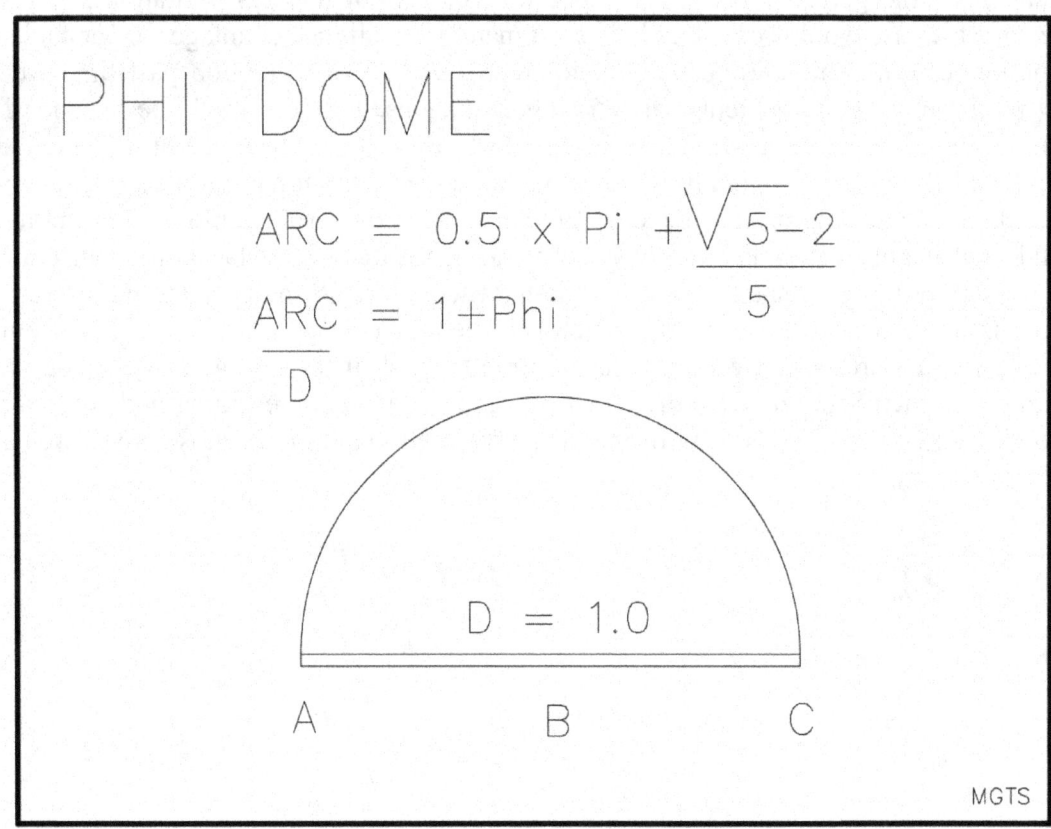

PHI DOME

$$ARC = 0.5 \times Pi + \frac{\sqrt{5} - 2}{5}$$

$$\frac{ARC}{D} = 1 + Phi$$

D = 1.0

A   B   C

MGTS

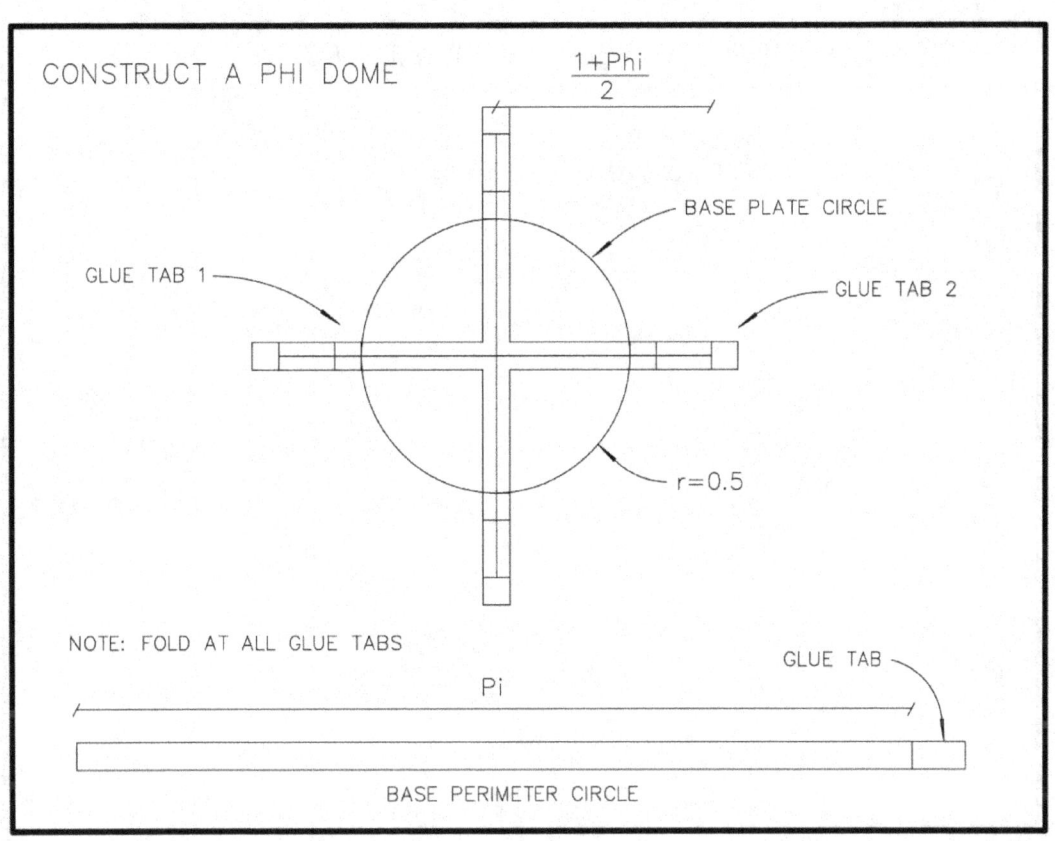

CONSTRUCT A PHI DOME

$$\frac{1+Phi}{2}$$

BASE PLATE CIRCLE

GLUE TAB 1

GLUE TAB 2

r=0.5

NOTE: FOLD AT ALL GLUE TABS

Pi

GLUE TAB

BASE PERIMETER CIRCLE

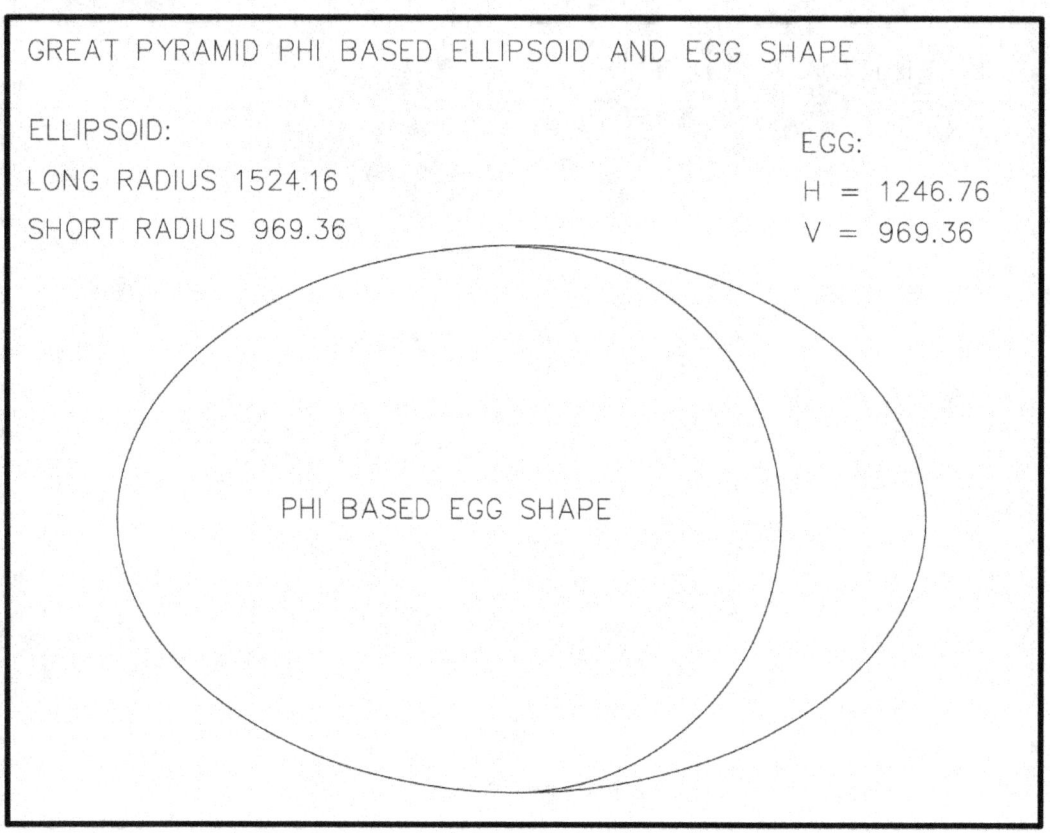

GREAT PYRAMID PHI BASED ELLIPSOID AND EGG SHAPE

ELLIPSOID:

LONG RADIUS 1524.16

SHORT RADIUS 969.36

EGG:

H = 1246.76

V = 969.36

PHI BASED EGG SHAPE

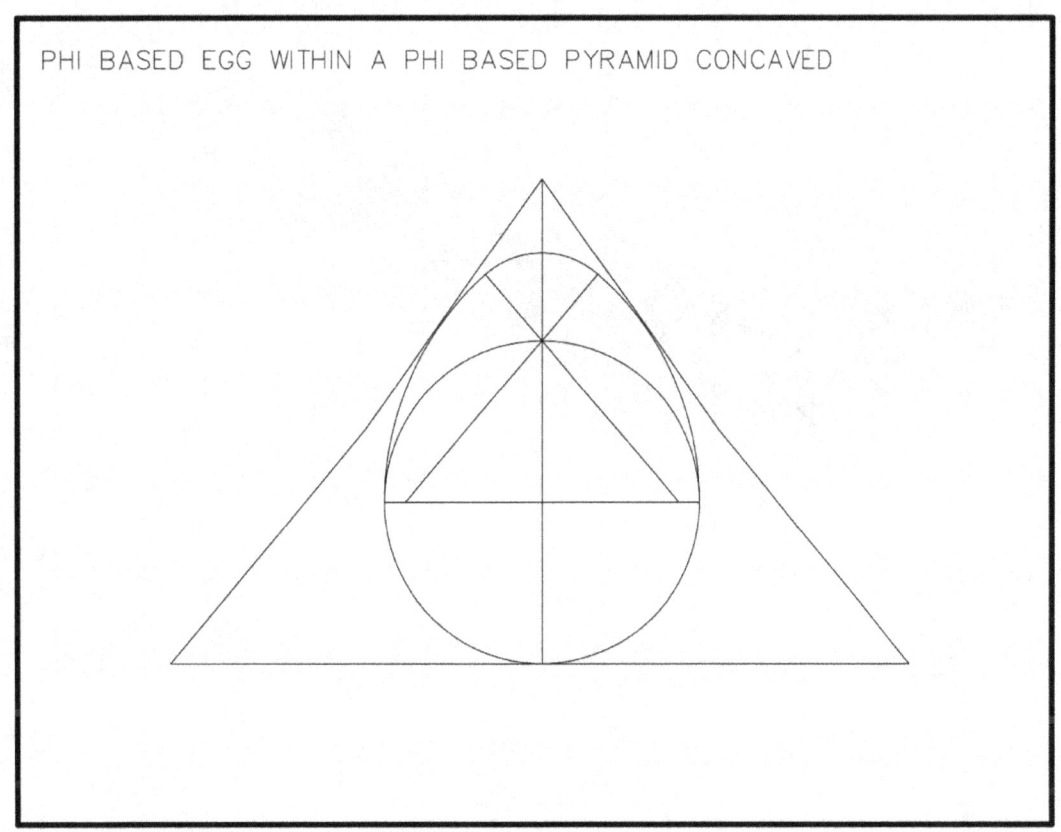

PHI BASED EGG WITHIN A PHI BASED PYRAMID CONCAVED

GEOMETRY OF AN EGG

SHOWING THE DIIFERENCE BETWEEN A DRAWING OF AN EGG AND A PHI BASED EGG

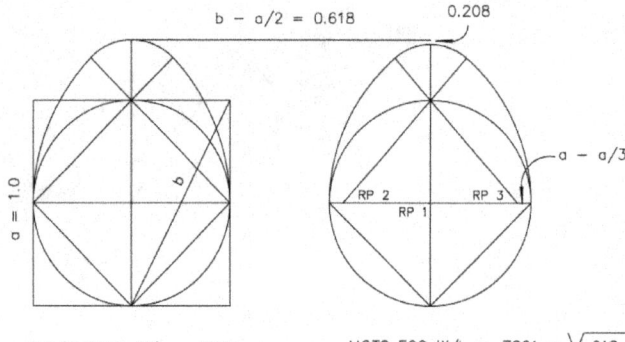

EGG DRAWING W/L = .7735

MGTS EGG W/L = .7861 = $\sqrt{.618}$

GOLDEN EGG THEOREM

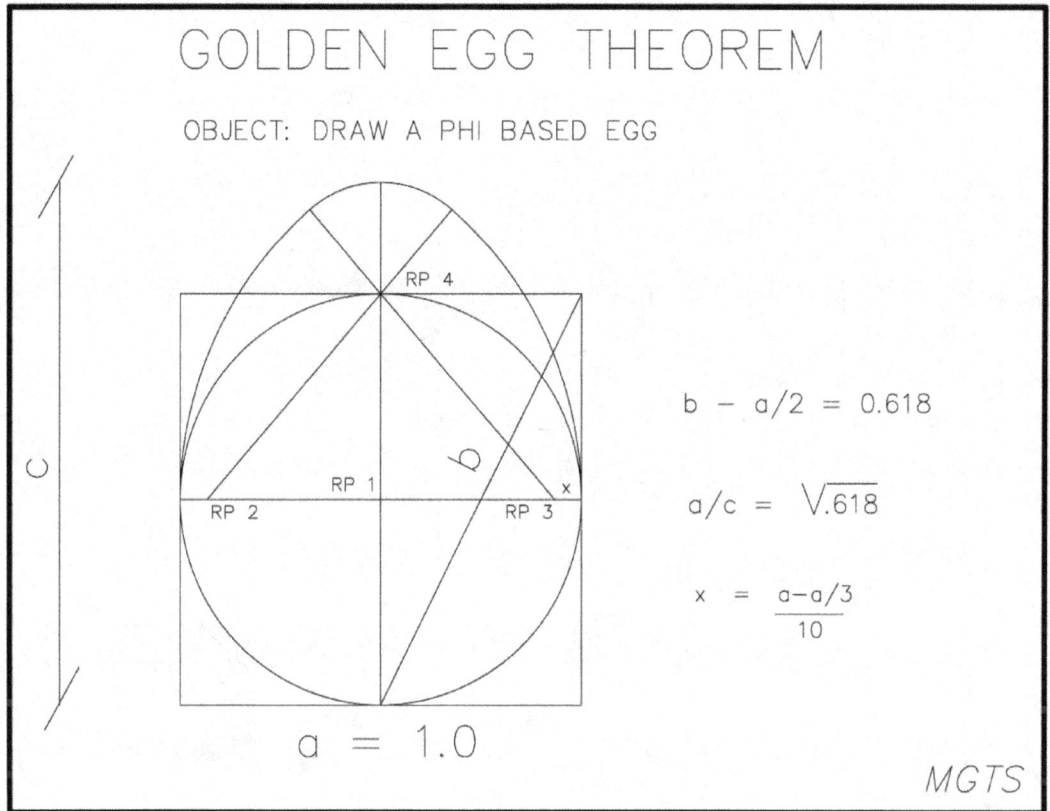

GOLDEN EGG THEOREM

OBJECT: DRAW A PHI BASED EGG

RP 4

RP 1

RP 2   RP 3

b

x

c

a = 1.0

$b - a/2 = 0.618$

$a/c = \sqrt{.618}$

$x = \dfrac{a - a/3}{10}$

MGTS

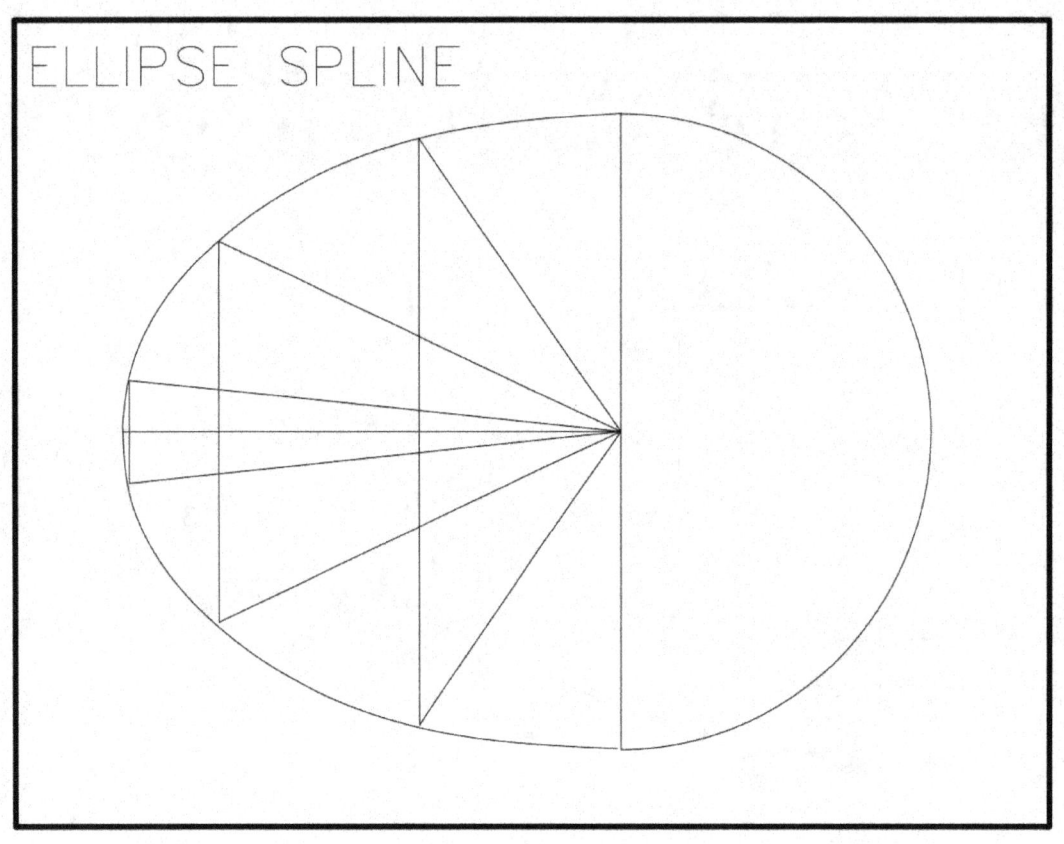

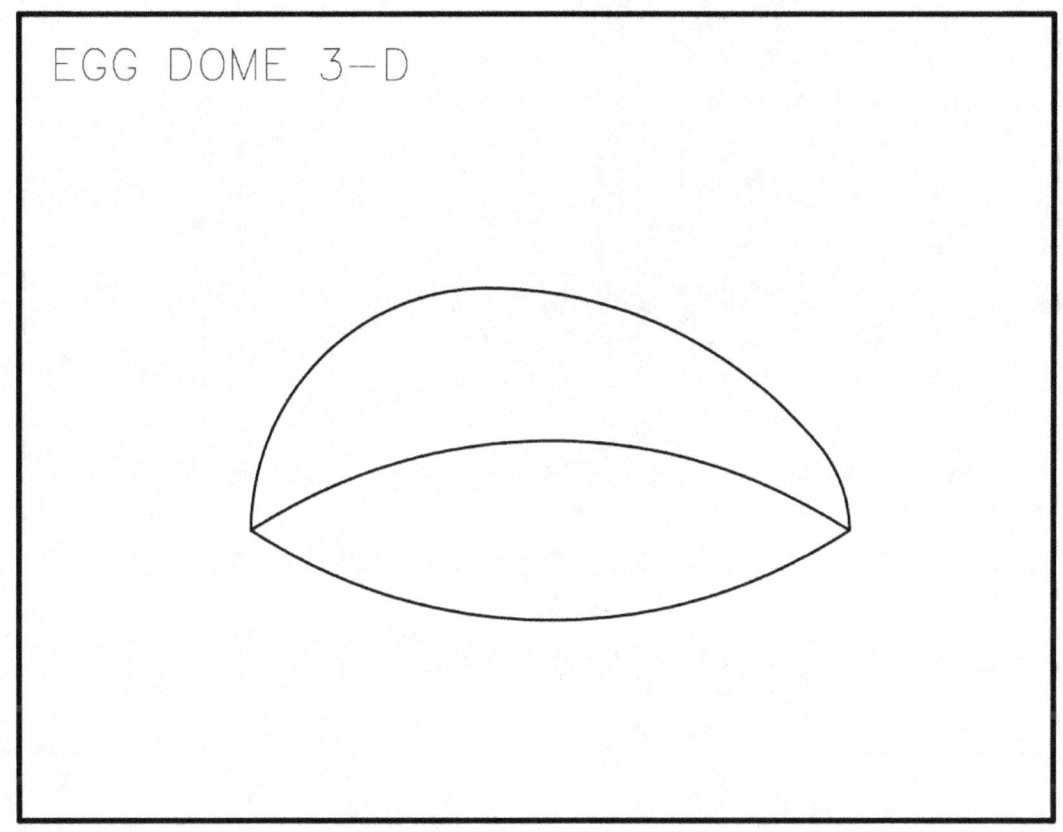

EGG DOME 3-D

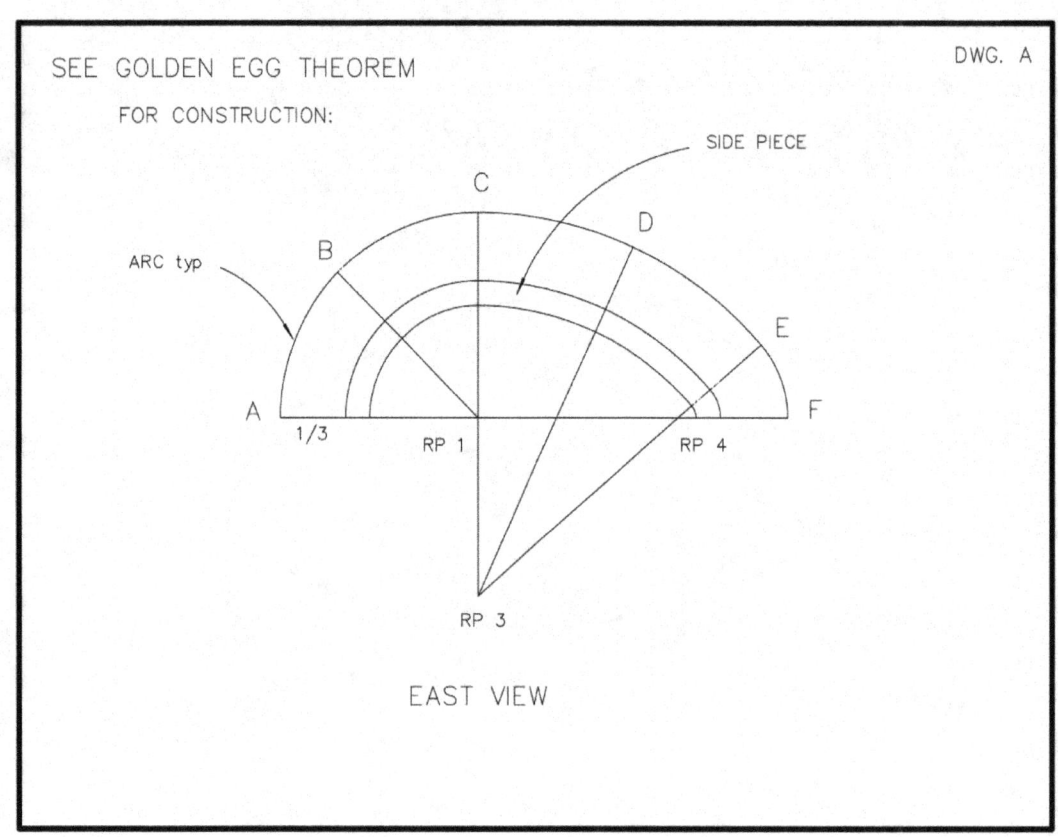

SEE GOLDEN EGG THEOREM

DWG. A

FOR CONSTRUCTION:

SIDE PIECE

ARC typ

C

B

D

E

A

F

1/3

RP 1

RP 4

RP 3

EAST VIEW

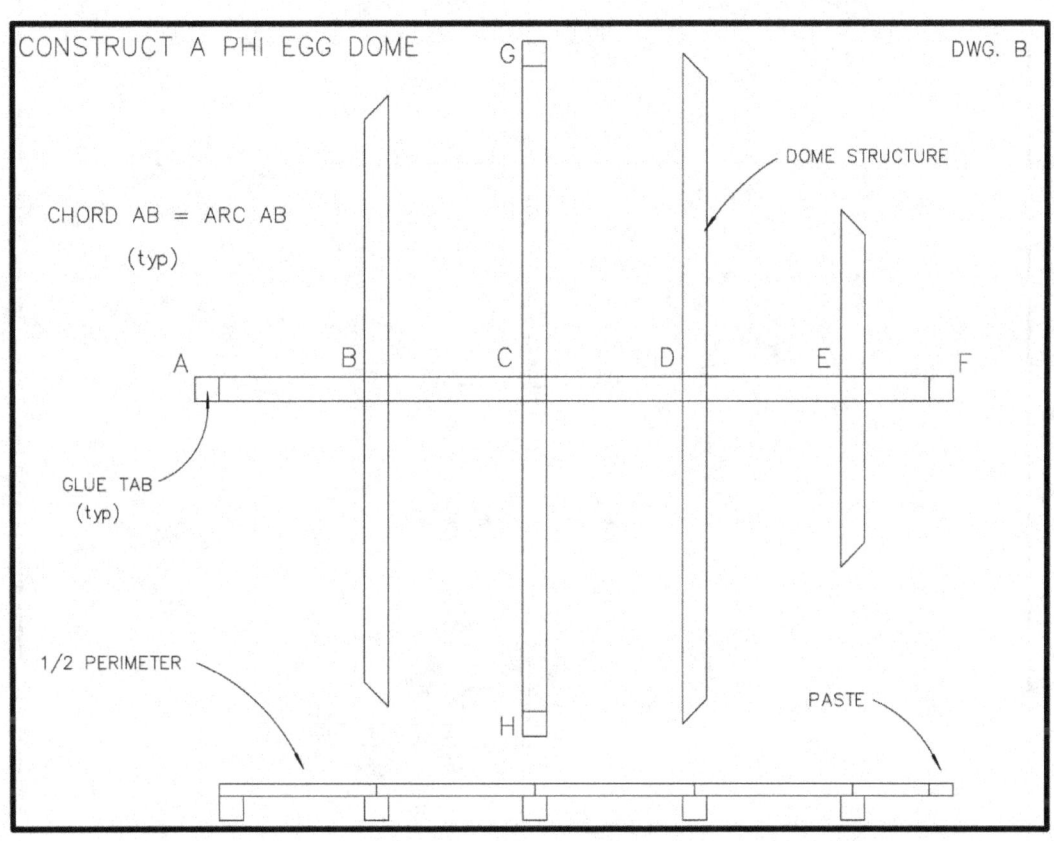

CONSTRUCT A PHI EGG DOME   DWG. B

CHORD AB = ARC AB
    (typ)

DOME STRUCTURE

A   B   C   D   E   F

GLUE TAB
  (typ)

1/2 PERIMETER

PASTE

G

H

THE COSMIC EGG

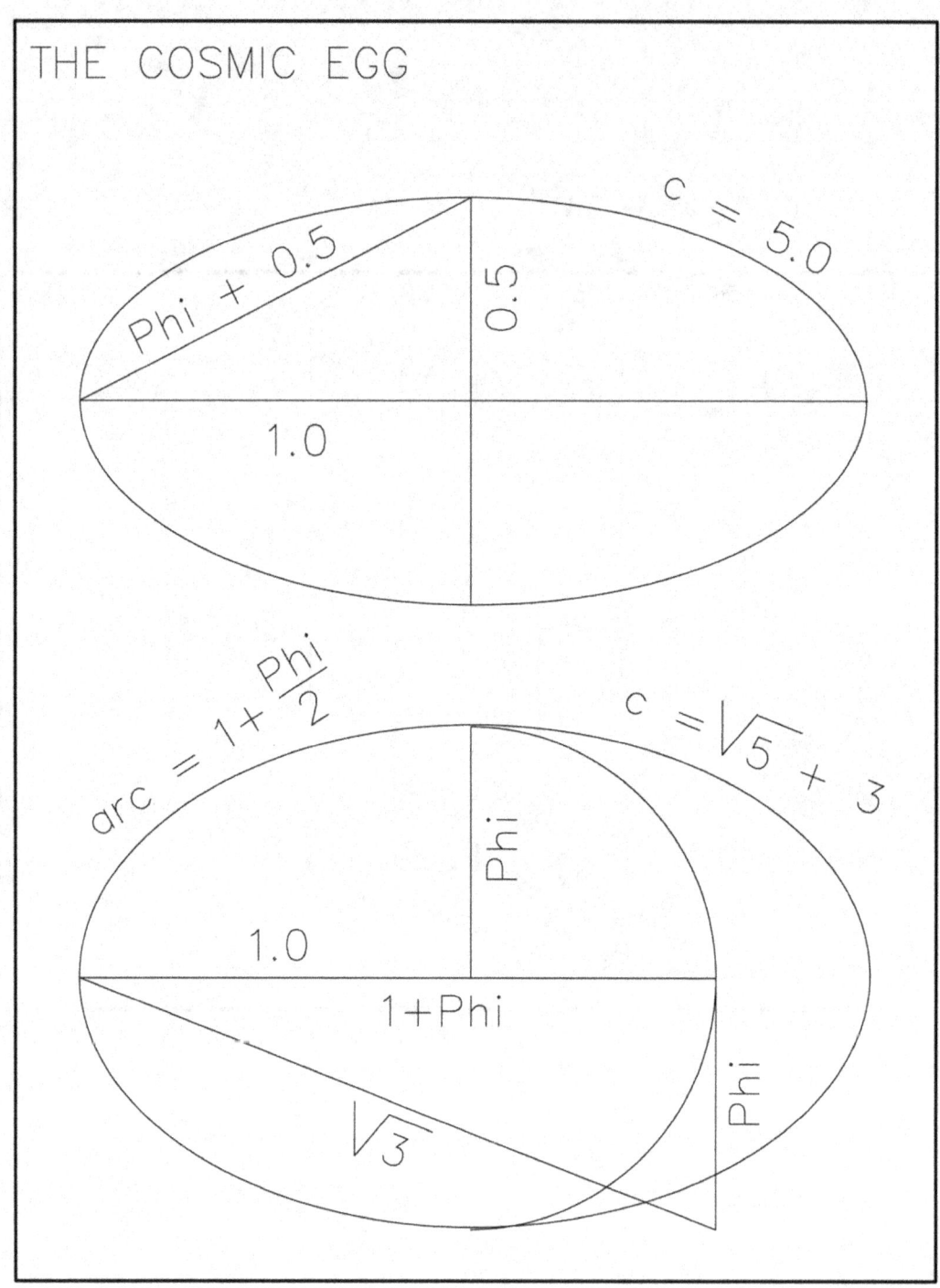

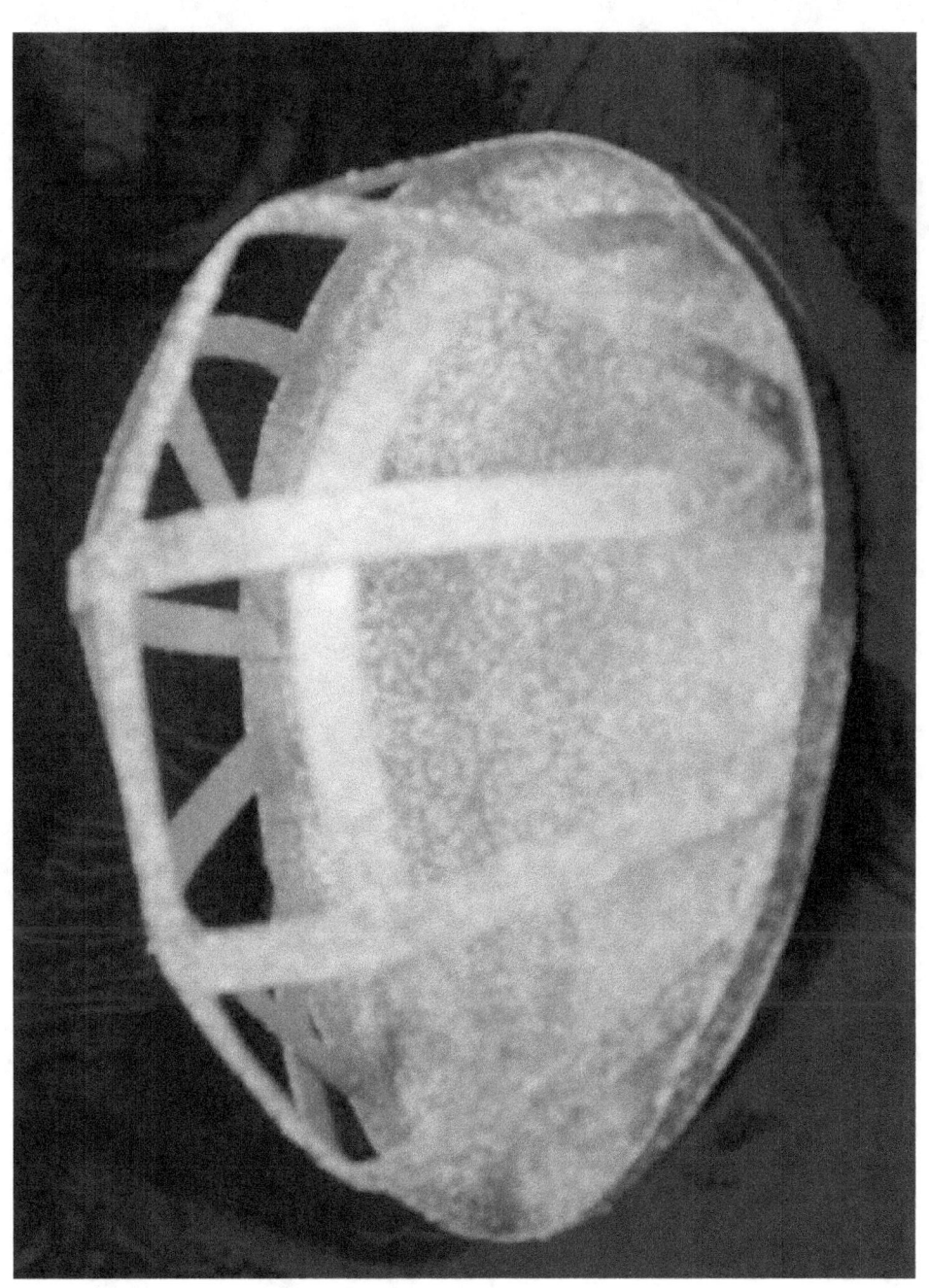

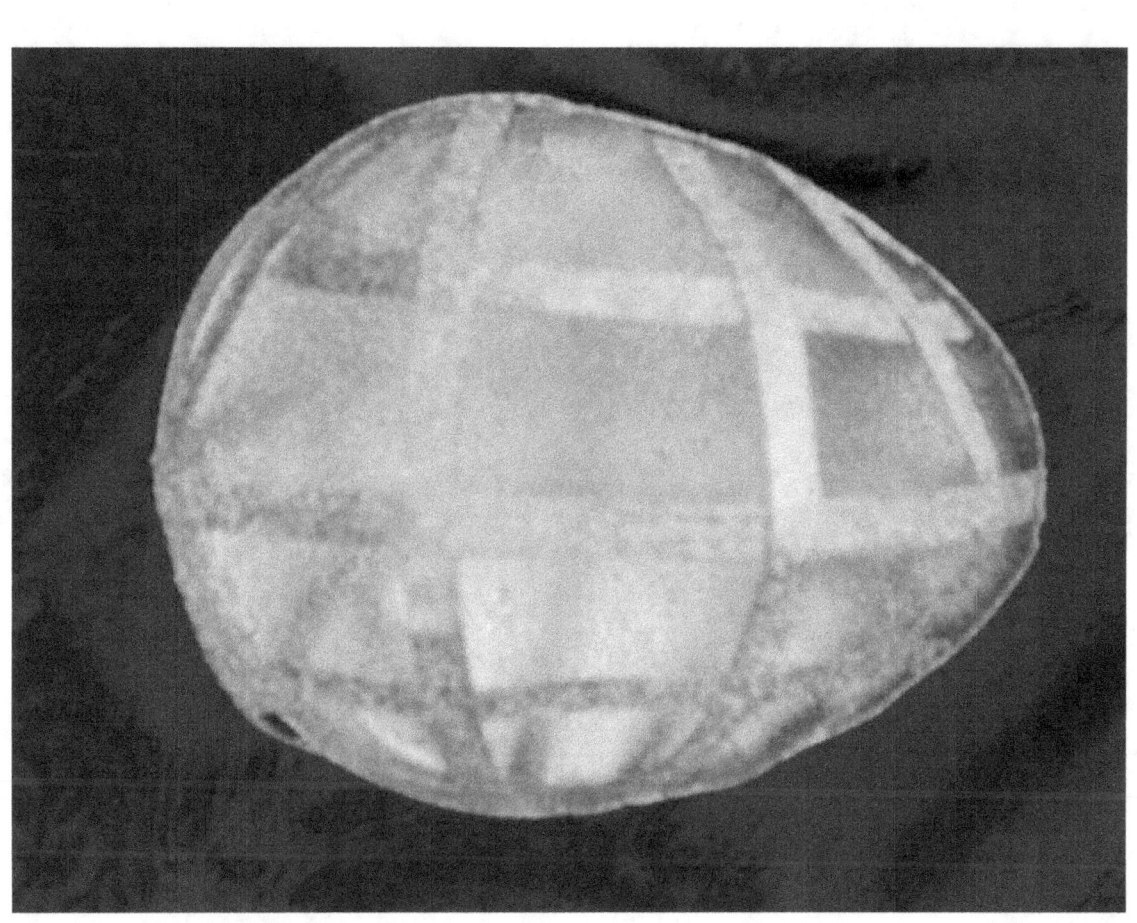

CHAPTER 7

Line and Angular Trisection and some Preliminary Closing Comments

I had studied mathematics well up into the post secondary level and had practiced application procedures amongst engineers, architects and land surveyors etc. to the point where I could give myself a private title: Professional Applications Geometrician, for what I had become to myself and others. The word geometrician isn't in the dictionary but it will do. In the old days the title would be "Geometer" but this isn't a common term used nowadays. At any rate I had developed a high level of competence, respect and expertise in the regular disciplines of surveying and engineering and that was all I really wanted to do in the first place. I was never introduced to the principles of the Golden Ratio at any time during my formal education years. I resent this and I am not alone in my thoughts on this matter. We know by now our education system has let us down in this department and western culture would be further ahead had this not been the case. The subject has been neglected long enough and it is time for us to deal with it in a thoughtful, constructive and creative manner. I had to discover its workings on my own and over a period of time, with research into its mysteries and assistance from available references I developed insights and overviews on life, its processes and terms. In a sense, because of what I do for a living and my disposition, my inquiries became not unlike a survey for a set of truths and rules I could live with by using the principles of this near forgotten science.

The one thing that eluded me in regular plane geometry problems was a workable theorem for angular trisection using compasses and straight edge. Line trisection is very simple but it doesn't help solve the angular trisection problem. It is an easy matter to measure an angle with a survey instrument or protractor and divide it into three equal parts but the real test and intellectual challenge is to do it by geometric construction analysis using only compasses and straight edge. Though, over the years, more than I care to admit to, I was determined to make it work, but try as I might I just could not develop a theorem for angular trisection except for the more simple angles. I came close a few times with more complex angles but no cigar. In more recent times I got caught up in a fascination working on an angular trisection theorem of my own for weeks using the principle of parallelism that involved Golden Section type angles such as 45, 36, 72, 90, 108 and 135 degrees using 1.0, the square roots of 2,3 and 5, and the Phi Ratio values in a unique way. Hence an introduction to The Golden Section Angular Trisection Theorem is made. Enclosed are six drawings that show it can be done this way, but only if the Golden Section Ratio is thoroughly understood. The square roots of 2,3, 5 and Phi including 1.0 didn't let me down in this inquiry and I was pleasantly surprised with the results because using this method proves to be a further assist in understanding these values. Note: in the case of the trisection detail for 45 degrees the Phi value is reduced by 0.994 % to make it work, meaning 0.618 becomes 0.614, but the solution is still quite reasonable. Using Phi and 1.0 for the trisection of 36 degrees this way appears straight forward and for 72 degrees it might seem a round about way to solve the problem but it does work. The trisection of 90 degrees using the square root of 3 and 2.0 this way is a joy because it offers a simple solution. With regards to the trisection of 108 degrees on these terms it is very simple using Phi. Reference to this angle was made in Tee Pee construction and the Number of the Beast in the earlier chapters and it is certainly a value to reckon with. And for the trisection of 135 degrees the easy way to solve is simply scribe an arc with center A having a radius of the square root of 2 through B but other solutions are offered as seen in the graphic and much is offered in it. This drawing aside the one for Elements of the Golden Section provides an even fuller relationship of these values in the arrangement of the geometry of the Comos. It can be seen that the angle of 18 degrees, 26 minutes, 06 seconds on the 135 degree trisection drawing subtracted from 63 degrees, 26 minutes, 06 seconds on the Elements of the Golden Section drawing equals 45 degrees. Aside from that the square root of the sum of the squares of the 3 x Square Root of 2 and 2 x Square Root of 5 sides of the triangle for the 135 degree trisection, x 100 = the slope side of the Great Pyramid within a thousandth. The above mentioned theorem is fair enough because compasses and straight edge are used to construct these important angles that demand special attention in the study.

The problem is, it is virtually impossible to use this method to trisect odd angles that have values in degrees, minutes and seconds. I was hunting for a more workable method to do this because, still ringing in my ears was my tutor's claim it couldn't be done but since that period in the 60's I was still stubbornly clinging to the belief it was possible. Then one day, January 1, 2007, to be precise, I was surfing the net on a geometry website fussing and fuming over the problem and, "Eureka!" there before me on the screen was a three step theorem for Angular Trisection by Archimedes. I could hardly believe my eyes. I put it on my drawing program for an angular value and saw to my delight that indeed the theorem proved out for that particular angle. Then I spent days trisecting every angle value between 0 to 360 degrees right to the second and went through a ton of ink and printing paper. My gosh, I said to myself, it took 40 years to find it but my hunch proved out and it was possible after all. Then some weeks later, I found another angular trisection theorem by Hippocrates. Now I look back on my efforts for angular trisection and come to the conclusion that I had learned quite a lot in the process and it had made me stronger in the end but I had indeed finally found it and besides that I had taught myself how to trisect Golden Section angles and this was also an important step in the process. Then, precisely on the 777 day in July this year I finally got around to drawing up the Royal Cubit Theorem and shortly thereafter determined the true dimensions of the Great Pyramid.

My tutor would be around 75 years old by now, his hair was very gray when I spent time with him in the 60's. He might very well be dead by now. If he is still around and we reconnected somehow I would have a few things to say about angular trisection to him all right, and I'm left wondering how many other students were told it couldn't be done. I think if I had an opportunity to speak with him and he still had all his faculties we might discuss angular trisection theorems that were available over 2000 years ago, and to my Golden Egg and Sign of Infinity theorems he might be somewhat attentive. I am not sure what he would say about my Royal Cubit Theorem, the Solar System and Milky Way Theorems etc., probably not much, as he would still be at the level of the Isaac Newton crowd. He might drop dead or run away once he saw what I have uncovered. We never at any time discussed the Great Pyramid or the Golden Section values and looking back I can say we were all robbed of this aspect of education and enlightenment. About all I would do is thank him for helping me to transform my stubbornness into the power of thought, and come to think of it a philosophical thought based on angular trisection is a reminder for the need of balance between spirit, mind and body. Not unlike a tree our feet are rooted on the ground, yet we are mentally and spiritually connected to the Heavens and the entire Universe whether conscious of it or not. By tri-viding the whole each piece retains its unique relationship to the whole. Here we see three that are two that are one, or one that becomes two, then three. At any time in my process I could have tossed the notion of the shape of the Universe aside and accepted what a man with far more education than I had said about angular trisection but I reached for the limits and am enriched by the outcomes in my own personal and private way.

Closing Comments:

I couldn't avoid using our favorite "8" letter word Geometry in the contents of this presentation because indeed, that is what we all intrinsically live with. Ultimately the investigation led to decoding the grand science behind the Great Pyramid when it told a story that rocked the foundations of the Establishment after which much substance was added to the inquiry. There it is before us, the most comprehensive and enlightening study on this subject, the world has ever seen before. After approaching it like any other survey / engineering type problem, or project I have ever encountered in my career I saw this one to be very special and deciphering its long lost code became a labor of love and joyous occasion. At times I was not unlike a hummingbird or a busy honey bee flitting and buzzing around in a limitless flower garden, sampling the sweet golden nectar of nature's mathematics. Wherever I searched there were more of those blossoms of ancient knowledge within reach. Apologies are made for any errors or omissions. These will be dealt with in the follow-up edition if necessary. Our fictional hairball in a space suit is on the way back home since modern science caught on by sending a return to earth signal many times the speed of light from the Great Pyramid to his space craft. In the process some thought provoking concepts and jolts of reality on the state of our affairs, sense of values and state of being within the Living Universe have been put forward because not all is what it appears to be. All is like the Theory of Everything and it is our challenge to come to an understanding with it and at least some of the way has been discussed in terms of the language of the Universe. Once the truth is separated from the false we can creatively and constructively return to the ways of the natural world in a manner more sensitive to the reason for living and in support of it. The truth of anything is available if one is willing to dig for it. For starters our education system needs to include instructions on the Golden Section at all levels in art and science programs starting at the primary level. The mathematics of today has the right idea but it is lifeless unless it is integrated with the Golden Ratio. It is as important as reading and writing and as we know, it has its fair share of arithmetic. Yesterday is gone, most of us have lost contact with the ancient ways, but we have today and tomorrow to get back on the learning curve. The hope is to release the spirit from its confines, make the leap and bridge the gap between the physical and the metaphysical, transcend the mundane and scale the rainbow's reason to find our Phoenix and explore the spiritual levels between the finite and the infinite, for it is then we can summon our true potential in order to reach the pinnacles of achievement we so desire. Life and existence is about far more than death and taxes or escaping reality by any abusive or false means and there truly is a lot more to life than getting old, sick and dying. We need to be sensitive to what we can pass on to our children and their children and be an asset to human society and life in general while in the state of living. As stated, the most powerful forces in the Universe which have no mass or substance are conscious thoughts. Other equally powerful forces in the Universe including the Life Force itself, which have no mass or substance are controlled will, desire, passion and emotion. These come with the life package which includes the state of Spiritual Being, that part of us that needs truth and enlightenment in real terms. The heaviest thing in the Universe which has no mass or substance is an uneasy or guilty conscience from a wasted life. If we are not happy with the way external conditions present themselves to us then internal changes such as the way we think and perceive need to be altered. Positive results are manifested by embracing deep convictions, respect for life at all levels, trust in your intuition, staying with your convictions based on truth, determination and perseverance that will place us in a good state for each tomorrow. The main theme of life is not just to survive but to merge our strengths, character and spirit with the divine. Again, the harmless yet intriguing notion of this thought is put forward, what else could life be if it is not an eternal state of consciousness, spirit and being within the Infinite Architecture of the Living Universe? Love is the freedom from all influences and forces which inhibit the well being of the Spirit and in knowing the God-Force we will never be alone.

Children of the Stars

Should someone ask me who I am and where I am from, the response would simply be; my origins are the same as yours, and because I could see beyond the limitations imposed upon us by the Establishment and made certain queries the answers to many questions were provided, especially with regards to those revelations that manifested themselves in chapters 3 and 4, from which we became blessed with knowledge that has been lost for thousands of years. The truth of it is, I merely let the Great Pyramid tells its story in the language of the Universe once its true dimensions were established. Our psyche is from a place somewhere and everywhere within the unlimited boundaries of the Infinite Living Universe, and we have existed in the state of consciousness and form for an incalculable period of time that has no beginning or end. This became evident when the code within the Great Pyramid was deciphered and the story will be the same once the teachings of the Emerald Tablet are revealed in the final chapter. It was an energy matrix satellite craft in the form of the Great Pyramid with its base fixed inside the hull of an enormous Ark. This was the space vehicle that travelled to its destinations in a dematerialized state at a velocity of 0.0866 light years per second in hologram / quasar form not unlike a directed laser beam that teleported the homo sapiens to their respective third planets at sun-stars in the Milky Way after relocating from the first galaxy to the west of the Milky Way that could no longer support life. Details of the distance and time it took for the journey are as follows: *1.732 x 2 = 3.464 x 1,000,000 = 3,464,000 x 1.333 = 4,618,000 light years x 0.0866 = 400,000 seconds, divided by 3600 = 111.111 hours, divided by 24 = 4.6 days. It took another 2.4 days to disengage from dematerialized state to material state while orbiting Earth, a total period of 7 days = the Solar number = one week in Earth time. This came about 360 Zodiac Cycles x 5,150 = 1,854,000 years ago. Note: 1,854,000 years divided by 100,000 x Phi = 3.0 and the square root of 3 = 1.732 = One Royal Cubit. The first humans to reach Earth had been teleported to the waters of life on the third planet from a sun-star in the north eastern limits of the Orion Arm where they became established at the Nile River Delta on Pangea from where the Empire of Atlantis began to spread to other key centers around the world at a time when there was travel to and communications with their fellow homo sapiens who inhabited the other third planets from their suns in the galaxy. The Ark design of the space craft was in the Golden Section Proportions but many times larger than the one mentioned in the scriptures. Its propulsion system was based on the principle of the central area of the Great Pyramid when it was fully functional that had the capacity to tap into the vast black hole electromagnetic energy fields of the galaxies it was navigating through. When this energy was combined in a 2:1 ratio with the piezoelectric forces generated from within the Great Pyramid it became possible to transport a human in a dematerialized state to the mother ship from the Sarcophagus in the King's Chamber through those parabolic concave features in the faces of the Great Pyramid that we will discussing shortly. It was unlike anything known in today's level of science. The giveaway on the velocity of space travel in an ancient time becomes known when it is learned that the volume of the ascending passageway divided by that of the Grand Gallery = 0.0866 = the pyramid inch, a ten millionth part of the sun's diameter. A distance that becomes a measure of time like the foot became a light year earlier in the discussion. It can be seen that the reciprocal of 0.0866 x 10,000 = the diameter of the Milky Way in light years. Therefore the velocity = 0.0866 light years per second. Some basic yet astounding proofs of this concept before the follow-up edition titled, Wings of Gold...Footprints in the Sands at Giza becomes available are explained in the following brief:

SOL / 186,216.56, divided by 0.0866 = 2,150,000, the amount this velocity exceeds the speed of light. Note: 2,150,000 divided by 5 = 430,000.

4 light years to Alpha Centauri
x 0.0866 = * 0.3464 seconds...this rings a bell.

Time it took to travel around and across the Milky Way...

1.0, divided by 1.732 x 100,000 x 100,000 = 57,736.72055 light years x (2 x Ancient Pi) = circumference 363,124.0286 x 0.0866 = 31446.54088 seconds, divided by 36 = 8.7 hours, divided by 24 = 0.363 days x (Phi/100 + 1.0 x 1000) = 365.24 days in a year. Also (363,124.0286.0286, divided by 365.24, divided by 24, divided by 100 + 1.0) = the square root of 2.

1.0, divided by 1.732 x 100,000 x 2 = diameter 115,473.4411 light years x 0.0866 = 9999.999999 seconds, divided by 3600 = 2.78 hours, divided by 24 = 0.1158 x 360, divided by 100 + 1.0 x 100 = 12.0 months in a year and 365.24, divided by 12 = 30.436666 days in a month.

The craft could stop on a dime when required and it could safely hover-vibrate through any type of atmosphere. It could sprout wings at the touch of a screen button when needed. The occupants inside the craft were totally protected at all times by an energy field that engulfed it. Included in the cargo below were highly specialized navigation instruments, a life support system, certain plant seeds and selective animal DNA that would be activated once they arrived at their desired location. It was an operation that had taken place countless times in the past. They knew all about earth, the sun, the solar system and its makeup because of their advanced knowledge of astronomy and frequent space travel episodes elsewhere in the Universe and previous excursions to this solar system. Hence the so called Garden of Eden, or its equivalent was theirs once the Ark landed at Pangaea, the only continent on earth at the time. It wasn't a weekend camping trip because they were there to stay, knowing death was the only alternative had they stayed where they were from. The early humans on Earth were skilled in the ways of the Ancient Sciences or Golden Ratio Mathematics and lived for many generations. After making some physiological adjustments with the atmosphere, gravity and climate they and their offspring flourished there at a home base for many thousands of years but since that very, very long time ago the people came to be restless and some came to be at odds with each other, therefore many of them migrated to other parts of the world after the new continents formed. The original family unit was breaking up because to be independent, venture out, explore, see what is on the other side of an ocean, what is on the other side of the mountains, see where a river comes from and where it ends has forever been an inbred trait in the human makeup. Eventually they became permanently separated after many horrific cataclysms while contending with savage earth creatures and, except for a few, no memories of their earlier training and origins existed and the process of degeneration set in with some groups who became hand to mouth nomadic hunters and gatherers while others who had the genetic strength became technologically advanced and built fantastic civilizations that have long since vanished with no trace of their existence remaining. Over tremendous spans of time these various empires rose to lofty peaks and fell into decline, then the pieces were picked up to start over again and in the end the human race became earth bound. The duration of western culture as we know it is but a mere fraction of a second compared to the time man has inhabited this planet. By now people from different parts of the world differ in appearance and race due to lengthy periods of climate, latitude and food sources. Language and customs vary amongst the peoples of the world but the human spirit has no color or is confined to any particular race. One huge difference between man and the other creatures on earth was, he could make fire, yet the history of human development on this planet became not unlike the case where it would advance one step ahead and fall back two. It might be surmised that some 5,000 years ago a certain group of humans knew the story of the Universe and the knowledge of it was stored in the ancient structures of stone like the Great Pyramid left behind for us to study so that we too could, possibly become as enlightened as them.

A thought to contend with is this; whenever gazing upon the canopy of stars above at night, not fully understanding them, yet an instinct like a genetic memory stirs, it might occur, ah yes, my origins are somehow, from somewhere out there, or in some mysterious way I can relate to them and it all makes sense because for some reason that is how I feel about it deep inside.

"For my part, I know nothing with any certainty, but the sight of the stars makes me dream" …Vincent Van Gough.

"The only real valuable thing is intuition", Albert Einstein.

When gazing up into the Heavens on a clear night it also occurs that the speed of vision which is connected to the mind, that stimulates the thought processes greatly exceeds the velocity of light and this provides an insight into the potential of human mental capacity. The reclamation of our true heritage was made when the Great Pyramid was decoded and now that the way has been shown by the wondrous science behind it, from which we will learn much more, there will come a time when we manage to explore the Universe efficiently again and be enabled to move on because there will be no choice but to relocate to other habitable locations in the Heavens from whence we came. The need to do this might be accelerated due to the global warming factor we have imposed on life and ourselves here on planet Earth. Today's space travel attempts will eventually evolve to the stage when this will be possible, it is a repetitive cycle that has taken place an infinite number of times in the past. Today we send signals to the constellations waiting for a response from intelligent life and search the night skies for the presence of extra terrestrial beings who want to visit, help or conquer us, not knowing we have been looking for ourselves all along. And now that we know who we are, we must come to terms with the enemy within. We are no longer alone but at one with the Universe as children of the stars. And though modern science and religion may scoff at such notions, what does the Establishment have on file that is more believable and even worth the paper it is written on?

Since year 2008 has begun, in terms of numerology, $2+0+0+8 = 10$ and this speaks of a new beginning. $10 = 1+0 = 1.0$, the number of all and in conjunction with $3138 + 2008 = 5146 = 5+1+4+6 = 16 = 1+6 = 7$, the solar number, the number of infinity makes it presence known in this year, 2008. Perhaps then, should mankind of today wish to take up the cause for change and the betterment of his mental, physical and spiritual state, our efforts on behalf of an improved understanding of the workings of God, Life and the Living Universe will be blessed by these influences. This is the war we should be collectively fighting, all we need to do is set our differences aside, get organized, be aware of our potential, learn that special language, get our heads on straight, roll up our sleeves and, Get to Work!…

Everything has hidden meanings when it comes to numbers and proportions it could be said, and there are reasons for all that transpires in life. The speedometer in my work truck just turned 577,000 and the word count of this edition is 109,000 rounded off. Well, $1 + 0 + 9 = 10$ and $1 + 0 = 1.0$, the number of all, the Universe and this is what we are coming to an understanding with. Perhaps not by coincidence alone a whopping phone bill for $314 came in and 314, divided by $100 = 3.14 = Pi$. Coincidences these may be, but the values relate to the details resulting from the study and are perhaps a further indication that the record of findings should be released to the public which is hungry for the answers they provide.

The Bottom Line on the most Astounding and Greatest Discovery of our Time:

The Great Pyramid, from which the ancient and sacred code of truth and light is echoed around the world and across the far reaches of the Universe teaches us what we need to know in life, and it has become our most reliable guide. With the completion of Living with Geometry close at hand we are no longer deaf to its words now that the Language of the Living Universe is understood. The way it communicates with us is in the proportions, number values and forms of the physical universe which is in conjunction with the laws and rules of the Divine Ratio, Golden Section Code. The problem has been, we and our forefathers have become deaf to its words, blind to its actions and dumb in the language not unlike mental cripples for thousands of years because, as stated earlier, somehow the link with the natural domain was lost and that has been a major setback in our development. There is to be rejoicing in the streets and dancing with the stars now that we are able to communicate with it. Possession of this knowledge is like the touch that heals when miracles take place. A reflection is, looking back on the theme, especially in chapters 3 and 4, it took only the details of three drawings, Elements of the Golden Ration / Phi Ratio, The Royal Cubit Theorem and The Great Pyramid, including the explanations thereof, which served well in the study in order to understand the workings of life, God and the Cosmos.

I trust the reader hasn't tired of number crunching and plowing through the details just yet...we are near the end of what has become an epic odyssey of enlightenment. Actually, there is no end to the story because literally, a vein of gold has been struck and it could only come to closure at the limits of infinity.

The answers to what has appeared to be the most complex and impossible to solve since time immemorial is seen in the simplicity of a twirl of the compasses to bisect a square of one unit and a line join where "all" is seen in the Geometry of the Masonic Symbol and Elements of the Golden Section drawings.

Please see the following: Recalling that the polar radius of earth, divided by its equatorial radius = * 0.997, or 0.9966 to be more precise. This constant applies to the elliptical composite of all planets and sun stars throughout the Universe...find the drawing, All Heavenly Bodies Theorem at the chapter's end where it is shown the equator diameter, divided by the slope from its end point to the top or bottom at the polar points = sq. rt. 2.0, 1.414, and the ratio of the angle to the pole, divided by the pyramid angle = 0.866 = 1.732, divided by 2. Re: the Royal Cubit Theorem.

When 1.0 becomes 2.0, then becomes 3.0 through association (+) (A ~ Phi, divided by Phi = *0.9966, divided by 10 = 0.09966.

then, 0.9966 - 0.09966 = 0.90 + 3.0 = 3.900.

then, 3.900 x .9966 = 3.8867, subtracted from 3.900 = 0.0133,

then, 3.8867 - 0.0133 = 3.873, divided by sq. rt. 5, or 2.236 = sq. rt. 3, or 1.732..., and 3.873, divided by sq. rt. 12 = 1.118 - 0.50 = Phi, 0.618 x 0.9966 = A ~ Phi...

Interestingly, 1 + 2 + 3 = 6.0, divided by 1.732 = 3.464, 3.464 squared = 12.0 and, 3.464, divided by 2.0 = 1.732, sq. rt. 3.0...

Simplicity is the key to the wonders of the ancient sciences that tells of God, life and the

Universe. It is this that has eluded Man for ages gone by, the details of which were not only forgotten thousands of years ago, but have been totally overlooked in this time era. That the so called modern era has missed the boat in terms of the true science behind life and the Universe is the understatement of our times. However, the grand opportunity has arrived to work with the fundamental principles of these subject areas and go on to learn much more. The time has come to pick ourselves up and start over again and by doing so we cannot go wrong. What has developed is, the keys to our future successes have been found in the sciences of the remote past in ancient times. Please have a look at the drawing titled, Enlightenment vs. Limitations. The Ancient Greeks who snooped around old Egypt before it went totally into decline did gain knowledge about the Golden Proportion, however, that is about as far as it went and the Romans didn't do any better. The drawing on the left sub titled, Geometry of the Infinite, Geometry of the Living provides an insight into the workings of God, Life and the Universe yet the one to the right sub titled Geometry of the Finite, provides only limitations to the material plane and such are the type of works by those early geometers who's influences continue to work on western culture with its immature level of science to this day. What happened was, they overlooked the true values of the elements of nature's geometry seen in the square roots of 2, 3 and 5 from which the secrets of life are revealed and the heavens can be charted out, and unfortunately, it has been this type of approach for the past three thousand or more years that has promoted the separation of Phi from Pi. It is fair to say then, the founders of western culture were about as uniformed as the scribes who wrote the scriptures and their thoughts on life and God were based incorrectly on a material interpretation.

The Golden Spinnaker Theorem…for sail boat enthusiasts and thinkers:

I was in doodle mode a while back, thinking how impressive the sight of a spinnaker sail is when the wind billows it out into that appealing arc. When Phi from A to B was drawn, then 1+Phi from A to C on a 90 degree azimuth, then 1+Phi again from C to G on the Great Pyramid angle, I found that arc A to G looked very much like this type of sail in action. This would be the case since the Golden Section is the geometry of nature, and the natural phenomenon of wind occurs in these terms, but then I found in this simple drawing a gold mine of information that relates to our favorite subject and my enthusiasm peaked out. Hence, the drawing titled, The Golden Spinnaker Theorem, is shown at the end of this chapter. Here is another example of how important information can be uncovered by simply experimenting. Have a look at the follow up drawing titled, Back Up to the Golden Spinnaker Theorem (Construction), see: Elements of the Golden Section. It is readily seen that bisecting a square of one unit and performing line join 1.) the sq. rt. of 5, divided by 2.0 - 0.5 leads to the values of Phi and 1+Phi, and line join 2.) provides the sq. rt. of 2. And 3.) The sq. rt. of the sum of Phi squared + 1+Phi squared = sq. rt. of 3.0. And not to overlook that the reciprocal of 1.732 - 1.414 = Ancient Pi. The triangle to the right with base 1+Phi and that its right side equals 1+Phi and the left side equals sq. rt. of 2, a simple and correct construction of this arrangement becomes available. Therefore the theorem stands on its own and is every part as important as the Elements of the Golden Section drawing which has served so well to get us this far in the discussion. It gets more interesting when xy = sq. rt. 3 and yz squared = 3+Phi which equates to 2.0 when 1+Phi is subtracted from it, and when 3+Phi is divided by 1+Phi the quotient = sq. rt. 5, 2.236. Also, the value 0.636 appears, and when this is divided into 2.0 the quotient is Ancient Pi. A further note in this drawing is, the arc AG squared = Pi - 1.0 and that is of interest. To be thorough about it another amazing feature of this theorem is, when Arc AG is divided by the full circumference from G to A having a radius of 1+Phi x 2 Pi the quotient is Ancient Pi - 3.0. The last drawing in this chapter, and aptly so, is titled, Set Sail for the Great Adventure…on the Wings of Time. In this drawing is seen the geometry of the spinnaker sail that can transport the mind on the wings of time to an ancient era on this planet. The point of the discussion is, very often information of great value is found when looking for something else, and that, it might be said, is when providence plays its hand in the face of ignorance and adversity. The same can be seen in a piece of driftwood down on the beach, in the fins of a fish, in a sea shell, the wings of a bird, the horns of certain animal species, in the physical body proportions of reptiles, insects, a human, and in the way a plant grows, alas, what has been obvious in the ways of life the intelligent chatter of the Living was missed, because mankind took a wrong turn in the study an eternity ago and hasn't been able to see for looking ever since. And now an awareness that the same is seen in the arrangement of a solar system, a galaxy and the distant star patterns, all of which express in the geometric terms of the Living Universe. Yet, all that is changing, now that our understanding of such matters has been elevated to a higher level by becoming knowledgeable about the mathematics of the Golden Section, the Geometry of Nature and its language.

Blue Prints for the Great Pyramid:

The question came up earlier in the discussion on how the blueprints for the Great Pyramid came into being, and by whom they were drawn? The answer to this question must be, its design and form is based on the geometry of the Living Universe itself, but how is this possible, and that of course leads to another very large question. In short, what is known as the works and wonders of God, the Great Spirit, the Creator, the Supreme Being, and the Grand Architect, a survey of the heavens was made possible from the long lost code and that is certainly no coincidence. It is therefore perceived that a more precise answer to the timeless question, who or what God is, has been made available. Therefore, it is believed that the Language of the Living Universe is that of the Eternal One, and Only One Spirit of Life, immortalized in stone, constructed by an intelligence on earth as a sign and witness to absolute truth and enlightenment for the benefit of Mankind and we are humbled in its presence because its code provides us with the power "to be", and the supreme, indisputable, grand knowledge of the ages is now ours to treasure.

Who, or What Group Built the Great Pyramid?

Hermes Trismegistus ~ the Thrice-Great, a god, or God in human form during a very ancient time on earth when the spirit, mind and body were at one. The study of the Kabbalistic Tradition, the Emerald Tablet, the Hermetic Philosophy and the Masonic Symbol finds the link and the connection is made. Hermes was the original Grand Master of a Lodge that had an understanding with God, Life and the Universe, from which the Craft of Free Masonry came into being. Therefore, the deduction is made, that it was the Ancient Hermetic Order who designed and constructed the Great Pyramid and the other grand structures at Giza and this is why by no coincidence, the Free Masons of the United States of America, its founding fathers, placed its image on the American one dollar bill. The true facts come in and a final proof of this contention will be made in the last chapter.

TRISECT A LINE

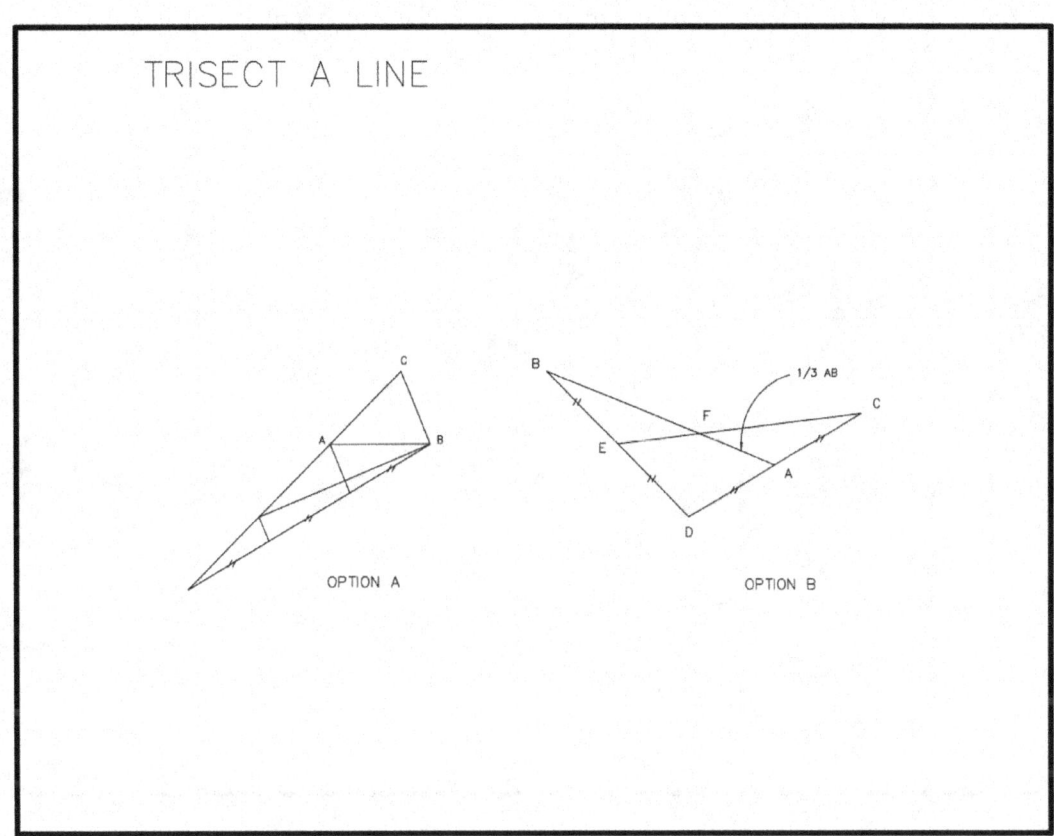

OPTION A

1/3 AB

OPTION B

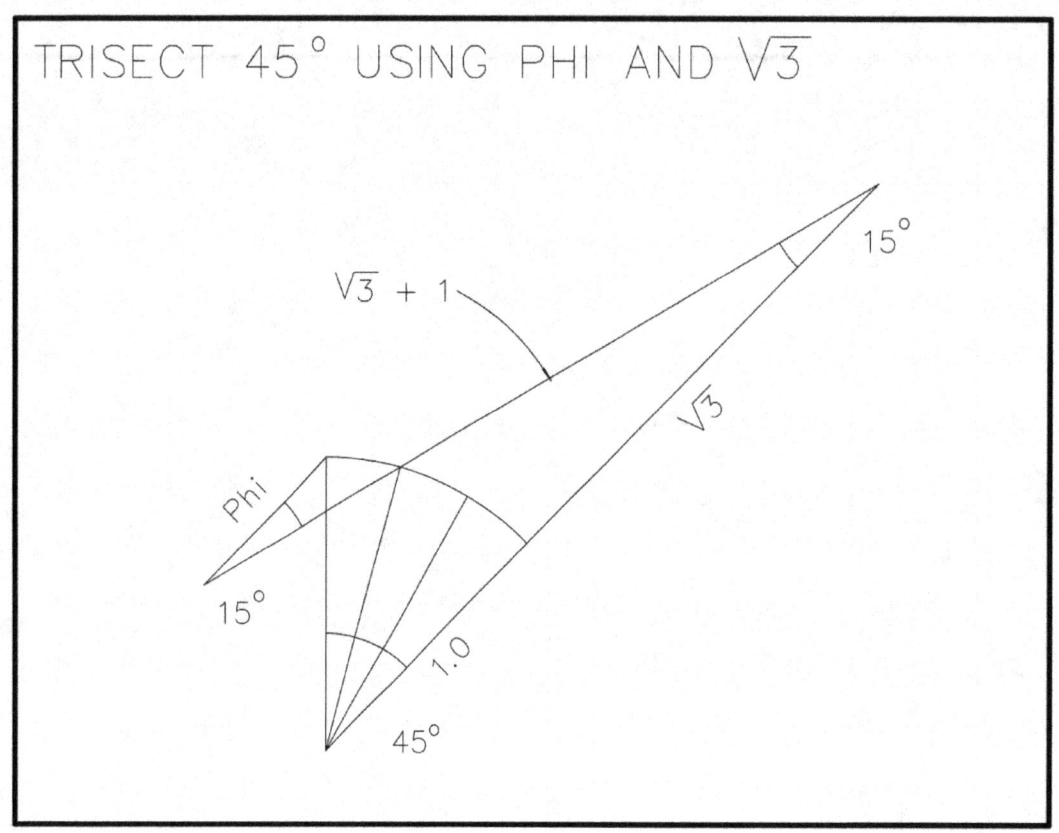

TRISECT 45° USING PHI AND √3

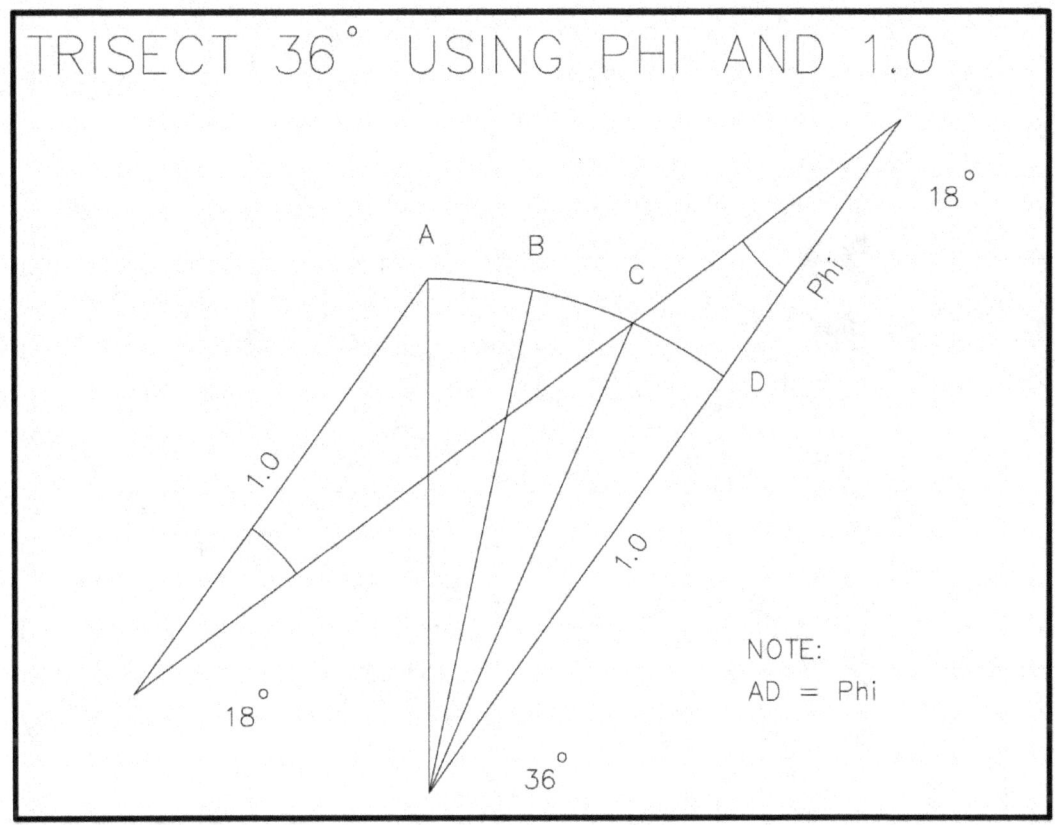

TRISECT 36° USING PHI AND 1.0

A   B   C   18°

Phi

1.0

D

1.0

1.0

18°

NOTE:
AD = Phi

36°

TRISECT 72° USING PHI

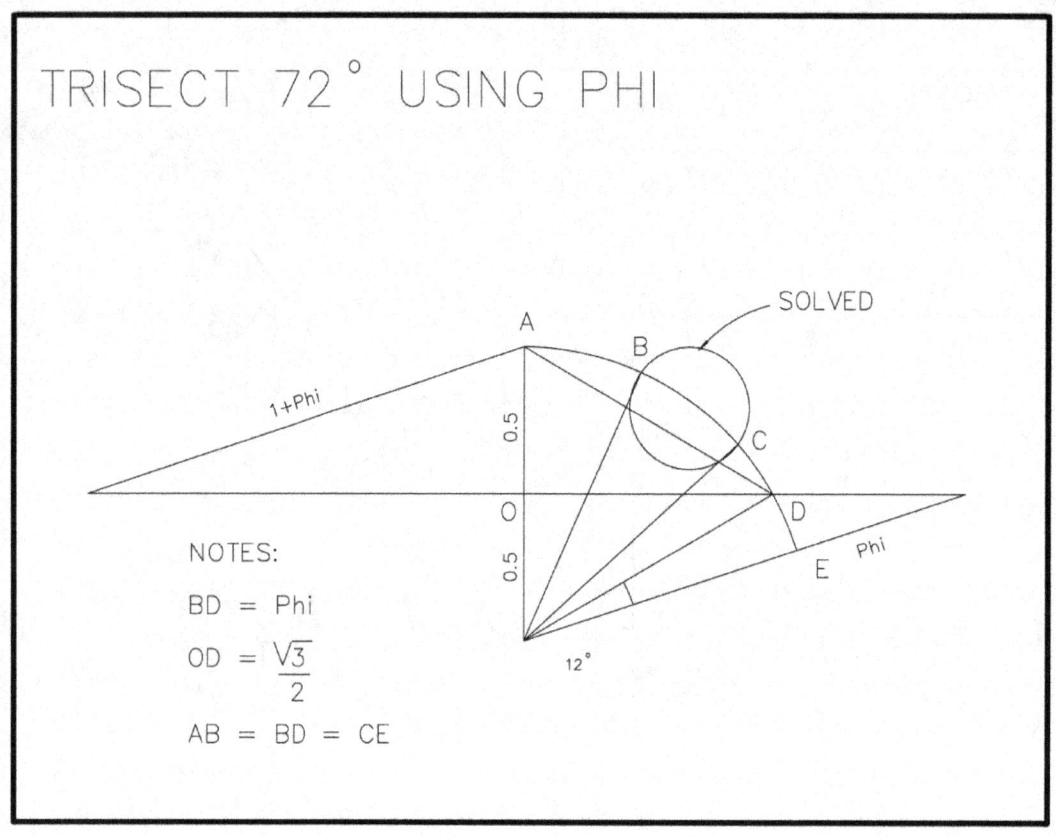

SOLVED

1+Phi

0.5

A

B

C

O

0.5

D

E   Phi

12°

NOTES:

BD = Phi

$OD = \dfrac{\sqrt{3}}{2}$

AB = BD = CE

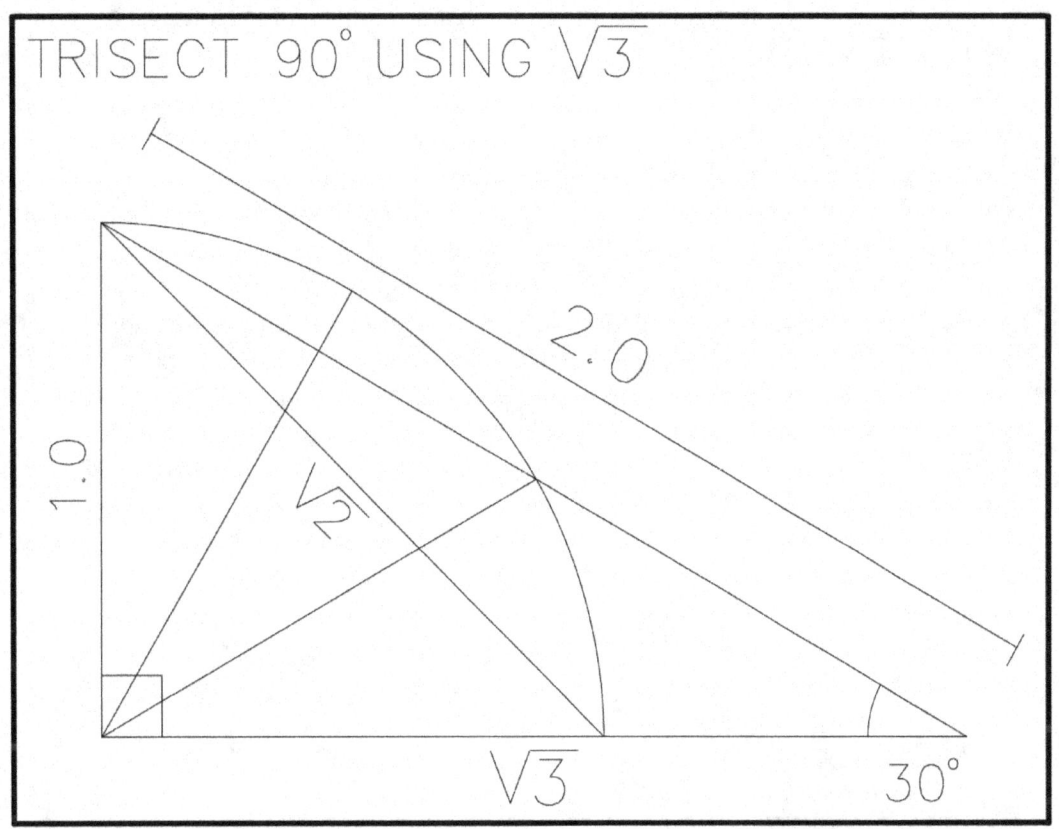

TRISECT 90° USING √3

1.0

√2

2.0

√3

30°

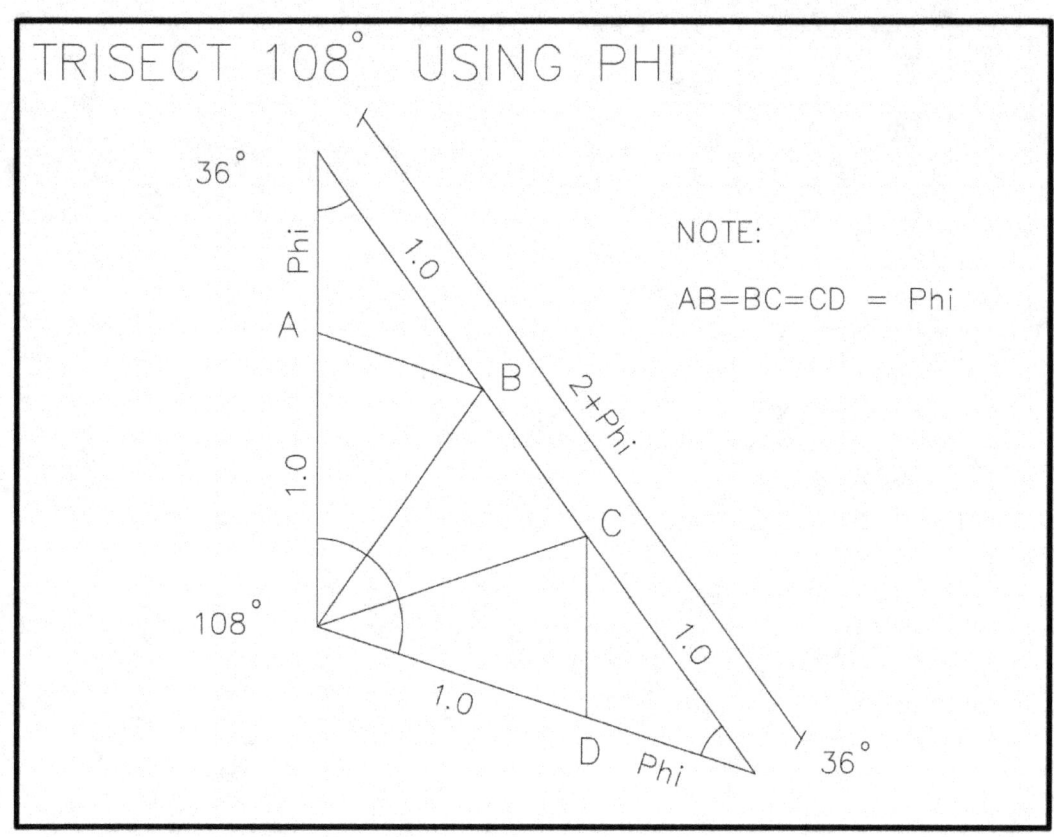

TRISECT 108° USING PHI

36°

Phi

A

1.0

1.0

B

2+Phi

C

1.0

108°

1.0

D Phi

36°

NOTE:

AB=BC=CD = Phi

GOLDEN SECTION ANGULAR TRISECTION THEOREM

TRISECT 135° USING $\sqrt{2}$, $\sqrt{3}$ And $\sqrt{5}$

MGTS

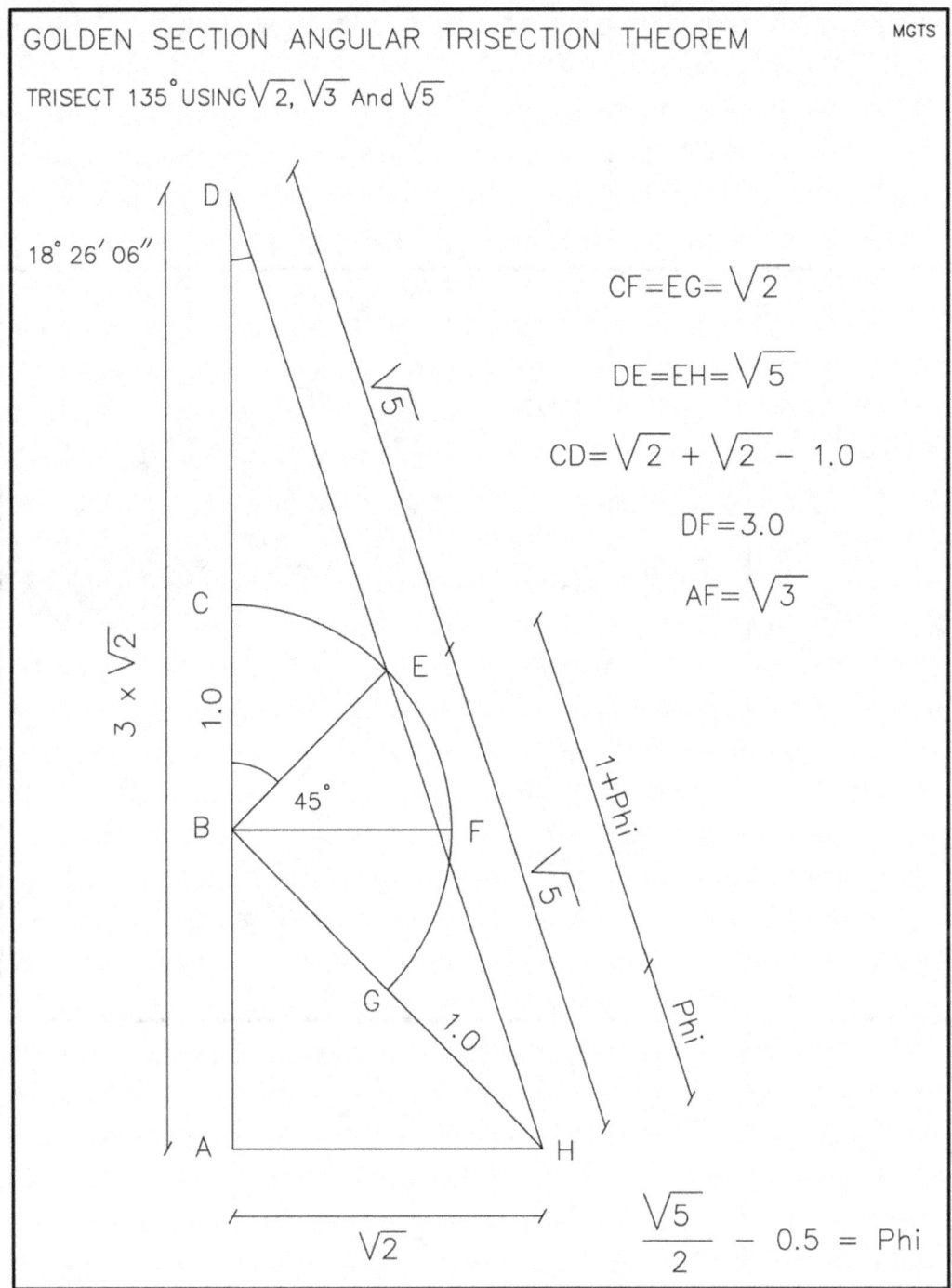

18° 26′ 06″

$3 \times \sqrt{2}$

1.0

45°

$\sqrt{5}$

$\sqrt{5}$

1+Phi

Phi

1.0

$\sqrt{2}$

$CF=EG=\sqrt{2}$

$DE=EH=\sqrt{5}$

$CD=\sqrt{2}+\sqrt{2}-1.0$

$DF=3.0$

$AF=\sqrt{3}$

$\dfrac{\sqrt{5}}{2}-0.5 = Phi$

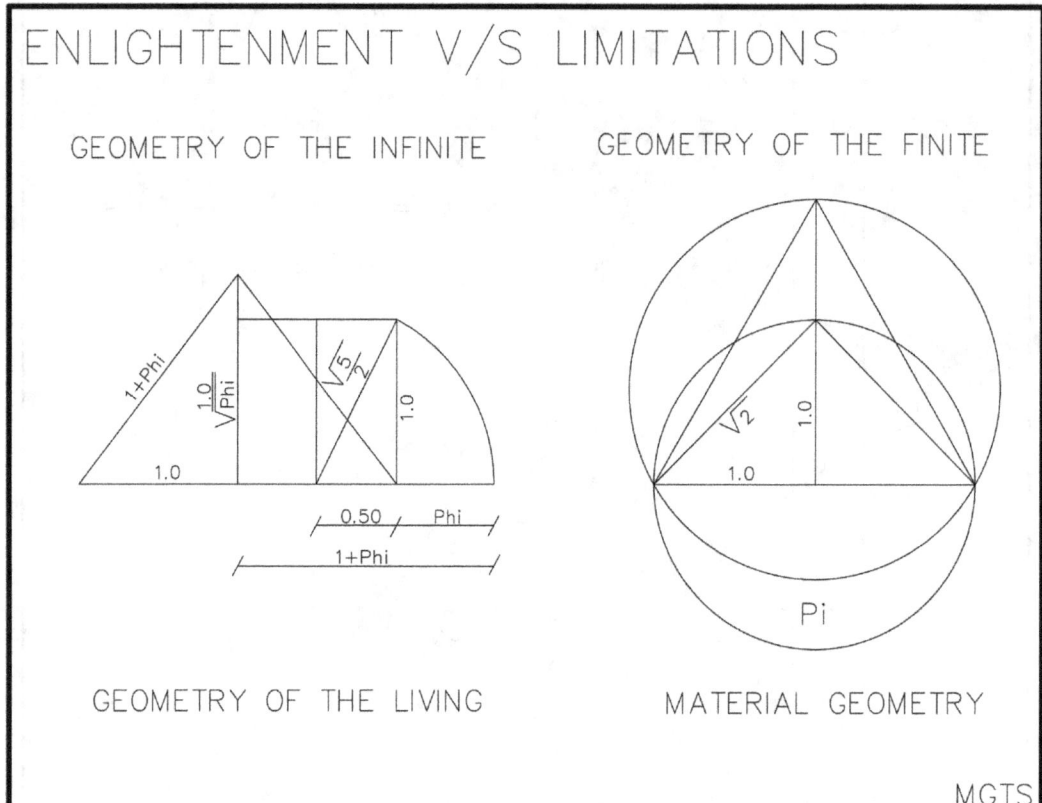

ENLIGHTENMENT V/S LIMITATIONS

GEOMETRY OF THE INFINITE

GEOMETRY OF THE FINITE

GEOMETRY OF THE LIVING

MATERIAL GEOMETRY

MGTS

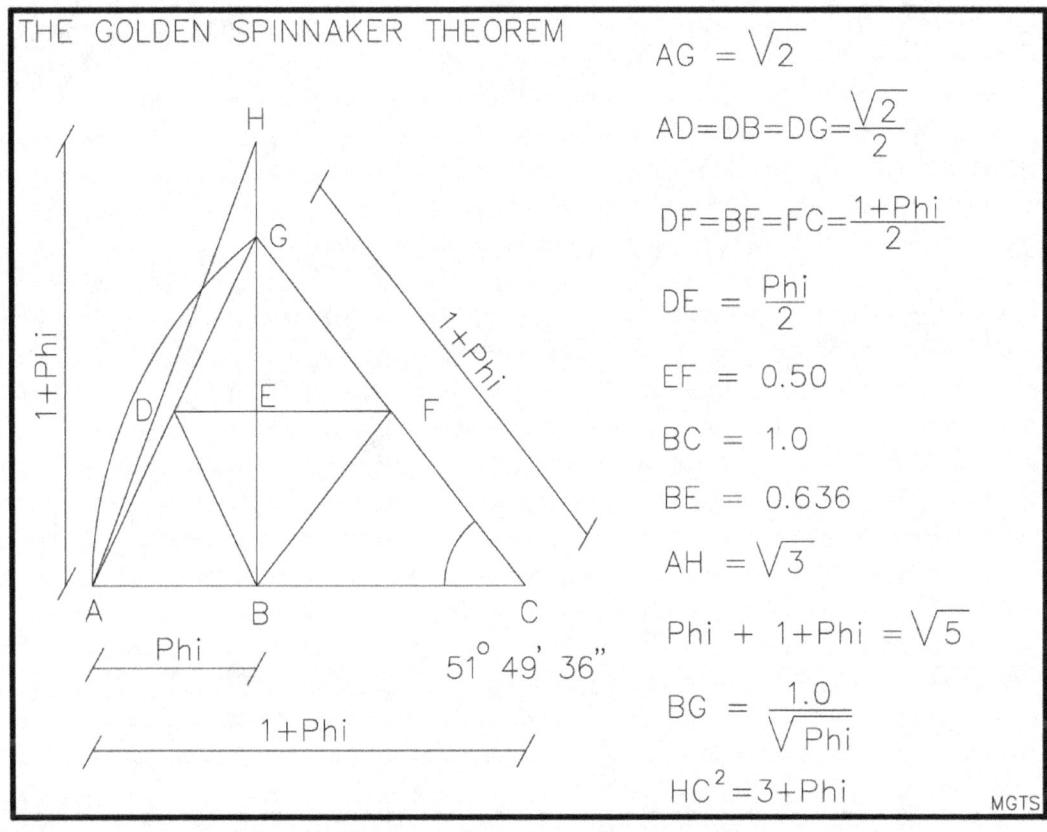

THE GOLDEN SPINNAKER THEOREM

$AG = \sqrt{2}$

$AD = DB = DG = \dfrac{\sqrt{2}}{2}$

$DF = BF = FC = \dfrac{1+Phi}{2}$

$DE = \dfrac{Phi}{2}$

$EF = 0.50$

$BC = 1.0$

$BE = 0.636$

$AH = \sqrt{3}$

$Phi + 1+Phi = \sqrt{5}$

$BG = \dfrac{1.0}{\sqrt{Phi}}$

$HC^2 = 3+Phi$

1+Phi

Phi

1+Phi

Phi

1+Phi

51° 49' 36"

MGTS

BACK UP TO THE GOLDEN SPINNAKER THEOREM (CONSTRUCTION)

see: ELEMENTS OF THE GOLDEN SECTION

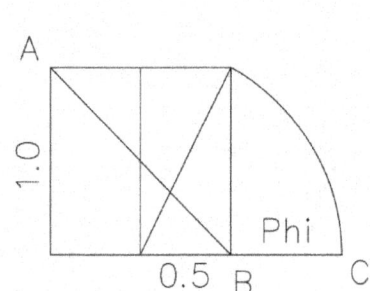

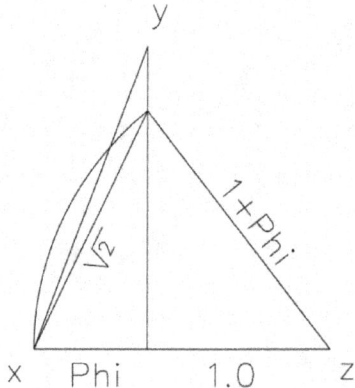

$$AB = \sqrt{2}$$

$$AC = yz$$

$$xy = \sqrt{3}$$

$$yz^2 = 3 + Phi$$

MGTS

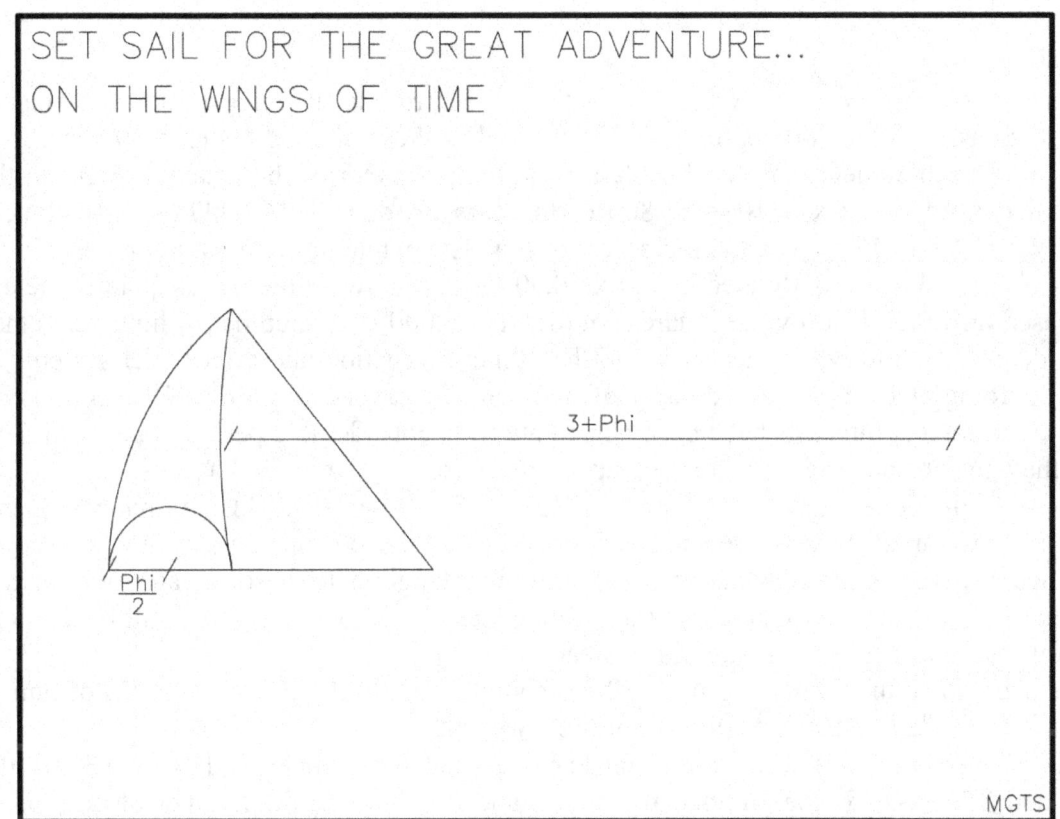

SET SAIL FOR THE GREAT ADVENTURE...
ON THE WINGS OF TIME

3+Phi

Phi
2

MGTS

CHAPTER 8

The Ancient Galactic Zodiac as it relates to Astrology and Numerology...
The 12 Planets of our Solar System, the final analysis of the Masonic Symbol,
The Big Bang Number Theorem, a proposed Labor of Love Project...

As discussed, study of the Ancient Galactic Zodiac provides the bigger picture of the heavens. It is based on the 360 degree circle of the Milky Way, divided by 12 = 30 degrees and the arc of each 30 degree interval cycle around the perimeter of the galaxy = 5150 earth years. The singular proof is, 12 x 5150 = 61,800 earth years divided by 100,000 = 0.618, Phi, or, 1.0 divided by 0.618 = 1.618, 1+Phi, the reciprocal of Phi. The magic in these values appears in the case when the reciprocal of 5150, divided by Phi x 10,000 = Pi. Knowing the radius and perimeter of our galaxy based on 1.0 divided by the square root of 3 x 100,000 = the number of light years made this a simple deduction, however, others have not had this information and their belief systems when it comes to astrology have been considerably off the mark for a very long time. What it boils down to is today's astrological forecasts have been inaccurate for a considerable period along with those who profess they understand the code locked up in the Great Pyramid. It has been determined that construction of the Great Pyramid took place 2012 – 5150 = year 3138 B.C., a time period that suits the rough approximations and preconceived notions of those who lacked an understanding of Golden Section Mathematics which provided nothing tangible to base its age on. Then it was concluded year 3138 B.C. was the beginning of the Pisces cycle and year 2012 will be the beginning of the Aquarius age. The Age of Aquarius is near.

It has been noted the number value 3138 has some meaningful qualities in terms of numerology and when it is divided by 1000. A quick review shows that

3 + 1 + 3 = 15 = 1 + 5 = 6, fire and water, the double trinity number, and 3 + 1 + 3 = 7, the solar number, and 8 by itself is the number of infinity, and 8 - 7 = 1.0, the number of all. Also, 3138, divided by 1000 = 3.138, rounded off to two places of decimal = 3.14, Pi, the cyclical number value. Then it was discovered there appears to be a correspondence between the number 3138 and the Gregorian calendar dates when it came to details such as major inventions, historical events and people of influence whose activities are recorded by history, as if the future or nature of events and incarnations of famous people was being predicted along the golden spiral of life and time. The analysis of such is provided by number interplay and their relationships with each other. The question that comes to mind is, if there are 12 Zodiac Signs and 12 months in a year, why would there not be 12 planets in ours, or any other solar system as well? We are told there are nine planets in our solar system, but in truth there are twelve and there are no names for three of them. A proof that there are indeed twelve planets in our solar system and all others has been put forward on page 117, chapter 3, therefore, modern science needs to do some homework in this area as well. Computations for the distances in miles from the sun to planets 10, 11, and 12 were carried out in a logical manner using Pi, the cyclical number beyond Pluto. In the process we are reminded that the Golden Section Proportions are the same as the nautilus shell, "As above so below, as below so above, and the familiar echo of an ancient philosophy rings true. This is seen in the drawings titled,

Planet Ellipse Orbit, Planet Orbits - Solar System and Inner Solar System - Golden Spiral, knowing the Golden Spiral energy that originates from the all powerful sun passes through all the planets. It is interesting to note that the distances from the sun to the inner planets is based on Phi, while from Mars to Pluto they are based on A ~ Phi, then from Pluto through to planets 10, 11 and 12 they are based on a multiple of 2, Pi and Phi. It is difficult to let go of the intuitive thought that those other three locations of interest in the galaxy that are identical to our solar system might possibly have animal, vegetable, mineral and quite possibly human life on the third planets from their sun / stars. Taking it a step further the probability of life existing on every third planet from its sun / star in the Universe has immense possibilities. This topic will be explored in the follow-up edition, please stay tuned. The problem is the distances between earth and those locations are too distant for communications or visitations using present day technology.

The problem with today's so called science of astrology is there are 12 Zodiac Signs but the Sun and Moon are counted in as bodies of influence when they are not planets, and Mercury is shared with Virgo to add up to 12, therefore, it becomes evident that all the flowery words and descriptions given by these pretenders for an astrology chart or horoscope reading are bogus. Planets 10, 11 and 12 have names and today's astrologists have it all wrong. Until the subject is more completely understood by us about what influences the heavenly bodies in the Universe have on earth beings the wild guesses by today's so called astrologers should be put on hold because the truth of it is, it has little to do with the divining of human life on this planet. What this study has brought to light is, the distances, locations and orbits of the planets and Zodiac constellations relative to the sun, earth and the center of the Milky Way are in the Phi / Pi Ratio, that which is in agreement with the proportions of all forms of life in the Living Universe.

Following is a list showing the Zodiac Constellation Signs and their associated Planets concocted by today's astrologers along with an attempt to assign three of the signs with the planets and constellations they belong to:

Aries = Mars (March 21 - April 19)
Taurus = Venus (April 20 - May 20)
Gemini = Mercury (May 21 - June 20)
Cancer = Moon * = Planet 11 / Cancer (June 21 - July 22)
Leo = Sun * = Planet 12 / Leo (July 23 - Aug. 22)
Virgo = mercury * = Planet 10 / Virgo (Aug. 23 - Sept. 22)
Libra = Venus (Sept. 23 - Oct. 22)
Scorpio = Pluto (Oct. 23 - Nov. 21)
Sagittarius = Jupiter (Nov. 22 - Dec. 21)
Capricorn = Saturn (Dec. 22 - Jan. 19)
Aquarius = Uranus (Jan. 20 - Feb. 18)
Pisces = Neptune (Feb. 19 - March 20)

The solution to the problem may appear simple on the surface but there is always a catch when it comes to understanding the full underlying truth about any subject. It needs to be considered, any person born between 3138 B.C. and 2012 re: the Gregorian calendar, irregardless of their birth sign, is under the influence of Pisces in some way. With more research it might be determined just how astrology really works. It needs to be determined just how the Sun, Moon and Earth fit into the picture with the other planets and physical Universe when it comes to divination of human events. That is, if there really is such a thing, because the truth of the matter is, any one can create a successful future by being diligently aware and knowledgeable about the processes of life. On the physical level the sun is the cosmic engine that is vital for life on earth in the form of light and electromagnetic energy, and certain events that take place on it such as coronal mass ejections, or solar flares and the formation of sun spots also have an influence on our planet, and of course, it helps us keep track of time. The moon influences conditions on earth in the form of ocean tide fluctuations by distorting it shape through the forces of gravity. It reflects sunlight to earth during the night for the practical purposes of life and its phases also help keep track of the time and it is the main force that moves the poles in the precessions of the solstices and equinoxes etc. Aside from these physical features of influence the proportions and distances between the solar system members and where it is located in the Milky Way with reference to the positions of the Zodiac constellations at their various distances from earth, the sun and the center of the galaxy provide the bigger picture when it comes to the true workings of astrology. It needs to be considered as well, since the Universe is a living organism, it must also have the capacity to think and sense.

Ever since the true dimensions of the Great Pyramid were established, which led to its lost code being deciphered, one thought provoking revelation after the other has come to light that challenges what the Establishment's belief systems are based on, leading to the conclusion that it is minimally informed about the most important issues man of this era should know about. This information is what the mystics, sages and holy men need to know about along with the rest of the world populace who seek enlightenment. The voice of the Great Pyramid has been heard around the world and across the unlimited expanses of the Universe, telling us what we need to know and who we are. To know is "to be" and spiritual strength is achieved by having the right type of knowledge, which is like medicine of the mind, and the intentions of the contents herein will do exactly that, mingle with the minds of the masses to find and release the "truth". The hope is, notes on these findings which belong in future text books and public libraries will be taken before they get lost for thousands of years more. The basis of an all knowing, all seeing set of truths, what might become known as a likeness to a religion that will unite mankind around the globe is in the workings. One might only need to explore the fundamental teachings of Free Masonry because it has been brought to light that the geometry of its symbol is a perfect match with that of the Great Pyramid from which many valuable truths were found. One is to the other as the mysteries of God, Life and the Universe are revealed. Please see the drawing titled, Great Pyramid and the Masonic Symbol which again serves as a reminder that it was the Science of the Hermetic, Greek name, for the Ancient Egyptian God Thoth Order from which Free Masonry originates, who designed and constructed the Great Pyramid and other wonders of architecture around the world during that unrecorded period of history on Earth. Following is the Great Pyramid ~ Masonic Symbol ~ Final Analysis.

There isn't enough room on this drawing to provide all the details but what is shown has value enough to carry on with the study, therefore, the following list will confirm the facts. The approach will be making use of the compasses and straight edge in terms of construction along with a basic knowledge of surveying.

Masonic Symbol Theorem ~ Seven Steps to Enlightenment

1.) The compasses are drawn first as outlined in the drawing titled, Draw the Masonic Symbol. The self explanatory drawing to the left of the 1.0 unit square, bisected and a line from D to K is the first step in learning how to draw the symbol. To the right, compass point at E, radius DE, scribe at D and F. The compasses form an angle of 60 degrees at the top when the intersection at J is achieved with the compass radii DF and FD, therefore, triangle JDF is equilateral. This angle plus 90 degrees = 150 degrees, the azimuth from J to F, the azimuth from J to D = 210 degrees, the supplement of the latter, and the dimension of these is 1.0. The dimensions from E to K and D to F are also 1.0 unit each. From E to J = sq. rt. 3, divided by 2. The position of K is the result of intersecting compass radii DK = FK on the Ancient Pi circumference at the top of the circle, or by producing line EJ to that point for a dimension of 1.0 unit. The corners of the square are acquired by a line intersecting the circle on both sides commencing at radius point I, parallel to line DF. Dimension EJ = 0.866, sq. rt. 3, divided by 2. Therefore, J to K = 1.0 - 0.866 = 0.134, that which is used for the width of the square to begin with. The significance of this value is, the sq. rt. of its reciprocal = sq. rt. 3 + 1.0. K to F is on azimuth 153 degrees, 26 minutes, 06 seconds and from K to D = 206 degrees, 33 minutes, 54 seconds. The dimensions work out to 1.118 = sq. rt. 5, divided by 2 and we are familiar with these values. Again, simplicity is in the workings here.

2.) Now to the drawing titled, Great Pyramid and the Masonic Symbol. The line EB is on azimuth 315 degrees and line EH is on azimuth 45 degrees. * AK = 0.136 = the width of the square. Note: * Bx + xy = BE / yz + zH = EH. BH intersects on the Ancient Pi circumference with radius 0.5, parallel to DF.

The Perimeter of the Square:

0.707 x 2 + 0.136 x 2 + 0.571 x 2 = 2.828 = sq. rt. 8 = the number of infinity...
* Inner length of square, 0.571 cubed x 1,000,000 = speed of light. The dimensions from B and H to the outside lines of the compasses = sq. rt. 5 - 2.0 = 0.236, which is the same value as Phi cubed.
* The circumference of the circle = Ancient Pi, divided by Phi, then divided by sq. rt. 3 + 1.0 x 100,000 also = the speed of light.

Note: AK - JK = 0.002, its reciprocal = 500, divided by 1000 = 0.5, the radius dimensions DE, EF, BI and IH.

3.) The triangle dimensions BC + CE + EB add up to sq. rt. 5, the same as HG + GE + EH on the right side. BC + CD + DB = sq. rt. 5 = right side. The symbol speaks totally of the dual nature of the Universe.
4.) When the intersection point at A is determined using radii 1+Phi, divided by 2 = 0.809, angles ABH and AHB = 51 degrees, 49 minutes, 36 seconds, the life angle slope of the Great Pyramid occurs. Line AB is projected to point C to intersect with line FED, and line AH is projected through to point G to intersect with line DEF. Angles ACG and AGC are also those of the life angle slope of the Great Pyramid as explained above. The dimension from C to G is 1.0 + sq. rt. Phi = 1.786151321. Another way of explaining it is, dimension EA = 1.0 over 1.136, divided by sq. rt. Phi = 1.272054628 x 2 = * 440, the number of royal cubits in the base. What becomes apparent in this discussion is, the reciprocal of 1.1363636...= 0.88 x 100 = * 88 (Aha! The number of constellations), then 88, divided by 2 = 44 x 10 = 440. To end up with the same answer an alternate option, 0.50, divided by 1.1363636...= 0.440 x 1000 = 440.

Therefore, to compute the dimensions of the Great Pyramid:

The base / 440 x 1.732 = 762.08 ft.
The slope / 220 x 1.6180339 = 355.967 x 1.732 = 616.53 ft.
The height computation / 440, divided by Ancient Pi, divided by 2...440, divided by 1.572327 = 279.84 Royal Cubits x 1.732 = 484.68 ft. / Review Chapter 3.

Hold on there, another gem of geometry has just come to light on the 9 th. day, of the 9 th. month of 2009. The brief story on an enormous revelation is as follows: I had long been intrigued by the way the last two digits in the value for Ancient pi, 3.1446540(88) happened to be a pair of eights, 88, the number of constellations and it didn't register with me right a way. First I divided 88 by 3.144654088 and the quotient was 27.984. Then history was made with the following computation : 88 x 10 = 880, divided by Ancient Pi = 279.84, the height of the Great Pyramid in Royal Cubits and its base dimension is 880, divided by 2 = 440 Royal Cubits. Shortly after it was determined 279.84, divided by 2 = 139.92 x (2 x Ancient Pi) = 880 and the height of the capstone = 1.0 Royal Cubit, divided by 2 = 0.5 x (2 x Ancient Pi) = Ancient Pi.. Then it occurred that 440, divided by 5 = 88 and 440, divided by 8 = 55, that pair of quintessential numbers mentioned earlier where 5 + 5 = 10 = 1 + 0 = 1.0 = the number of the Universe. It also occurred that 2 x Ancient Pi = 6.289308(1760) ends with the value for the perimeter of the Great Pyramid, 4 x 440 = 1760 Royal Cubits. Then it occurred that Ancient Pi, divided by 2 = 1.5723270(440) a number that ends with the base dimension of the Great Pyramid in Royal Cubits. In addition it was discovered that 8, divided by (2 x Ancient Pi) = 1.272 squared = 1+Phi, reciprocal = Phi, and it was then that I more fully realized that 8, the number of infinity, and Ancient Pi indeed have magic qualities when it comes to Golden Ratio Mathematics. Please see the self explanatory drawing titled, Great Pyramid – Royal Cubits at the end of this chapter. Refer again to the drawing titled, Elements of the Golden Section / Phi Ratio (a) and we are reminded that a square of one unit, the number of the Universe, bisected provides the answers needed. There will be a drawing titled, Elements of the Golden Section / Phi Ratio (b) in the follow-up edition which will be the beginning of a new adventure into how the Royal Cubit was dealt with during those ancient times. Two additional drawings in Royal Cubit dimensions titled, Great Pyramid Interior Detail and Great Pyramid / Planetarium of the Milky Way are introduced at the end of this chapter. These serve as a summary of what lays ahead in the study. To carry on then, it will be seen that the earlier values shown above can be equated with the true dimensions of the Great Pyramid in what at first appears to be an uncommon way.

1.0 divided by (5.00 - 1.25) = 0.2666 + (2 x 2) = 4.2666. In simpler terms, 5.00 - 1.25 = 3.75, the reciprocal of which = 0.2666 + 4 = 4.2666. A proof, since the base length has proven to be correct, 762.08, divided by 100 = 7.6208, divided by sq. rt. 1+Phi (1.786151321) / 7.6208, divided by 1.786151321 = 4.2666. Explanation: The reciprocal of 4.2666 = 0.2344 x 1.0, the number of the Universe + 7.0, the solar number, divided by 1000 = Phi cubed. The numbers in action...0.2344 x 1.007 = 0.236 = Phi cubed. The Royal Cubit Theorem makes its presence known when 1.732, divided by 1.1363636 = 1.52416 x 2 = 3.04832 x 1000 = 3048.32 = the perimeter of the Great Pyramid.

Therefore: 4.2666 x 1.786151321 = 7.6208 x 100 = 762.08 ft., the base.
Side slope: 0.809 + 0.636 = 1.445 x 4.2666 = 6.165 x 100 = 616.5 ft.
Height: 1.136 x 4.2666 = 4.8468 x 100 = 484.68 ft.
Height of King's Chamber:
B to I / I to H = 0.50 x 4.2666 = 2.1333 x 100 = 213.33 x 9.0 = 1919.97, divided by 100 = 19.1997, or 19.2 rounded off to one place of decimal. This is good to know because 484.68, divided by 213.33 = 2.272, the reciprocal of the square root of Phi + 1.0 and it appears to satisfy the question about this back in chapter 4.

Note: Ex + Ez = AC and AG x 2 = 1.44 x 100 = 144.00 = 12 squared...

* A proof of those oh so valuable correct dimensions of the Great Pyramid are at hand through the simple geometry of the Masonic Symbol.

5.) AK = 1.136, the radius of the smaller circle at the top, the compass twirling device x 2 = 0.272 = 1.0, divided by sq. rt. Phi - 1.0 = 0.272 x 4.2666 = a one millionth part of the diameter of the Milky Way.

6.) AK = 0.136 x Ancient Pi x 2 x 4.2666 = 3.6524 x 100 = 365.24 days in a year.

Also, AB + BE + EA = 2.6524 + 1.0 = 3.6524 x 100 = 365.24 days in a year.

The 12 divisions of the square on both sides x 2 = 24 hours in a day. The 8 th. division of the square, hidden by the compasses divided by 12 = 0.666... is that proportion of the height of the Great Pyramid from its apex to the bottom of the King's Chamber.

7.) It was established earlier that angle KDP has a value of 63 degrees, 26 minutes, 06 seconds, or in terms of its decimal value, 63.435 degrees. Back in chapter 3 it was determined the arc value squared is a one billionth part of the number of miles in a light year. The review is as follows: Ancient Pi, 3.144654088, divided by 180 = 0.0174703 x 63.435 = 1.108 x KD, sq. rt. 5, divided by 2 = 1.118 = 1.239. The square root of 1.239 = 1.113. The interesting results from playing around with this value are 1.113, divided by *4.26 = 0.26 x 100 = 26 and 2+6 = 8, the number of Infinity. Also, 0.26 x 5, the quintessential number of life = 1.3 x 10 = 13, the number total of the sun and the twelve planets. Furthermore, when 1.113 is multiplied by (2xAncient Pi) the result is 7.0 and the solar number comes into focus. Try this one, 1.0 x 1.113 x 2 = 0.318 = reciprocal of Ancient Pi. Last but not least 5x7 = 35, divided by Ancient Pi = 11.13, divided by 10 = 1.113 and when 7.0 is divided by Ancient Pi the result is 2.226, divided by 2 = 1.113 and the fascination with the subject grows when it is realized the reciprocal of 7.0 = 0.143. Therefore 1.0 + 0.143 = 1.143 x 100 = 114.3, the dimension from the base of the Great Pyramid at center to the floor of the Grotto. Seen here is that elusive value half way between Pi - 2.0 and Ancient Pi - 2.0.

Since the dimensions of the Great Pyramid are in agreement with the earlier input which were evaluated independently from the Masonic Symbol details, but using the same Golden Number values, not a word, phrase, sentence or computation needs to be changed except for that part which refers to the concave features of the faces. With reference to the Great Pyramid and Masonic Symbol drawing, when line yx is intersected with line AC and line yz is intersected with line AG the value between the two = 0.01618 = 1 + Phi, divided by 100. In other words the depth of the concave feature in feet = 0.01618 x 4.2666 x 100 = 6.903 ft., divided by the slope dimension = 0.011196535 and the reciprocal = CG, divided by 2 = CE and EG. The reciprocal of 6.903 = Ancient Pi - 3.0. The distance from A to x and z = 287.569 leaving a difference on the slope of 616.53 = 287.569 = 328.961, and the reciprocal of 287.567, divided by 328.962 = Ancient Pi - 2.0. This arrangement appears to have mathematical significance, therefore, there is a strong possibility this is the detail needed to begin understanding why there are concave faces on the side slopes, but the reason for these remain a mystery until this day. As stated earlier, the belief or thought still lingers that it is possible this feature, not unlike one of our parabolic dishes must have had something to do with the workings of a sophisticated communications system which operated on an interstellar basis far beyond today's understanding. Please view the drawing titled, Great Pyramid - Concave Feature. It is a little more than an artists rendition as it involves the very values which have been under discussion in this inquiry. It might be speculated that the faces of these parabolic faces were comprised of thousands of gold inlaid quartz crystals but the truth on this may never be known. Who or what were they communicating with is a good question. Then view the drawing: Masonic Symbol = Great Pyramid = 1.0, the number of all, the Universe = Infinity. Therefore the image of the Great Pyramid from the American one dollar bill seen on the book cover is in this respect a coded way of showing all of these. Also, it occurs, since the workings of the Great Pyramid has been determined to be the greatest teacher we will ever have, the same can be said about the study of Free Masonry and additional closure on the topic becomes available.

A stunning revelation is at hand when it is discovered the Masonic Symbol has had a simple and accessible code locked up within it for thousands of years from which the design of the Great Pyramid was manifested. Only one conclusion can be made, and it explains what the science of Free Masonry was based on during ancient times. To this day its modern version continues to have a profound and postive impact on its members and society in its own humble way because its belief system embraces the essence and existence of the true God Force, Life and the Language of the Living Universe. The question of what did the blueprints look like that the Great Pyramid was based on is answered and that thousands of years ago a road map of the Cosmos was made available by way of the ancient sciences. Beyond doubt, this was the elite and mysterious group who were in possession of those closely guarded secrets that went missing a very long time ago. These are what Bro. Manly P. Hall refers to in his edition, The Lost Keys of Free Masonry. The descending and ascending passages, the King and Queen's Chambers, the Sarcophagus and Grotto were used as part of the initiation processes we, today, can only guess and theorize about. All along there has been far more to this elementary looking symbol than first meets the eye. What was lost has now been reclaimed and returned to us by way of dedicated research, focused attention on the details and a form of providence, or divine guidance while conducting a survey in search of the truth. My involvement with Free Masonry surely has lead to finding the answers needed. I would often see the Masonic Symbol on the face of what would be the likeness of the Great Pyramid and one day I was motivated to go to the drawing board to perform an experiment on paper. It was simple enough to determine point A using the compasses because AB on the left side which equals AH on the right side = DK - 0.5 + 1.0 = 1 + Phi, divided by 2 = 0.809. Then after drawing in lines AC, CD and GC as discussed above I got into fine tuned computations mode and there before me was an astonishing gem of geometry that captivated my full attention. The solution came when it was concluded the symbol represents the vertical plane of the Great Pyramid. I have had intuitive thoughts, or premonitions while engaged in the study from the beginning and now it has been proven, that Modern Free Masonry, to say the least, truly is a living link to a wondrous ancient period in history. Another detail I came across in the compasses and square arrangement for the Lodge of Perfection, within the Masonic Scottish Rite Order of Free Masons has an enneagram in it and this resolves the question back in chapter 2 of where it comes from. Ancient Egypt of course. A drawing of this symbol is made available at the end of the chapter. What is interesting about a detail in the drawing is, the circumference of the circle around the enneagram, radius. 0.183 x Ancient Pi x 2, divided by Phi x 100,000 = the speed of light. The radius 0.5, divided by 0.183 = sq. rt. 3 + 1.0. This serves as another example of just how serious the philosophy and science of Free Masonry was in ancient times, and it still is very much so today.

I am left to wonder if it is my distant Celtic heritage, choice of education style, my career activities in survey / engineering practices and general perceptions of life that has enabled me to come to terms with the ways and means of the ancient mind set. The simple truth of it is, anyone who would trouble themselves to find the keys to the geometry of nature and life could do the same. Perhaps the greatest challenges for Modern Free Masonry will be to restore the Great Pyramid. It is possible this ancient group were the survivors of the Atlantis cataclysm, and this almost appears to fit in with the Noah's Ark story but so much time has gone by, the true story is left entirely up to speculation. They and their blood line might well have been the first human inhabitants on earth, from a distant location elsewhere in the Universe as discussed in the previous chapter. The assumption made is the Pharaohs of earliest times in Egypt are the descendants of the ancient Free Masons but no hard evidence on such a claim exists, however the thinking on this does seem quite reasonable.

Whatever the case might be, it occurs that the image of the Masonic Symbol has been available for a long while right under the nose of Western Culture since its beginnings in this era, and its full meaning was missed by those who have been parading around in our history books for centuries claiming to be at an advanced level in the scientific disciplines and passing themselves off as authorities in the subject areas of astronomy and how it relates to the Great Pyramid. It is apparent this information was not available at the time Pythagorus was instructed by the scribes and educators in Egypt when he was initiated into Free Masonry of that time, because by then, in all probability, the Language of the Universe had been long forgotten at a time, some 2500 years after the Great Pyramid had been constructed. To be certain, Isaac Newton and his following would never have given this symbol a moments thought because they were influenced by the dark age religion of the time that was totally ignorant about the ancient sciences. The scriptures do mention the Great Pyramid and God's Ratio but the details were never explained because the scribes had no real understanding of the subject. They were unaware that reference to the Great Pyramid was an acknowledgement of the ancient masons who left behind for us the true story of God, Life and the Universe, carved in stone so that we could follow their way. This Alter to God, the many obelisks and temples are the remnants of a far greater legacy, near forgotten in the sands of time.

A drawing titled, The Masonic Trowel is presented at the end of the chapter and it has great meaning to all members of that persuasion. Associated with Wisdom, Strength and Beauty, on the esoteric level the Masonic Trowel is used to spread the cement of brotherly love and affection that unites the Oldest Brotherhood of Man in the history of the world, among whom no contention should ever exist… uniquely, the geometry of the trowel shown inside a drawing of the Great Pyramid is by no coincidence the inner perimeter of the compasses and square. When the perimeter is divided by Ancient Pi, its sq. rt. = the sq. rt. Of 3.0 - 1.0. The azimuth of the square on the right side = 150 degrees, subtracted from 180 = 30 degrees. That azimuth subtracted from the azimuth of the right side of the Great Pyramid = 8 degrees, 10 minutes, 24 seconds + 30 degrees = azimuth 38 degrees, 10 minutes, 24 seconds, that of the azimuth of the Great Pyramid on the left side. It is the same case working from left to right.

The last of these type of drawings, Signature of the Grand Architect explains why it is difficult to end this story. That is because the pathway has been along the Golden Spiral which travels to the infinite reaches at both ends. This helps to describe the link between the grand capacity of man's mind and the Universe when it has the right type of information to work with. The compasses and square represent the power of thought and the ways and means of drawing on our mental resources and intelligence to strengthen the fabric of our "state of being" and that is why the study of the Golden Section, the Geometry of Nature, of Life and the Living Universe is so very important. It provides the freedom and joy of learning. Now that we have become acquainted with the ancient way of thinking it is seen, the answers to the most complex are based on simplicity itself and this provides the means to better understand ourselves while that material entrapment is overcome and our spiritual connection becomes improved under the guidance of that which is the "Truth and Light of God". In closure on this section, it can be said, the mysteries of the Great Pyramid are more fully completed with more answers than questions.

About Numbers…The Big Bang Number Theorem

In solving the mystery of where numbers come from, the answer is based on a relatively simple and natural sequence of order involving the following analysis:

9.0, the triple trinity number, the full number before the recount takes place
8.0, the number of infinity
7.0, the solar number
A ~ 9 x 9 = 81, divided by 10 = 8.1
B ~ 7, divided by 100,000,000 = 0.00000007

A + B = 8.10000007

The reciprocal of 8.10000007 = 0.1234567891 x 10,000,000,000 = 1234567891

…Not unlike the revelation beheld when 1.0 was divided by the square root of 3 x 100,000 to determine the dimensions of the Milky Way in light years. To be sure then, number is in all, all is in number, geometry is the expression of their form and proportion in the eternal " Living Universe. Numbers, like the Universe have always existed.

Note: The reciprocal of 0.111... to infinity = 9.0
The reciprocal of 0.999...to infinity = 1.0

Further input on the Golden Numbers:

2.236 x 1.414 = 3.162 squared = 10 / 5 x 2 = 10

1.732 x 1.414 = 6
1.732 x 2.236 = 15
1.732 x 1.414 x 2.236 = 30

More on the AE dimension in the Masonic Symbol Drawing:

1.136363636...x 5280 = 6,000, divided by 440 = 13.636363.

3950 + 13.636 = 3963.636 = earth radius...

3950 x 5,280 = 20956000, divided by 484.685 = 43030. 186,216.56, divided by 43030 = 4.33, divided by 10 = 0.433 = 1.732, divided by 4.
The Big Bang Theory
    The word infinity implies, that which is limitless, boundless, endless and eternal…It is not a number but an important concept to mold the mind with, and the only thing that suits this definition is the unlimited expanse of the Universe itself, yet modern science has been insisting for a lengthy time that there must have been a beginning to matter, energy, space, time and life in the Universe and at one time very long ago when it was the size of a pea, there was a Big Bang, a cosmic explosion of some sort that started it all. The basic flaw in this argument is, modern science doesn't comprehend the true meaning of infinity or eternity. Thoughts in these terms coupled with an understanding of the Language of the Universe and the above number theorem reveal the truth in order for the spirit to merge with the divine and this is a "state of being" quite beyond that which rational science or any religion has to offer. It is about who you are and why you are incarnated as a

human being. In fact it is the other way around from what the establishment thinks. Within the living Universe, matter, energy, space and time exist, and the grand, ancient science behind the Great Pyramid that has been tapped into has been eternally in operation long before there was a Milky Way in energy form.

Review of the Mystical, or Metaphysical Meanings of Numbers:

1.) The meaning of 1.0 is seen in the Magician, One is the whole, One is all. One is the most fundamental part of all numbers. 1.0 = the Universe.

2.) The prime feminine number, duality, change, all pairs, oppositions, conflicts, complements, partners, man's awareness of his existence and therefore the number of intellect as distinct from ego, perception. 1.0 = 2.0, the mirror image of itself.

3.) Trinities ~ Child birth, life and death. Beginning, Middle, End. Childhood, adulthood, old age. Heaven, earth, the underworld. The frost world, the earth, the fire world. Heart, mind and spirit. One becomes two, becomes three as a result of union.

4.) The World, Earth, the Establishment, rational thinking based on truth tempered by wisdom. The number of the elements, earth, water, fire, air. A totally safe number, the square, and by implication the cube and polished act. A feminine number.

5.) The quintessential, most holy and lucky number, or the number of uncertainty and discouragement. The pentagram, Man, the five senses, Life, the four cardinal points plus the center, the four alchemical elements together with that which encompasses them all. Five frequently occurs in the natural domain and is associated with eroticism. As most odd numbers it is male in it orientation.

6.) Symbolic of the union between Fire and Water and therefore of the human Soul, ambivalence and effort, the ending of effort. A double trinity.

7.) An indivisible number and therefore compared to the God Force, Cycles of Time, periods of growth, the chakra centers, the number of music. Colors in the rainbow, the solar number, the number of completion.

8.) The intermediary between terrestrial and eternal order, infinity, with the balancing of forces, or with the equilibrium of different forms of power. The number of regeneration, genius, strength, inspiration, evolution and justice.
9.) Consciousness and psychism, efficiency, renewed energy. The triads of triads, the triple trinity and therefore a number of great power associated with the occult and with triple synthesis or disposition on the corporal, intellectual and spiritual levels simultaneously. A male number.

10.) Perfection, beginning a new cycle, completion, over completion, decadence, death to the past, future new life.

The human psyche is at all times under the influence of the mysticism of numbers…

To go along with this are the Seven Hermetic Principles, or Laws of Nature…

"The Principles of Truth are Seven; he who knows these, understandably, possesses the Magic Key before whose touch all the Doors of the Temple fly open.

1.) The Principle of Mentalism - "The all is mind; the Universe is mental."

2.) The Principle of Correspondence - "As above, so below, as below, so above."

3.) The Principle of Vibration - "Nothing rests, everything moves; everything vibrates."

4.) The Principle of Polarity - "Everything is dual; everything has poles; everything has its pairs of opposites; like and unlike are the same; opposites are identical in nature, but different in degree; extremes meet; all truths are but half truths; all paradoxes may be reconciled."

5.) The Principle of Rhythm - "Everything flows, out and in; everything has its tides; all things rise and fall; the pendulum-swing manifests in everything; the measure of the swing to the right is the measure of the swing to the left; rhythm compensates."

6.) The Principle of Cause and Effect - "Every cause has its effect; every effect has its cause; everything happens according to Law; Chance is but a name for Law not recognized; there are many places of causation, but none escapes the Law."

7.) The principle of Gender - "Gender is in everything; everything has its Masculine and Feminine Principles; Gender manifests on all planes."

Other input from ancient times...The Emerald Tablet

Also known as the Smragadine Table, Tabula Smaragdina, or Secret of Hermes, a large green hued tablet, dimensions unknown, made of crystalline material, perhaps of emerald gemstone itself, with an acient script engraved in it. First found by Alexander the Great in 230 B.C. when Eygypt was conquered. Then it was hidden away and refound in a cave-tomb on the Giza Plateau. Its age is estimated to be at least 10,000 years and its author is allededly Hermes Trismegistus, known by some as an Egyptian sage, a messenger of the Gods to humans in a very ancient time. Considered to be one of the greatest treasures of the Western World it was more fully deciphered as late as 1969. During the middle ages the alchemists saw it as the answer to their quest for making gold out of base metals such as lead, or finding the gold of the spirit in man from the earth upon which he stood and its relation to the cosmos. Their main problem was, they and others of that time didn't understand the workings of the Golden Ratio. The interpretation of the Golden Tablet was worked on by many notable individuals including Isaac Newton. It reads as follows:

True without falsehood, certain and most true, that which is above is as that which is below, and that which is below is as that which is above, for the performance of the miracles of the One Thing. And as all things are from One, by the mediation of One, so all things have their birth from this One Thing by adaptation. The Sun is its father, the Moon is its Mother, the wind carries it in its belly, its nurse is the Earth. This is the father of all perfection, or consummation of the whole world. Its power is integrating, if it be turned to earth.

Thou shalt separate the earth from the fire, the subtle from the gross, sauvely, and with great ingenuity. It ascends from earth to heaven and descends again to earth, and receives the power of the superiors and of the inferiors. So thou hast the glory of the whole world; therefore let all obscurity flee from thee. This is the strong force of all forces, overcoming every subtle thing and penetrating every solid thing. So the world was created. Hence were all wonderful adaptations, of which this is the manner.

Therefore am I called Hermes Trismegistus, having the three parts of the philosophy of the whole world. What I have to tell is completed, concerning the Operation of the Sun.

Knowledge of the Golden Ratio offers this interpretation:

The basis of the Golden Section Ratio possesses the three primary parts under the influence of Nature's Geometry. These are 1.0, Phi and its reciprocal 1+Phi = the three parts of the Philosophy of the Whole World concerning the Operation of the Sun. The Astronomic Unit between Earth and the Sun = 91,796,000 miles x 5,280 = 484,682,880,000, divided by 1,000,000,000 = 484.68288 feet = the height of the Great Pyramid which was designed, constructed and overseen by Hermes Trismegistus, the Thrice Great. Its construction was overseen by him. The dimensions and positions in the galaxy of Earth and the Sun become known. From these ingredients of knowledge a road map of the Universe is created...and thou hast the glory of the whole world, and all obscurity flees from thee. (in other words, one is no longer ignorant, but becomes aware of the true story of God, Life and the Universe.)

1.0 = the Living Universe = Phi / 1+Phi = Spirit of God and Life, free of its confines.

1.0 = the element number for hydrogen, the lightest, most abundant chemical in the Universe. 1.0 to 0.5 = a 2 to 1 ratio = 2 molecules of hydrogen to 1 molecule of oxygen = H_2O = Water = the main element that sustains and promotes the existence of life, an operation of nature that has been on going for eternity. This appears to speak of hydrogen combustion power of which the byproduct is Water.

My first Grandson...A study of his makeup employing the Golden Ratio approach in terms of realistic astrology and numerology as they relate to his birth date and those oh so meaningful numbers.

Julian Douglas ~ Male ~ Born: January 12, 2008...Astrological Sign Capricorn

Julian ~ Greek origin ~ Love's Child ~ variant Julius

Douglas ~ Scottish and Gaelic origin ~ the powerful Red and Black River Clans of Scotland

Numerological values attached to his names:
Julian ~ 10 + 21 + 12 + 9 + 1 + 14 = 67 = 6 + 7 = 13 = 1 + 3 = 4

Douglas ~ 4 + 15 + 21 + 7 + 12 + 1 + 19 = 79 = 7 + 9 = 16 = 1 + 6 = 7

Number influences related to his birth year re: the Galactic Zodiac

3138, divided by 1.618 - 5.0, divided by 100 = 0.5 - 5.0,
Divided by 1000 = 0.005 = 0.055 = 1.563, 3138, divided by 1.563
= 2008, birth year.
2008 = 2 + 0 + 0 + 8 = 5145 = 5 + 1 + 4 + 5 = 15 = 1 + 5 = 6.0, spirit number

In terms of the mystical, or metaphysical meanings of numbers and numerology which need to be fully understood in the process it can be seen that the quintessential number 5 energy of love and good fortune surrounds him and he is blessed with the numbers 1.0, the number of all, along with the energies of 4.0, the number of stability and 7.0, the solar number and 6.0, the double trinity of fire and water, making a very healthy combination. This is who and what you are about Julian Douglas, a very powerful soul and much, much more.

Welcome, Julian Douglas to a challenging life filled with adventure, love and enlightenment.

** We Love You **

According to the analysis he is well blessed with favorable numbers and being a Capricorn the hope is, his life will be filled with grand opportunities and spontaneous activities. Of course we want only the best for our children and theirs. I believe the above numbers have meaning and I trust that I will be a positive influence on him as he grows and matures. At this stage we cannot discuss that special language but for now he is surrounded by many Phi creations I have made for my daughter over the years. I made a Milky Way Pyramid Crystal for her to use while she was pregnant with him and she has success with it to help him get to sleep or when he is fussing. I predicted his sex to be male before he was born using my Gold Masonic ring suspended by a thin thread over his mothers tummy area when she was around three months along. Every one was amazed when the announcement came in, " It's a boy "! Predicting the sex of a child is quite simple, when the ring makes a circle the unborn child is a girl, when it goes back and forth, it's a boy.

Sand Box in a Pyramid ~ a Labor of Love Project

I live and work100 miles north from Julian and it isn't very often we get to visit. To help with this problem I have a plan in mind that will be a favorable influence on my new grandson when he is at the play stage. Please see the following conceptual drawing titled, Sand Box in a Pyramid and my intentions are explained in the graphic sense. While playing inside it he will be affected by the energies that are generated by the proportions of the structure just like the one I made for myself. I won't have to say a word to him about the language of the Universe because it will be happening for him before he learns how to talk. Spending time in my own pyramid has proven that that it is a worthwhile venture because, simply stated, its energies promote good health when the human metabolism harmonizes with the life force of the living Universe. The schedule for completion of this labor of love project is the Spring Equinox of 2009, by which time Julian will be 15 months old, just about the time a tot is ready to play in a sand box. His will be a rather unique one and it will become affectionately known as Julian's Space, where like a seedling the workings of his mind and body will spontaneously sprout under the influence of the Universal Mind. Inside there will be sea shells, especially of the snail type and crystals on the sand with star fish hanging down from fish netting suspended from above and these will surely catch his attention. While playing with his toys in his space his mom will be with him and they are talking about family overnight camp outs in it, and this will be a benefit for them. It was decided to construct the 2:1 access slope outside on the east side so as not to limit the play area inside. The upright support members on the outsides which will be fixed firmly to the sloping members with 6 inch screws will need to be adequate and secure because dry sand weighs 100 pounds per cubic foot and the volume of the sand box will be around 29 cubic feet and this equates to 2.9 tons. Because the most common constituent of sand is silicon dioxide, usually in the form of granulated quartz there is no better material to use for reasons that have been previously discussed. As a safety precaution the area below the sand box will be skirted in with plywood and access doors will be installed so that an adult can use the area below for storage. Violet hued plastic sheeting will be placed over the upper four faces in order to filter out harmful UV rays and it will be an impressive sight to be sure. I am thinking now that the pyramid should have a 12 foot base in order to provide more room since his parents and playmates will be spending time in there with him. This means the dimensions shown on the drawing will increase 1.2 times, or by 20%, including the volume of sand needed. The sand box will then be 8.0' x 8.0' with an area equal to the number of infinity squared and that sits well with me. It occurs that a facility such as this could be put to good use at all daycare centers in order for children to get a good start in life. To take it a step further the entire daycare facility center could be constructed in the form of a Phi Based Pyramid. It could have a 40 foot base with a height of 25.44 feet with a slope dimension of 32.36 feet. It is a thought at least.

Beijing Olympics 2008

That impressive structure, the Wall of China can be seen from the moon and during the 2008 Olympics the world outside got a glimpse of what this land of wonders is about. Approximately one sixth of the world population, China is on the move. In less than a generation it has transformed into a highly industrialized state with needs the same as western culture. In its main centers like Beijing it suffers pollution problems just like us. No doubt we will get to mingle and trade with China on a greater scale in future with opportunities to learn more about their fascinating culture.

The Bird's Nest Stadium looks somewhat like an elongated wrens nest with its opening on top. Its design was conceived of by one of their own artist / architects and is based on a series of overlapping ellipses. Its flowing lines and natural looking shape is certainly appealing but I am not acquainted with it proportions to give an opinion on whether or not they involve the Phi Ratio. Aside from the competitions the opening and closing ceremonies were very enjoyable. During the opening ceremonies which occurred on the eighth day of the week, on the eighth day of the month in the eighth year of the 21 st. century an unforgettable presentation was made that showed what China's contributions to the world have been with regards to gun powder, paper, printing and use of the compass. During the closing ceremonies was the first time I saw a light wheel bicycle and I was dazzled when these appeared with their lights on in the darkened arena area. These are very clever as the pedals drive a sprocket that engages with a chain fixed in place inside the inner circle and the frame consists of three fused ellipses that connect with the outer wheel at 60 degree intervals. I couldn't resist generating a drawing of one to show my appreciation for it.

I have studied the images of their dragons and in my opinion a thorough knowledge of the Golden Ratio is needed to draw them and I took note that the Olympic Torch was adorned with Golden Spirals. Some years ago I became engaged in the study of Tai Chi Chuan, which means, "The Supreme Ultimate Force." It is about natural movements, a meditative exercise for the body. Tai Chi activity, a meditative exercise for the body, has been practiced throughout China for countless centuries and this could be considered another gift to western culture where there is now a large following. Other gifts from them in these times are in the form of acupuncture and herbal remedies. It does no harm to study Taoism which features the working forces of Yin and Yang, the dual forces of the Universe. During the closing ceremonies the Olympic Flag was turned over to the Mayor of London, where the 2012 Olympics will be held. It was then I had a flash of inspiration along with some intuitive thoughts on what this stadium would be about with regards to the climate conditions, customs and history of England. Hence, another drawing came into focus:

Umbrella Dome ~ London Olympics 2012

I borrowed from the drawing titled, Masonic Symbol = Great Pyramid = the Universe = Infinity. There are of course a number of proposals in for this facility and plans for it most likely have already been decided but it will do no harm to play around with the Language of the Universe awhile and see what develops. This accompanying design would provide a grand opportunity to participate in a project that deals with the golden numbers and geometry of the ancient sciences. It works out that the 6 factor of the inner field dimension, 660 x 2 = *12 and the number of feet in one imperial mile, divided by *12 = 440, the number of royal cubits in the base of the Great Pyramid, and the square root of *12, divided by 2 = 1.732 ft. x 440 = 762.08, the number of feet in its base. Also, 5280 feet, divided by 660 = 8, the number of infinity., and 66 feet = the old surveyors chain length. The air supported dome material would be made of Teflon, which was used for the construction of the highly successful B.C. Place in Vancouver, Canada which has been functioning very well since 1983 in a west coast marine climate . A hue of translucent green for the dome seems an appropriate color.The height from center field to the top of the Umbrella Dome would be 619 feet. What is shown in this drawing is a seating capacity of 110,600 based on the area in square feet, divided by 6. An additional gallery above the ones shown in this drawing would provide seatting capacity for 193,300 spectators.

The drawing titled, Umbrella Dome - Plan View shows a proposal for three of 12 aisles, like a clock circle, and the three tiers, or spectator galleries encircling the inner field where more than enough room is provided for the pentathlon and decathlon events to take place. The perimeter of the dome will be securely fastened to the substructure and the air flow generated from the lower levels within the structure will billow it out and support it without difficulty. The electric fans will have backup power sources during severe weather conditions such as generators, propane, natural gas, temperature differential air flow intake and body heat when it is full to capacity. The slope of the galleries are those of the Great Pyramid and this attribute along with the Phi Dome will promote a healthy environment for both athletes and spectators. It has been noted that a play field area with a challenging diameter of 660 feet lends itself to a professional baseball park as the statistics on the home run distances to the left, right and center field fences are similar. The air-supported dome has become a popular feature for all outdoor sports activities. There is a possibility for an additional egg-shaped dome that could be included in this proposal but for now the circular coliseum concept will do. Having a circular umbrella dome for the 2012 Olympics makes some sense because it reflects on the west coast marine climate and the custom of using one in the British Isles. To take it a step further in order for the architecture to relate to local history the granite stone work on the outer face will be reminiscent of the castles of olden times with a mote established around it with eight iron gate openings at 45 degree intervals. On each entranceway of the draw bridges various sculptures which reflect on the history of the land and its people could be established. Turrets at the top of the outer wall above the gates with lookout pads on them would make its appearnce even more appealing and realistic. There would be plenty of room beneath the galleries where athletes can prepare for events and for shops and displays of all sorts. Beyond the 2012 Olympics the facility would be active for generations to come for professional soccer, dancing, music festivals and many other activities. Resistant to high winds and snow loads the future of this revolutionary type of dome construction is most promising.

Note: This design proposal is in feet because it has been proven beyond doubt, imperial units relate to the laws of nature and the ancient sciences, whereas the metric system does not, therefore, it is suggested that the Olympic and World Scientific Community face up to the reality of it and return to the imperial system, because 1.0 foot by way of the vesica pisces analysis provides the values for the square root of 2, 5 and 3, the true royal cubit.

One Last Tango with Numbers...

The decoding process of the Great Pyramid began during the Spring of 2007, close to the 777 day, but the final information as to who, or what group built it and the reason why came to light in August 2008, about the same time the Beijing Olympics got underway. On the triple 8 day, therefore, year 2008 will be used in the analysis to see if there are any special meanings attached to the outcome:

2008 ~ the year when the Great Pyramid was decoded.

3138, divided by 2008 = 1.563 / 1.618 - 1.563 = 0.055

Therefore:
1.563 + 0.05 = 1.613 + 0.005 = 1.618

A pair of quintessential numbers are at play in the analysis and justly so. Five is the number of life and certainly the use of its square root in the study cannot be overlooked. In simple metaphysical terms:

5.0 + 5.0, or 2 x 5.0 = 10 = 1 + 0 = 1.0, the number of all, which identifies with itself and becomes 2, 5.0 x 5.0 = 25 = 2.0 + 5.0 = 7.0, the solar number, 7.0 + 1.0 = 8.0, the number of infinity, 7.0 + 2.0 = 9.0, the full triple trinity number before reset. In the case of 5 x 5 = 25, squared = 625, divided by 5 = 125, divided by 100 = 1.25, square root = 1.118 – 0.5 = 0.618 = Phi, and 1.118 x 2 = 2.236 = the square root of 5. The following interplay of these numbers confirms beyond doubt that indeed, the decoding process matches with the triple 8 day mentioned above. A pair of quintessential numbers = 55 x *8, (the eighth year of the 21 st. Century) = *440, the base length of the Great Pyramid in Royal Cubits and 440, divided by 5 = *88, the number of constellations. By no coincidence it is as if it was meant to be. The dimensions and proportions of this monument of wonders provide the Golden Keys that unlock the Mysteries of the Living Universe. Its numbers are revelations that never lie.

In closing, I wish to thank the reader for allowing me to share these findings and insights into the workings of the Living Universe with them. In returning to my regular survey / engineering career I would dearly love to have a project that involved an ancient site to work on, or a huge piece of land where I would be free to apply the principles of the Golden Ratio and Geometry of Nature for landscaping purposes and various structures. In the interim I'll be busy connecting with the stars and planets in the various constellations at the speed of thought in order to manifest the follow up edition to Living with Geometry. I believe this presentation has provided a sane approach for all fellows and the world scientific community to work on and I am grateful to have had the opportunity to do so. One morning around the time of the Autumnal Equinox in 2006 I got started on it while tripping over triangles in my consulting office, when suddenly I was overwhelmed with thoughts on how the quality of our state of being could be improved. It was perceived that it would help if we knew more about that which is important to us and what the meaning of life is, rather than spinning aimlessly around on a globe somewhere in the vastness of the Universe like a misguided space vehicle going no where with no direction planned. The Universal Consciousness allowed me the opportunity and time to put it together, and the hope is that spin off information and other benefits will evolve from these contents and other meaningful breakthroughs in all types of endeavors such as medicine, engineering, athletics, horticulture and the arts and sciences in general are made in the process in order to stay on track with the true purpose of education and the meaning of life, because freedom is having right knowledge such as this in order to deal with, understand and improve the state of being. I don't know where else to go in order to have an exchange with my fellows about it except make a presentation in the form of a book such as this, that might, or might not be looked over and studied by others.

At times I jokingly remind myself of a movie called Good Will Hunting I took in some years ago, starring Matt Damon who played the part of an MIT janitor from the blue collar neighborhood in South Boston. He was discovered to be a mathematics genius by its professors after providing brilliant solutions to a number of problems on their wall boards when no one was around during the night shift. When asked how he knew of such things his response was, the only way I can explain it is, I see something and I just understand it. The particulars of what he saw and his solutions were never explained in the movie, but there are such things in the real world when it comes to the bisection of a one unit square and the square root of 3 x 440 and where it led to. Nevertheless there is a worthy moral to the story. Never judge a book by its cover, or underestimate a person by their appearance or station in life. A prime example of this scenario was the case with Susan Boyle, a comely looking unemployed middle age Scottish spinster who sang " I Dreamed a Dream " from les miserables in the April, 2009 Britain's Got Talent Contest. To everyone's astonishment hers was the voice of an angel that brought tears of joy to our eyes and rocked the cynical judges off their chairs, taking the world by storm to score first place in the hearts of millions. Who is to say that the innocent thoughts of a child couldn't manifest wonders of science. Members of the animal, vegetable and mineral kingdoms on this planet might have a better handle on the workings of life and the cosmos than we could ever hope to have, but now that the Language of the Living Universe is understood by us we can gain a better interpretation of life from other perspectives. I don't need to be a janitor on the night shift at one of these highly acclaimed institutions of learning to get my messages across, though it might be fun to draw the Royal Cubit Theorem on the black boards in the science and math rooms. I find myself pondering over what the responses will be from our astronomers and physicists of today once they get through the contents of Living with Geometry? No doubt they will find it eye opening that 1.0, divided by the square root of 3 x 100,000 x 2 equals the diameter of the Milky Way in light years and the orbit of Mercury around the sun equals the square root of 5 – 2.0 x one billion, that there are 12 planets in the solar system and the correct speed of light would become available to them. I'm left wondering if they might put this knowledge to good use or carry on providing false information to the future generations of students. The hope is they will in time also discover that the geometry of the Great Pyramid and Masonic Symbol are the keys to coming to an understanding with God, the workings of life and the Universe. There is no need to employ such dramatics for teaching what is meaningful to the populace as the New Age of Aquarius draws near. It was my passion for geometry and a form of divine providence that put me in this position. I was searching for the truth and became a student of the Universe while traversing along that Golden Spiral Road of Life and Enlightenment. The story can be easily compared to the Wizard of Oz tale about following the Yellow Brick Road. The only difference was I became linked up with the Wizards of Ancient Times and was provided with the opportunity to apply my experiential learning processes as I had done on many other projects. The result of which was composing a record of the survey details that are hereby made available in "Living with Geometry".

Michael Green

DRAW THE MASONIC SYMBOL

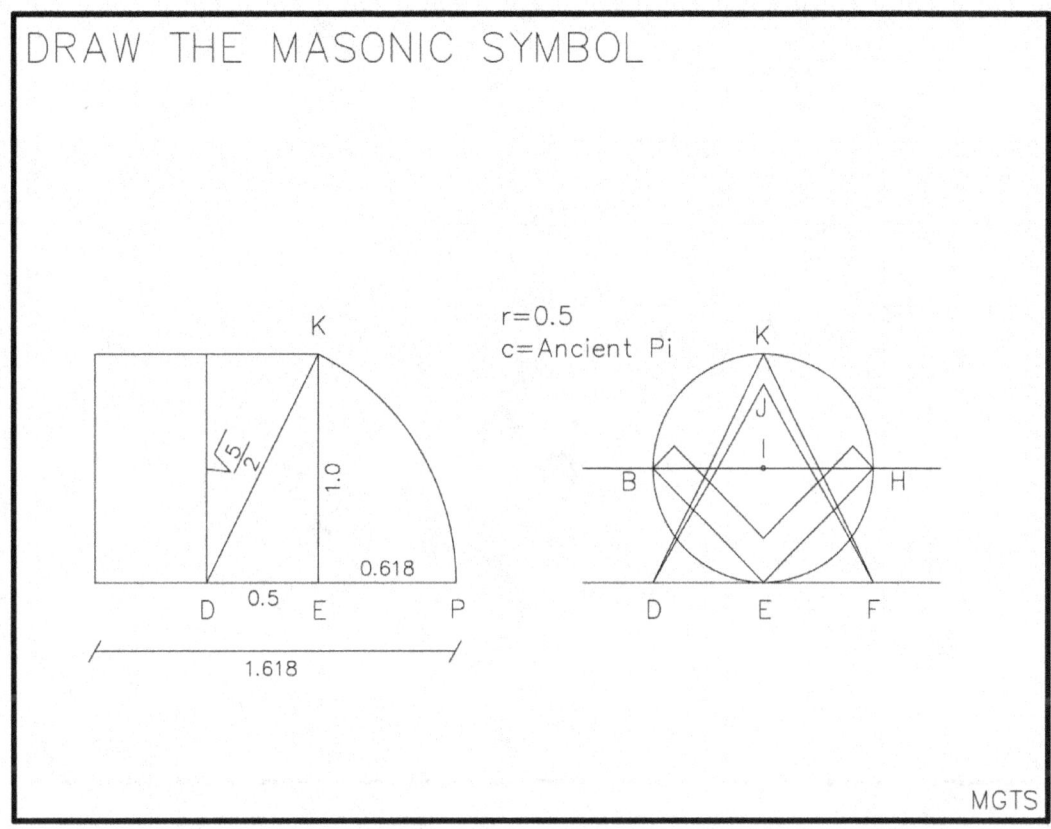

r=0.5
c=Ancient Pi

MGTS

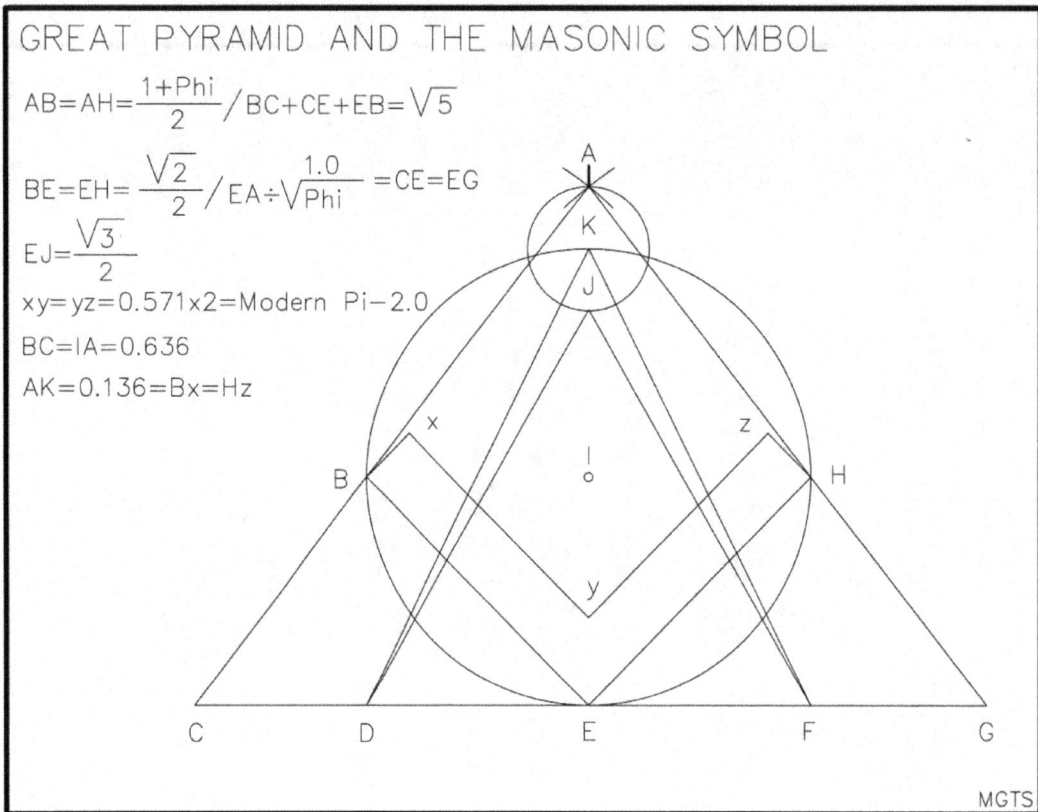

GREAT PYRAMID AND THE MASONIC SYMBOL

$AB=AH=\dfrac{1+Phi}{2}\ /\ BC+CE+EB=\sqrt{5}$

$BE=EH=\dfrac{\sqrt{2}}{2}\ /\ EA\div\sqrt{\dfrac{1.0}{Phi}}=CE=EG$

$EJ=\dfrac{\sqrt{3}}{2}$

$xy=yz=0.571\times 2=$ Modern Pi-2.0

$BC=IA=0.636$

$AK=0.136=Bx=Hz$

MGTS

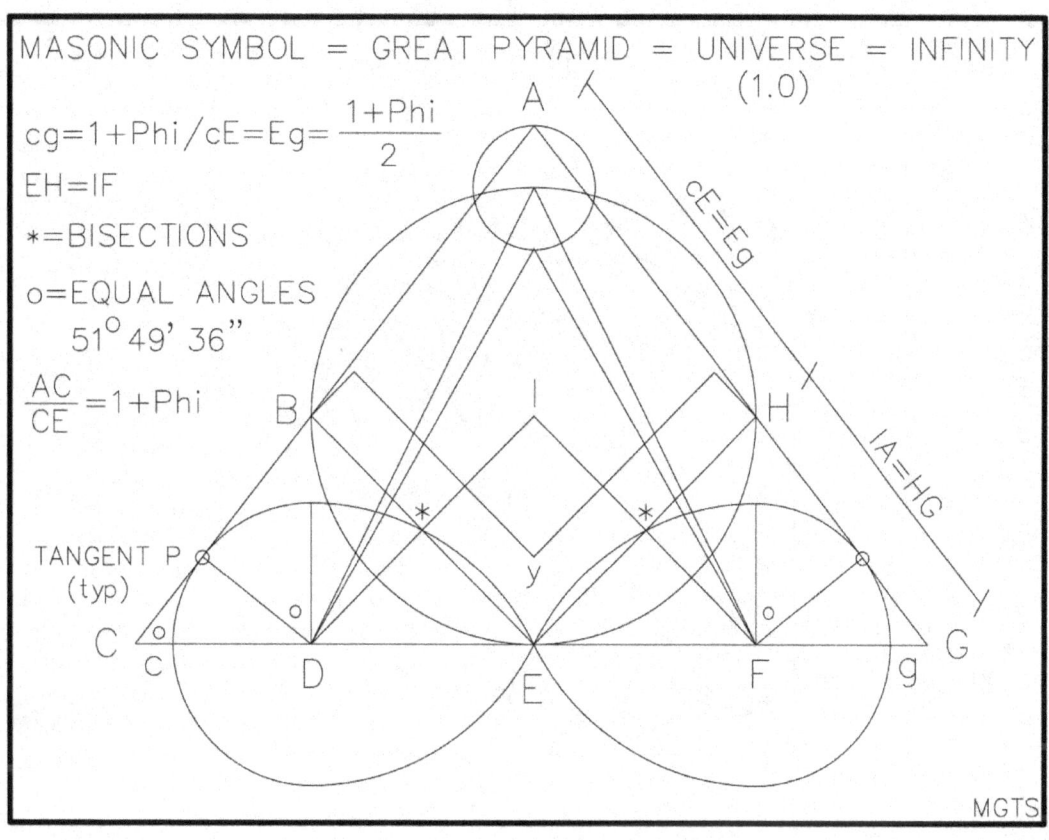

MASONIC SYMBOL = GREAT PYRAMID = UNIVERSE = INFINITY
(1.0)

cg=1+Phi/cE=Eg= $\dfrac{1+Phi}{2}$

EH=IF

∗=BISECTIONS

o=EQUAL ANGLES
    51° 49' 36"

$\dfrac{AC}{CE}$ =1+Phi

TANGENT P
(typ)

cE=Eg

IA=HG

A

B   I   H

∗   ∗

y

TANGENT P
o   o

C   c   o   D   E   F   o   g   G

MGTS

A = 1:0 = the UNIVERSE...

1.0

1.0

SYMBOLS OF THE UNIVERSE

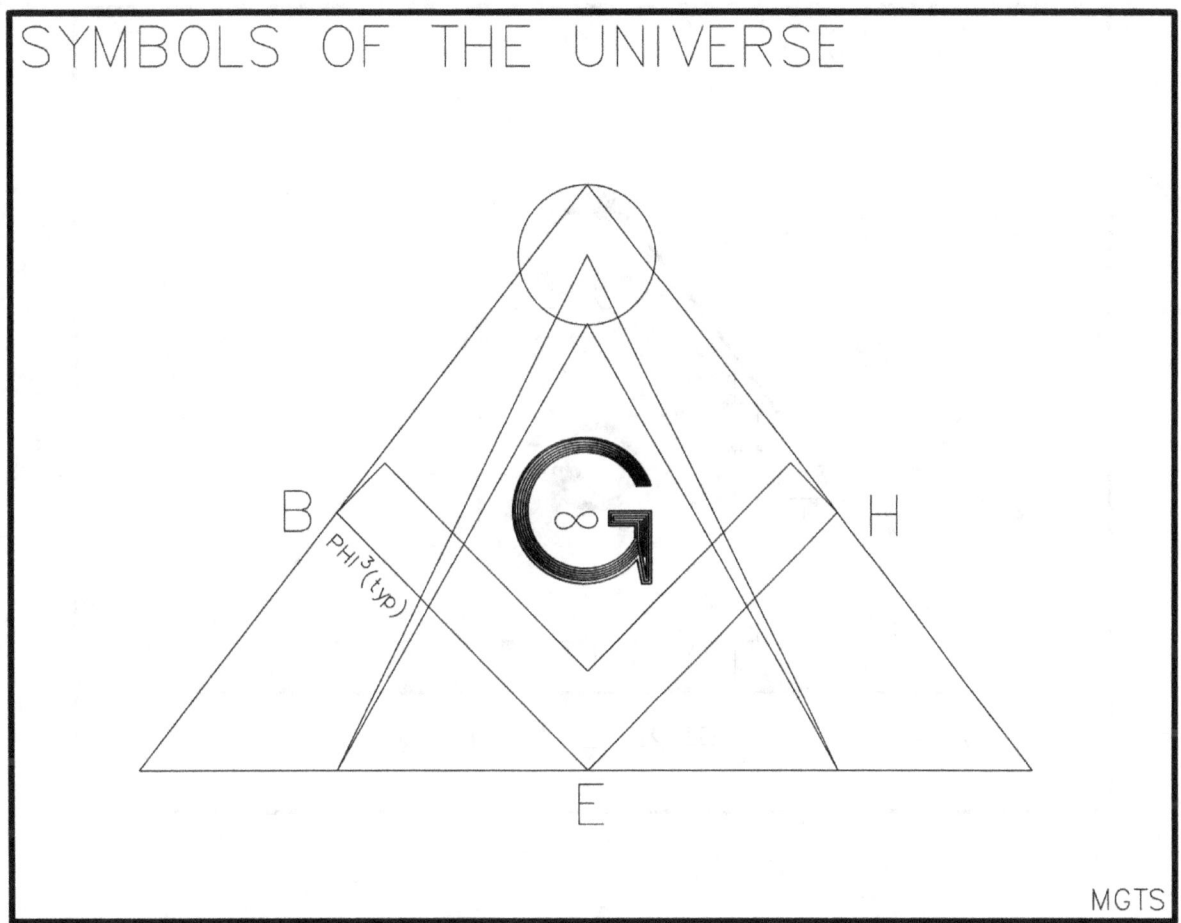

MGTS

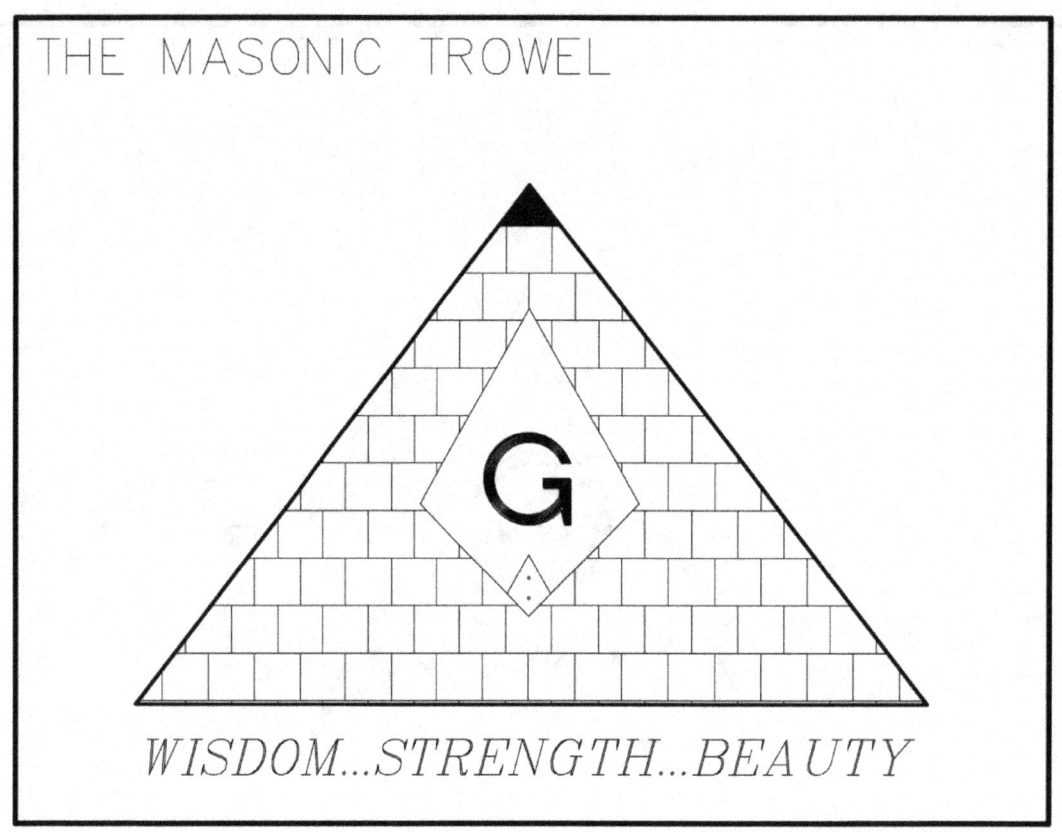

THE MASONIC TROWEL

WISDOM...STRENGTH...BEAUTY

LODGE OF PERFECTION

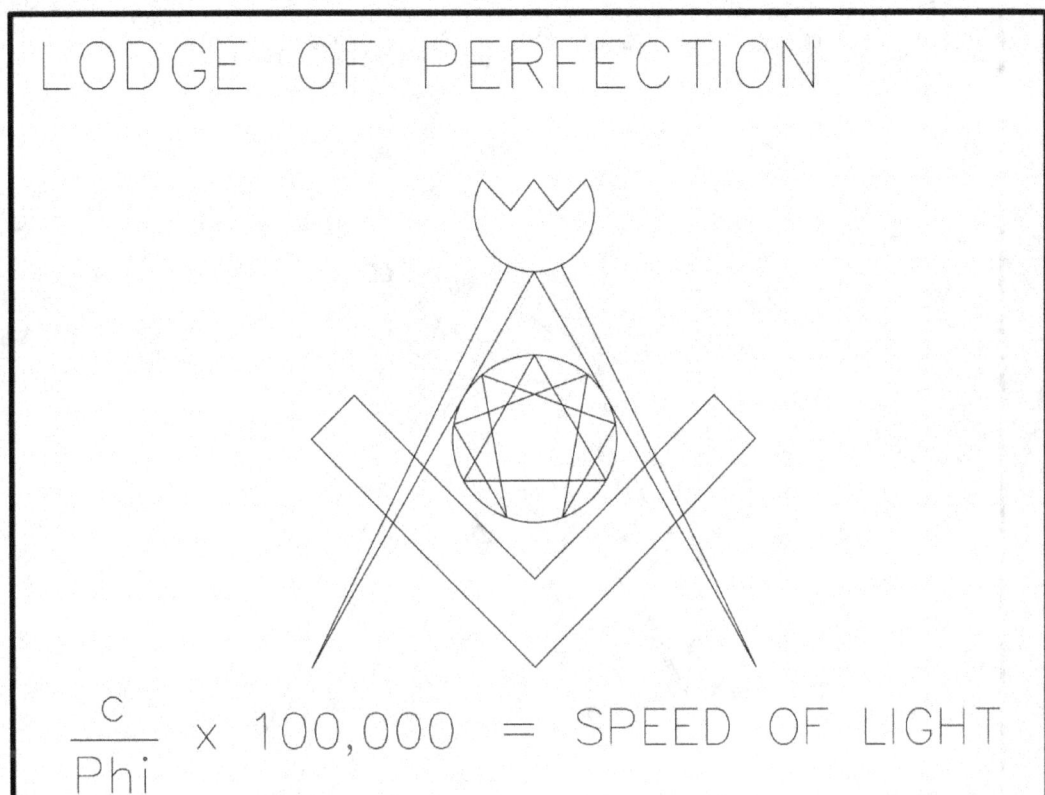

$$\frac{c}{Phi} \times 100,000 = \text{SPEED OF LIGHT}$$

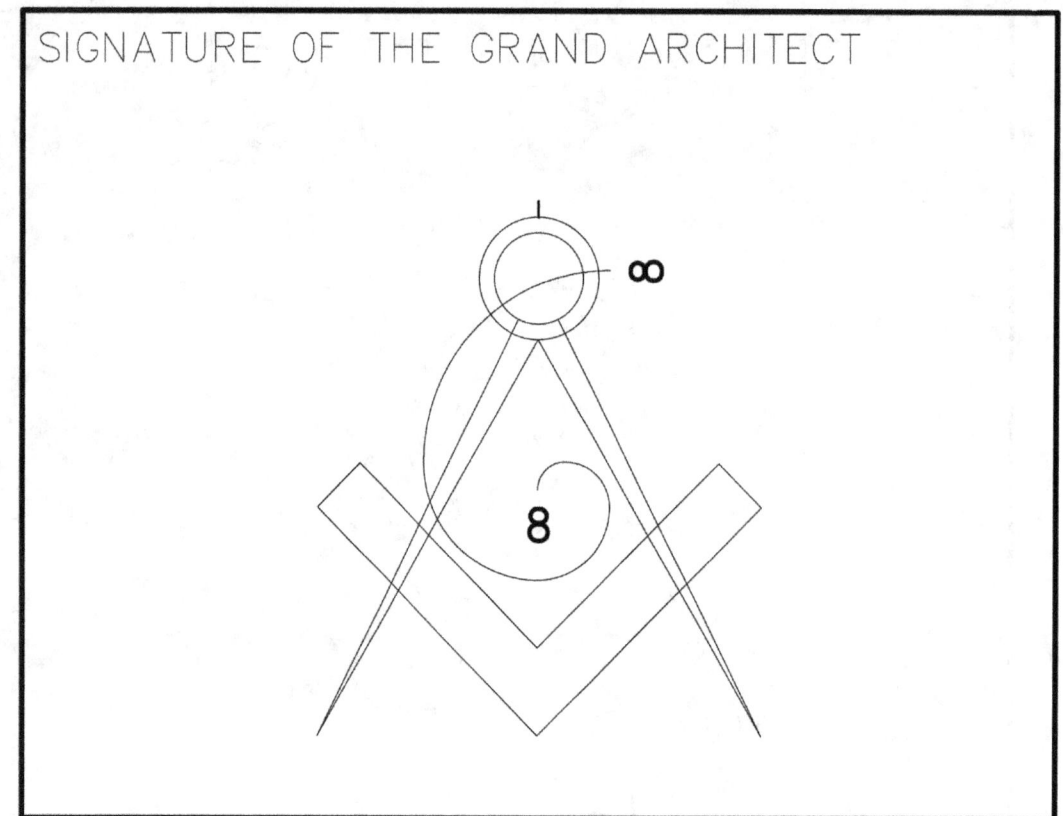

SIGNATURE OF THE GRAND ARCHITECT

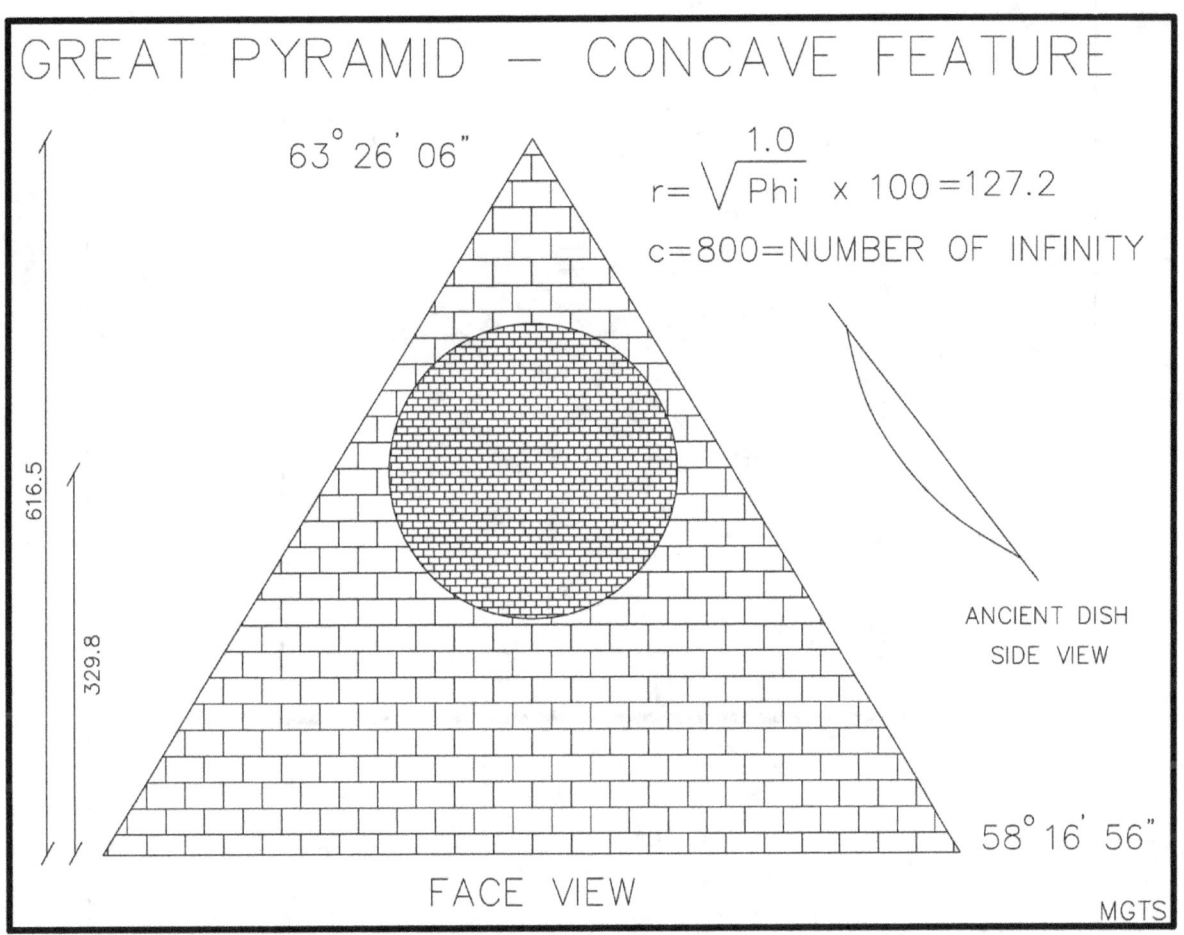

GREAT PYRAMID — CONCAVE FEATURE

63° 26' 06"

$$r = \sqrt{\frac{1.0}{Phi}} \times 100 = 127.2$$

c=800=NUMBER OF INFINITY

ANCIENT DISH
SIDE VIEW

616.5

329.8

58° 16' 56"

FACE VIEW

MGTS

UMBRELLA DOME — LONDON OLYMPICS 2012
CIRCULAR COLISEUM — FIELD DIAMETER 660 FEET
DOME — AIR SUPPORTED

$$AB = 1 + Phi$$
$$Ab = Ba = 1.0 + Phi^2$$
$$\frac{AB}{arc} = \frac{1 + Phi}{2} \ / \ Phi \ DOME$$

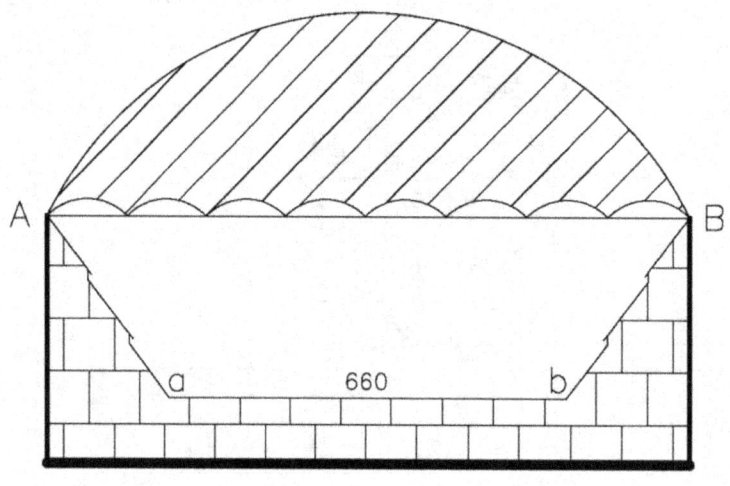

SIDE VIEW
SCALE: 1=660

OUTSIDE DIAMETER — 1068 FEET
HEIGHT FROM CENTER FIELD — 619 FEET
SEATING CAPACITY 110,600

MGTS

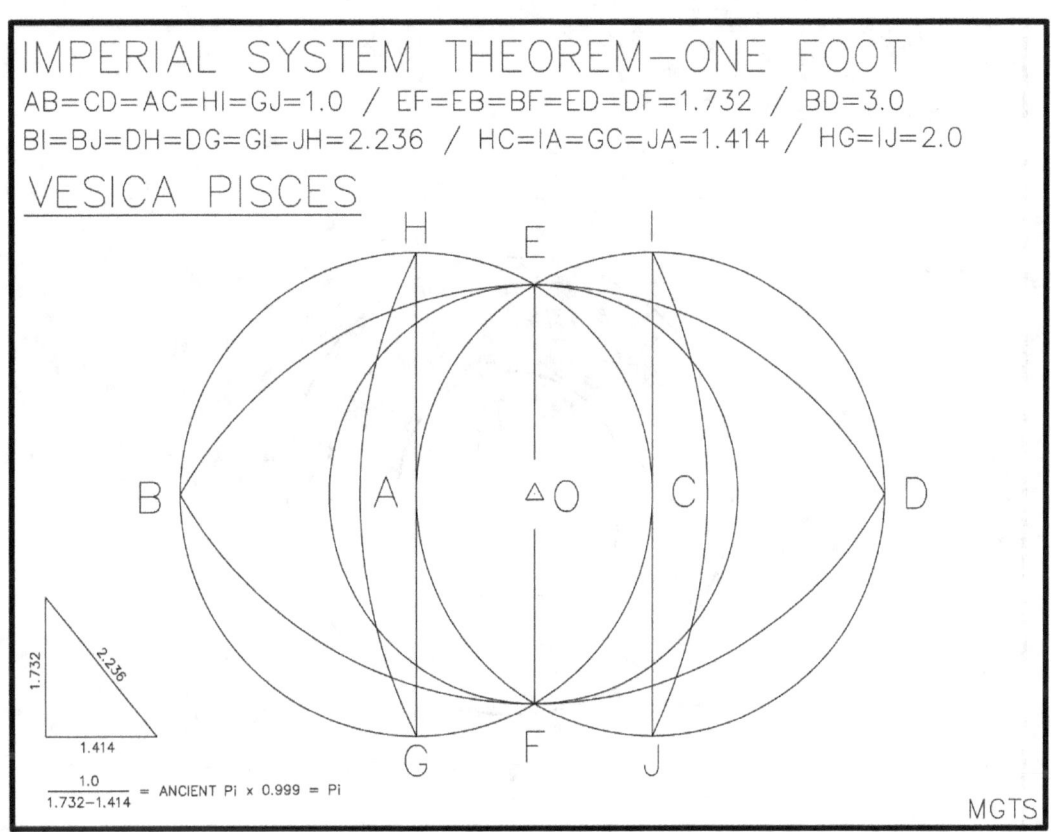

IMPERIAL SYSTEM THEOREM—ONE FOOT
AB=CD=AC=HI=GJ=1.0 / EF=EB=BF=ED=DF=1.732 / BD=3.0
BI=BJ=DH=DG=GI=JH=2.236 / HC=IA=GC=JA=1.414 / HG=IJ=2.0

VESICA PISCES

$$\frac{1.0}{1.732-1.414} = \text{ANCIENT Pi} \times 0.999 = Pi$$

MGTS

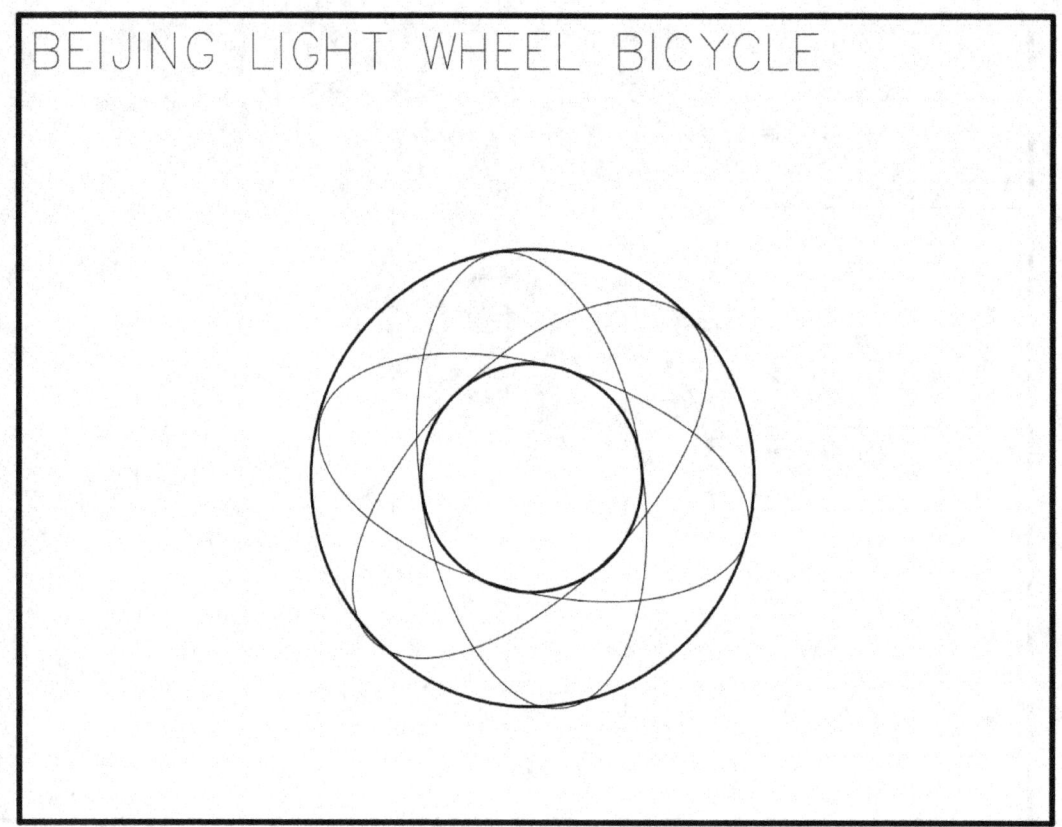

BEIJING LIGHT WHEEL BICYCLE

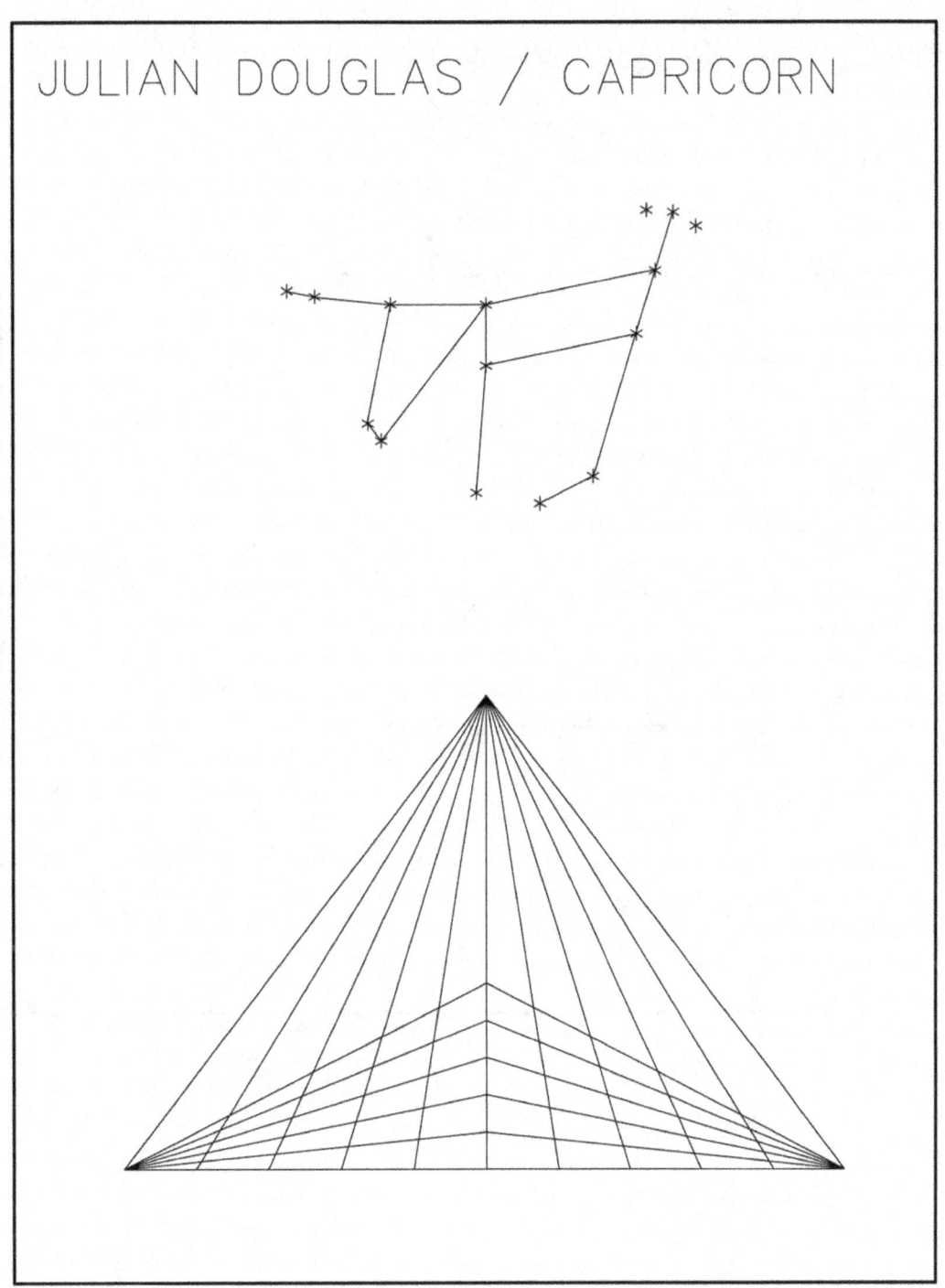

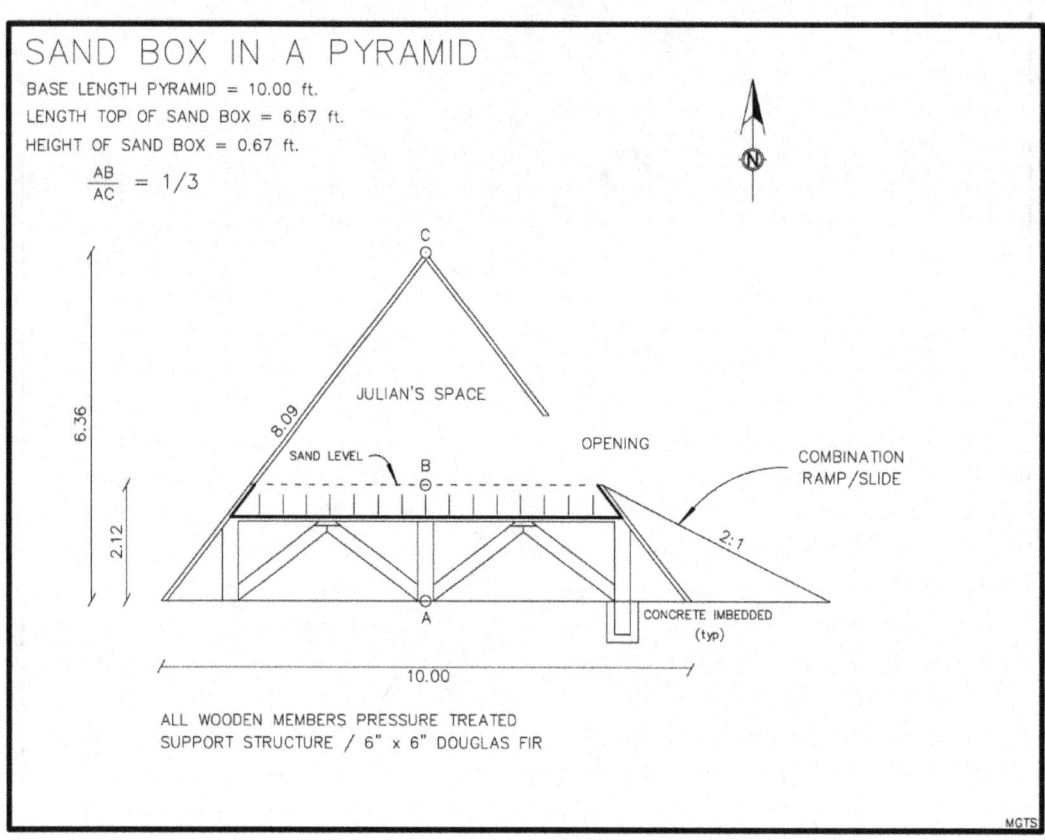

SAND BOX IN A PYRAMID

BASE LENGTH PYRAMID = 10.00 ft.
LENGTH TOP OF SAND BOX = 6.67 ft.
HEIGHT OF SAND BOX = 0.67 ft.

$$\frac{AB}{AC} = 1/3$$

C

JULIAN'S SPACE

8.09

SAND LEVEL

B

OPENING

COMBINATION
RAMP/SLIDE

6.36

2.12

2:1

A

CONCRETE IMBEDDED
(typ)

10.00

ALL WOODEN MEMBERS PRESSURE TREATED
SUPPORT STRUCTURE / 6" x 6" DOUGLAS FIR

MGTS

BLUEPRINT FOR THE GREAT PYRAMID

M=Top King's Chamber / Height / $\frac{IH}{100}$ x 9.0 = 19.2

$\frac{AM}{AE}$ = 0.666

y = flr. Queen's Chamber + 1.0 ft.

ox = zp = 6.928 ft.

BH = DF = 426.6 ft.

$\frac{P}{C}$ = 2.272 / $\frac{1.0}{2.272}$ x 1000 = 440

$\frac{AE}{CE}$ = 1.272

$\frac{LM}{LI}$ = 0.272

$\frac{CG}{AU}$ = 1.272

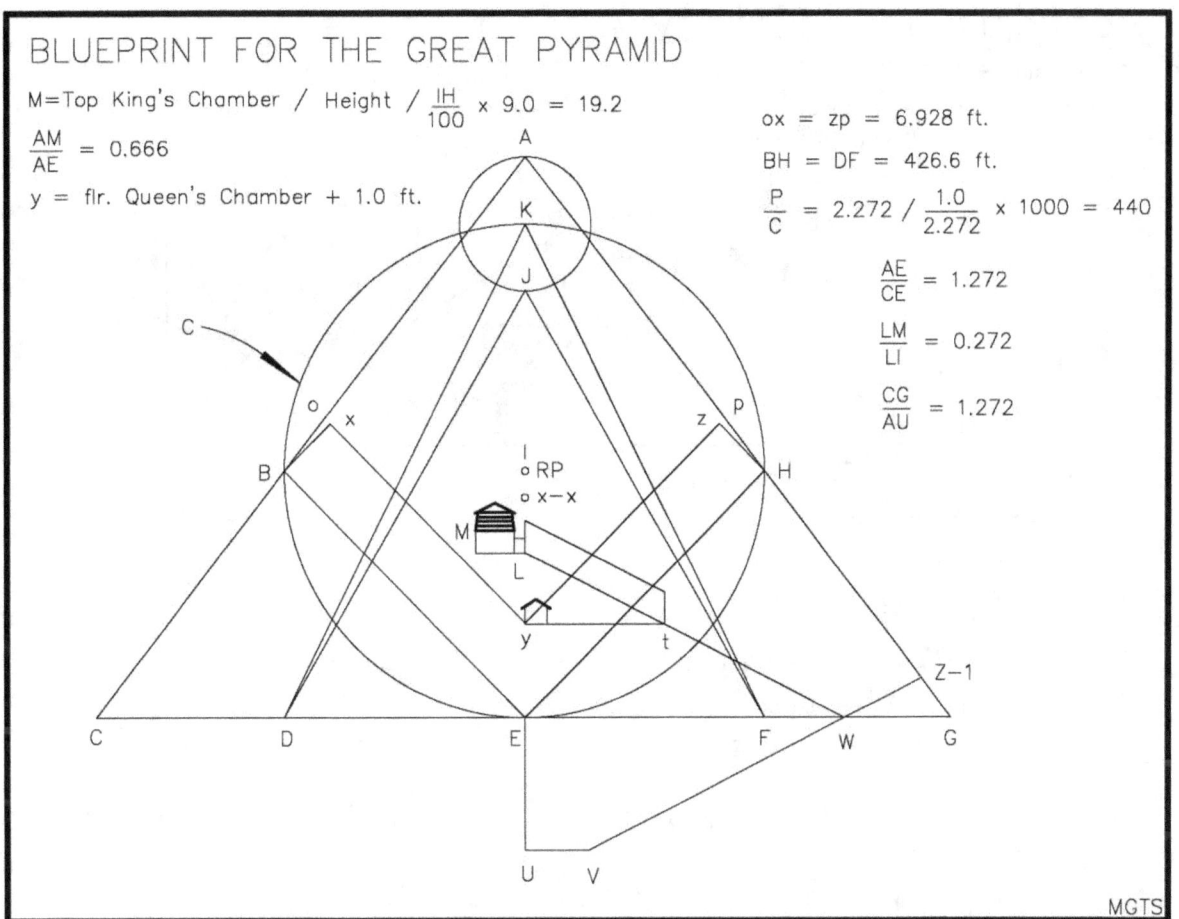

MGTS

Pyramid Inch — Velocity Ancient Space Travel

One Royal Cubit = 1.732 ft.

SOL = 186,216.56 mi. per second

Number of Planets in all Solar Systems = 12

Galactic Incline = 2:1 Ratio

Number of Infinity = 8

Quintessential Number of Life = 5

Note: $\dfrac{5 \times 8}{2}$ = 20

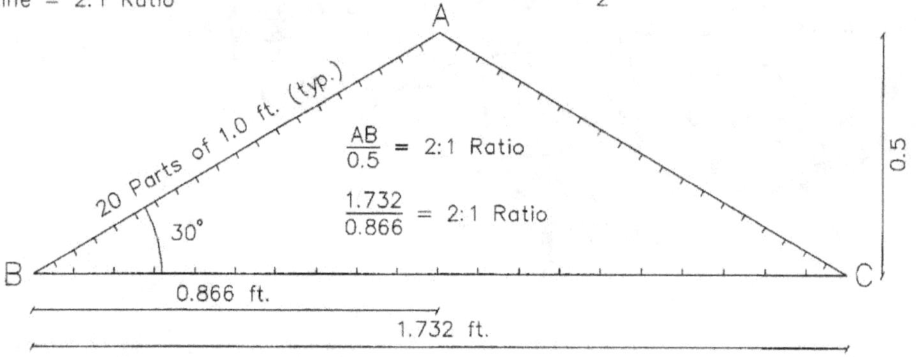

20 Parts of 1.0 ft. (typ.)

30°

$\dfrac{AB}{0.5}$ = 2:1 Ratio

$\dfrac{1.732}{0.866}$ = 2:1 Ratio

0.5

0.866 ft.

1.732 ft.

Pyramid Inch:

$\dfrac{20}{1.732}$ = 0.0866 ft.

Velocity Ancient Space Travel:

0.0866 lt. yr. per second = 2,150,000 x SOL

or, 12^2 x 1.0 E15 = 1.44 E15 m.p.h.

Width of King's Chamber: 20 x 0.866 = 17.32 ft. $\dfrac{34.64}{17.32}$ = 2:1 Ratio

Length of King's Chamber: 20 x 1.732 = 34.64 ft.

Diameter Milky Way:

$\dfrac{AB + BC}{BC}$ x 100,000 = 115,473.4411 lt. yr.

MGTS

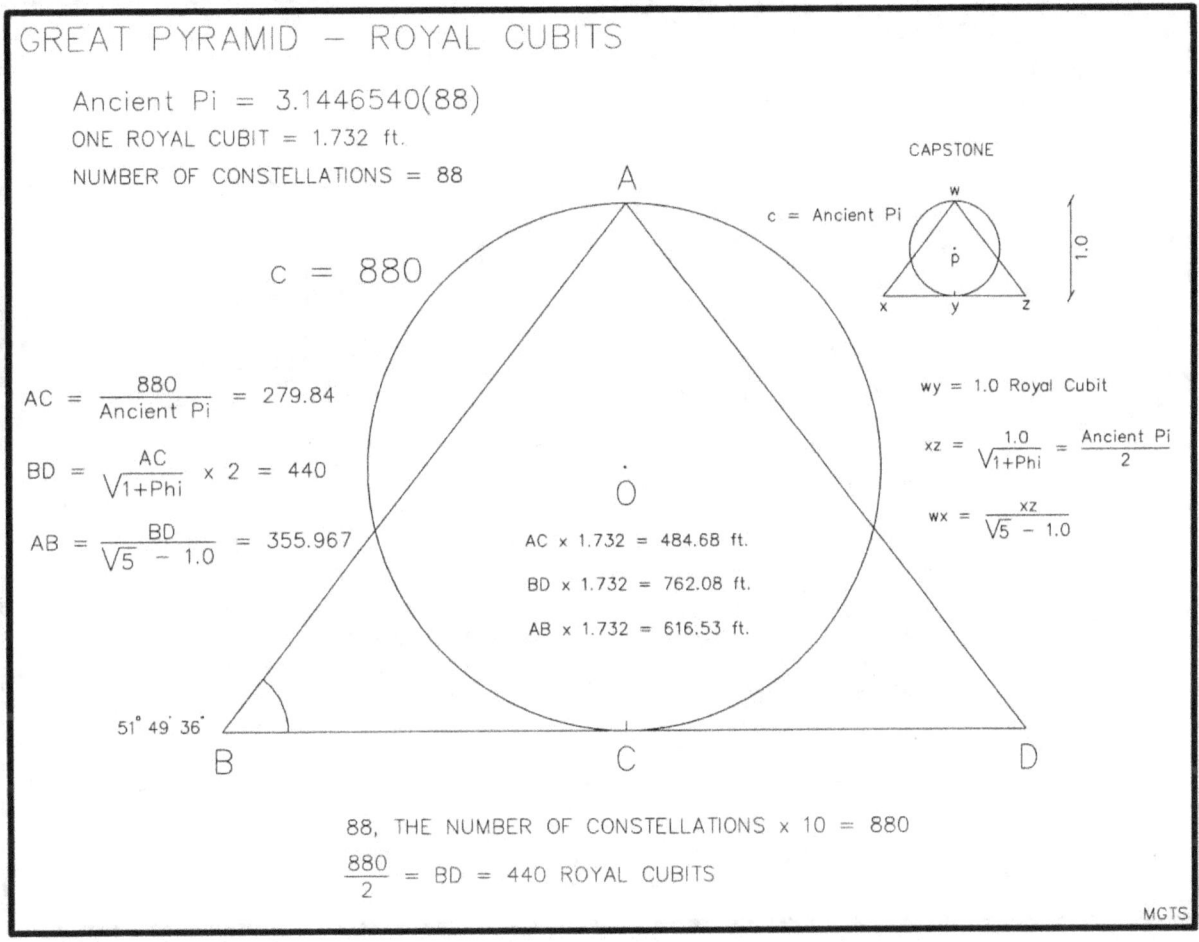

GREAT PYRAMID — ROYAL CUBITS

Ancient Pi = 3.1446540(88)
ONE ROYAL CUBIT = 1.732 ft.
NUMBER OF CONSTELLATIONS = 88

c = 880

CAPSTONE

c = Ancient Pi

A

$AC = \dfrac{880}{\text{Ancient Pi}} = 279.84$

$BD = \sqrt{\dfrac{AC}{1+Phi}} \times 2 = 440$

$AB = \dfrac{BD}{\sqrt{5} - 1.0} = 355.967$

O

AC x 1.732 = 484.68 ft.

BD x 1.732 = 762.08 ft.

AB x 1.732 = 616.53 ft.

wy = 1.0 Royal Cubit

$xz = \dfrac{1.0}{\sqrt{1+Phi}} = \dfrac{\text{Ancient Pi}}{2}$

$wx = \dfrac{xz}{\sqrt{5} - 1.0}$

51° 49´ 36˝

B   C   D

88, THE NUMBER OF CONSTELLATIONS x 10 = 880

$\dfrac{880}{2}$ = BD = 440 ROYAL CUBITS

MGTS

GREAT PYRAMID INTERIOR DETAIL

$\dfrac{BC}{AD} = 0.75$

$\sqrt{.75} = \dfrac{\sqrt{3}}{2} \times 10 = 8.66$

$\dfrac{y-1}{x-1} = 0.866$

$\left(\dfrac{K\ h-2}{K\ h-1}\right)^2 = 0.0866$

$\dfrac{1.0}{0.0866} \times 10{,}000 = d$ Milky Way

$\dfrac{y}{x} = \sqrt{1+Phi}$

$c = 440$

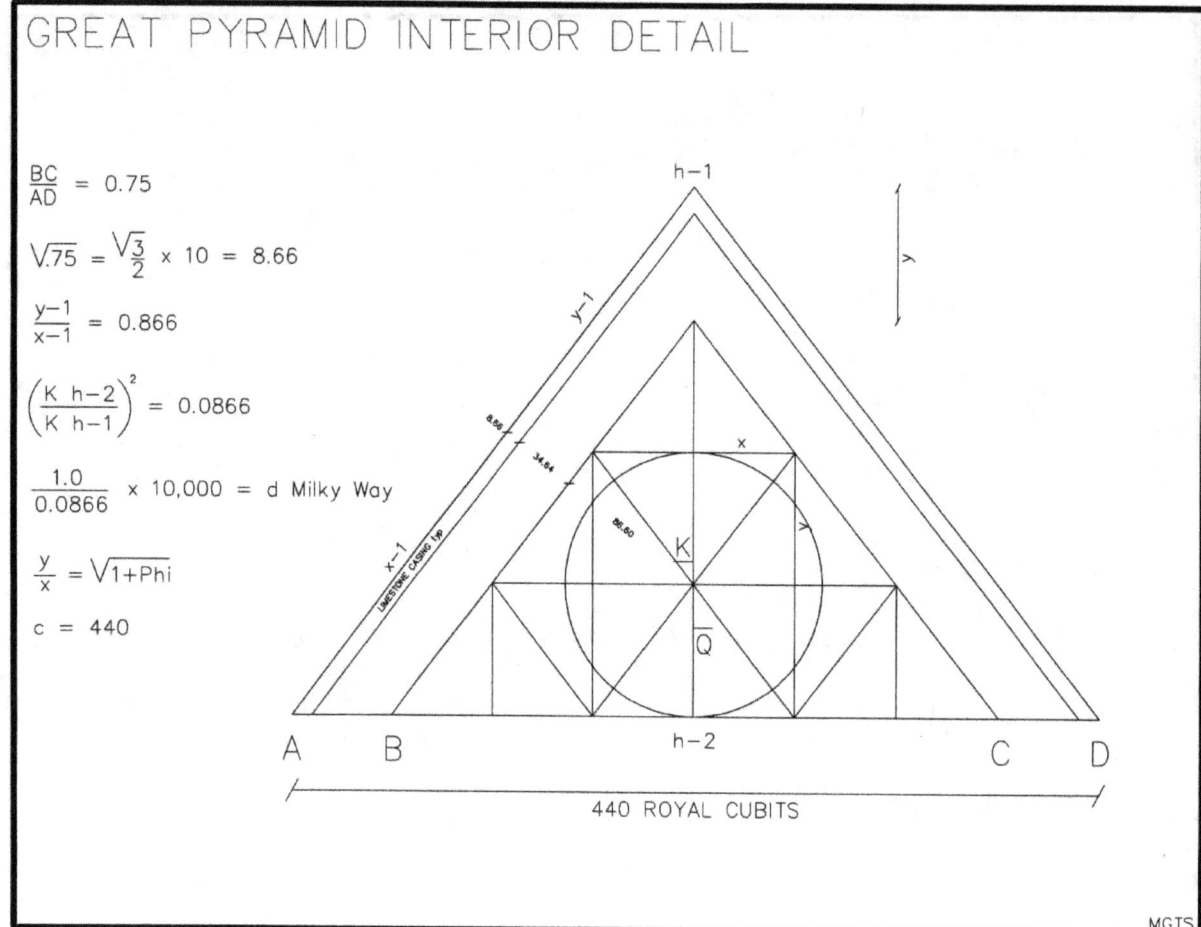

440 ROYAL CUBITS

MGTS

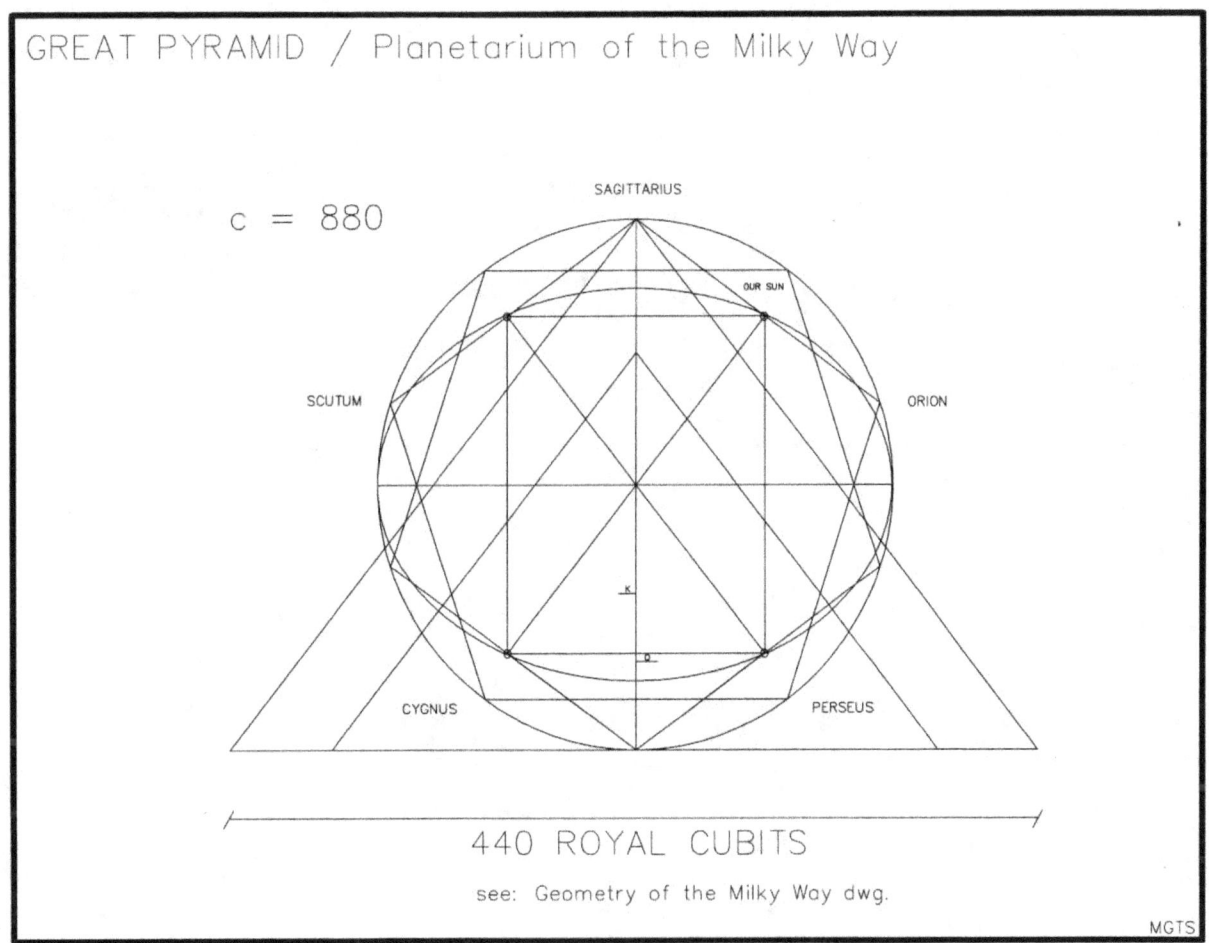

GREAT PYRAMID / Planetarium of the Milky Way

c = 880

SAGITTARIUS

OUR SUN

SCUTUM

ORION

K

D

CYGNUS

PERSEUS

|←———————— 440 ROYAL CUBITS ————————→|

see: Geometry of the Milky Way dwg.

MGTS

www.ingramcontent.com/pod-product-compliance
Lightning Source LLC
Chambersburg PA
CBHW081106170526
45165CB00008B/2343